EVOLUTION AND EXTINCTION

EVOLUTION
AND
EXTINCTION

PROCEEDINGS OF A JOINT
SYMPOSIUM OF THE ROYAL SOCIETY
AND THE LINNEAN SOCIETY
HELD ON 9 AND 10 NOVEMBER 1989

ORGANIZED AND EDITED BY
W. G. CHALONER, F.R.S., AND A. HALLAM

CAMBRIDGE
UNIVERSITY PRESS

Published by the Press Syndicate of the University of Cambridge
The Pitt Building, Trumpington Street, Cambridge CB2 1RP
40 West 20th Street, New York, NY 10011–4211, USA
10 Stamford Road, Oakleigh, Melbourne 3166, Australia

First published in *Philosophical Transactions of the Royal Society of London*,
series B, volume 325 (no. 1228), pages 239–488.
Hardback edition published by the Royal Society 1989
Paperback edition published by Cambridge University Press 1994

Printed in Great Britain at the University Press, Cambridge

British Library cataloguing in publication data

Evolution and extinction.
1. Organisms. Evolution. Theories
I. Chaloner, W. G. (William Gilbert)   II. Hallam, A. (Anthony)
575.01

Library of Congress cataloguing in publication data available

ISBN 0 85403 391 2 hardback
ISBN 0 521 40646 3 paperback

CONTENTS

CONTENTS

PREFACE

The year 1988 marked the bicentenary of the founding of the Linnean Society, the world's oldest active biological society. The bicentennial celebrations, which took place over the period 1987–1989, included a series of scientific meetings held jointly between the Linnean Society and several other learned societies, and dealt with a wide range of biological topics.

The Royal Society was already over a hundred years old when the Linnean was founded, and James Edward Smith, the founder President of the Linnean, was a Fellow of the Royal Society. Over the intervening years the two societies have had many shared Fellows. As a continuation of that historical link, this joint Royal Society – Linnean Society two-day Discussion Meeting was held towards the end of 1988, as a part of the series of bicentennial meetings.

In the past decade, two lines of thought have come to play a major role in our understanding of evolution. One deals with the actual rate at which the evolutionary process occurs, and questions whether the pace and the very nature of evolutionary change is gradual and more or less constant, or whether it is of an erratic, pulsed nature. This is linked to the separate but related question of whether so-called 'mass extinctions' seen in the fossil record are merely the end-members in a wide range of extinction phenomena, or represent processes of an entirely different order of magnitude. Those who favour the latter interpretation generally see mass extinctions as a result of catastrophic events, of either terrestrial or extra-terrestrial origin rather than the result of random fluctuations in the fortunes of survival. The bases of current thought go back over 100 years; yet the main source of the present controversy is new evidence apparently linking the rate of change seen in fossil populations to evidence of physical phenomena seen in the geological record. Although Darwin considered that biotic competition was the dominant cause of extinction, the fossil record is now interpreted by many as suggesting that change in the physical environment was the more important factor.

As M. Nitecki has observed, it is as inappropriate to consider evolution without extinction as to study demography without awareness of the process of death. We aimed at bringing together in this symposium a wide range of scientists interested in different aspects of these two phenomena. These reach in time from the earliest (Precambrian) record of extinction of microfossil taxa, to those documented – or merely threatened – extinctions associated with human activity. Our sources range from work on living populations to evidence derived from palaeobiology. The papers collected here extend from the deliberations of an evolutionary biologist and geneticist, Professor John Maynard Smith, F.R.S., to those of a palaeobiologist concerned with extinctions of ancient phytoplankton, Professor Andrew Knoll. In the last paper, the scientific broadcaster Mr Colin Tudge reviews mass extinctions and the human species, in the provocative style that we have come to expect of him. This final contribution may not fit the mould of a conventional scientific paper, but none the less presents a most appropriate message with which to end the volume.

March 1989

W. G. CHALONER
A. HALLAM

INTRODUCTION

In the past decade an enormous interest has been generated in the subject of extinction both among scientists and the general public. A prime stimulus to this interest has been the Alvarez hypothesis of mass extinction, about 65 million years ago, of dinosaurs and many contemporary marine and terrestrial organisms as a consequence of the impact of an asteroid. This has triggered a vast amount of research by scientists in several research fields on the end-Cretaceous and other mass-extinction events recorded in the stratigraphic sequence, leading to much ensuing controversy about the cause and possible periodicity of such events and the extent to which they were genuinely catastrophic. The other major stimulus is the growing awareness of conservation issues, arising from the realization that we live on a threatened planet as a result of the activities of an ever-increasing population of our own species, an issue that is dramatized in a spectacular fashion by the destruction of tropical rain forests.

Despite the burgeoning research activity several important questions have arisen which cannot as yet be regarded as satisfactorily resolved, such as the following:

(1) Are mass-extinction events catastrophic or not?
(2) Do they occur periodically or episodically?
(3) Is there a fundamental difference between mass extinction and background extinction?
(4) To what extent does taxonomic practice affect the recognition of extinction events?
(5) Is evolution driven primarily by biotic factors such as competition (as Darwin believed) or by changes in the physical environment leading to extinctions, thereby vacating ecological niches for occupation by survivors that subsequently radiated?
(6) Do relevant changes in the physical environment involve marine regressions and/or anoxia, climate, massive volcanism or impact of extra-terrestrial bodies, or an interaction of these various factors?
(7) Were the survivors fitter or just luckier than those that went extinct? In effect, what is the role of chance factors in evolution?
(8) Are we living in a catastrophic mass-extinction phase today as a result of the activities of *Homo sapiens*? If so, what can or should be done about it?

It was with such questions in mind that a symposium was organized jointly by the Royal Society of London and the Linnean Society and held at the Royal Society in November 1988. This volume contains the published versions of all the lectures given, together with some of the ensuing discussion, with the exception of one by Professor N. J. Shackleton, who obligingly substituted at short notice for Professor S. M. Stanley.

Maynard Smith addresses the gap in evolutionary studies between population genetics theory and palaeontological observations, and challenges palaeontologists to produce more statistical support for the patterns in the stratigraphic record that they recognize. He thinks that for many extinctions biological causes may have been equally as important as changes in the physical environment, and accepts only a limited role for species selection in the promotion of evolutionary change, one that allows the individual characteristics of particular species that prevail not necessarily to correspond to those that determine survival.

Hoffman notes that there has been a long-standing debate between supporters of catastrophic and gradualist schools of thought, dating back to Cuvier and Lyell, and suggests that many mass-extinction events are in fact clusters of minor extinction episodes accidentally aggregated in time over periods of several million years. He doubts that there is ever one overriding single cause as opposed to a coincidence of independent processes, but thinks that the biggest extinction event of all, at the end of the Palaeozoic Era, could be an exception. Illustrating his point especially with respect to the end-Ordovician event, Holland shares Hoffman's view that mass extinctions are complex events extended through time, and decidedly not catastrophic as claimed by some supporters of bolide impact. He is also sceptical about Raup and Sepkoski's claim of extinction periodicity and warns about the imprecision of available radiometric timescales.

It has been necessary, because of the limitations of the fossil record, to restrict attention about evolution and extinction to the Phanerozoic, but Knoll points out that, by the time animals first diversified in the Cambrian, planktonic protists had already undergone at least one radiation and one or two major extinctions. By and large, Phanerozoic radiations and extinctions of these organisms correspond quite closely to invertebrate events, but the massive plankton extinctions that occurred at the end of the Cretaceous were avoided by groups that had a capacity to resist increased environmental stress by the adoption of a non-planktonic resting stage. Knoll stresses the dominant role of the physical environment, as opposed to biotic interactions such as displacive competition, as the major engine of evolutionary change in the plankton, a conclusion at variance with what he inferred several years ago for terrestrial plants.

Fortey points out that systematics and the study of extinction are inextricably linked. If one accepts cladistics then a supposed extinction may merely record the disappearance of a paraphyletic taxon, in other words a pseudoextinction. In the early phases of cladogenesis there will be a tendency towards a temporal concentration of paraphyletic groups, and Fortey demonstrates how this reduces the importance of the mass-extinction event at the end of the Cambrian. On the other hand the end-Ordovician event is a genuine mass extinction, with deep-water taxa being adversely affected by oceanic anoxia.

Certain taxa are relatively resistant to extinction and the biology of extinction resistance is a research area that demands particular attention. Jablonski argues for a fundamental contrast between mass extinction and background extinction. Following mass-extinction events, evolution can be channelled into directions not predictable from situations established during more normal times. Based on his research on Cretaceous molluscs, he argues that high species richness and the possession of widespread individual species imparted extinction resistance among genera during background times but not during the end-Cretaceous mass extinction. Like Knoll, Jablonski sees physical environmental factors as significantly more important than biotic in controlling the course of evolution.

Dealing with the record of terrestrial plants across the Cretaceous–Tertiary boundary, Spicer reminds us that extinction rates based on isolated organs give a poor reflection of the evolution of the whole plant, because of the predominance of mosaic evolution. He considers that a more complex environmental scenario is demanded than has been assumed by some, with any short-term catastrophe having its effect against a background of longer-term climatic change. Whereas the record in North America is consistent with the kind of drastic environmental deterioration that one would expect to be associated with a bolide impact, there is no hint of such an event in the Southern Hemisphere, and he finds no support for the claim of a spectacular global wildfire.

The ammonoids, a marine cephalopod group that flourished through the late Palaeozoic and Mesozoic, provide good subject matter for radiation and extinction studies. House perceives a good correlation between marine regression–transgression couplets and associated anoxia, and extinction events. Evolutionary innovations are associated with the subsequent diversification and spread of the survivors. In House's view the evidence fails to support extinction periodicity and he dismissed the bolide impact that has been claimed for the late Devonian. The principal conclusions drawn by House are supported by Hallam for Phanerozoic marine invertebrates in general, with a strong correlation perceptible on both large scale and small scale. In contrast, he finds little convincing evidence for a major role for climate in pre-Cainozoic times, and virtually no evidence for bolide impact apart from at the Cretaceous–Tertiary boundary. Raup was invited by us to review the evidence for extra-terrestrial causation of mass-extinction events but chose instead to undertake a methodological exercise based on stratigraphic range truncation of fossil taxa. He argues persuasively that a hiatus in fossil preservation often produces an appearance of sudden extinction where none exists, that sudden extinctions may appear gradual and that so-called stepwise extinctions can be an artefact of the placement of fossil occurrences.

The vertebrate record, which is dominantly terrestrial, is considered by Benton, Charig, and Jaeger and Hartenberger. Benton notes that, as regards extinction analysis, this record is much more limited than for marine invertebrates, but has compensating advantages for ecological and phylogenetic studies. Major mass-extinction events appear to coincide with those among marine invertebrates, though the rapid radiation of 'modern' groups such as mammals and lizards was little affected by the end-Cretaceous event. Like Fortey, Benton pays attention to the problem of paraphyly, and he shares the scepticism about extinction periodicity expressed by Holland and House. The limitations of the record are also stressed, with regard to the end-Cretaceous dinosaur extinctions, by Charig, who observes that the only good record comes from western North America. It is indeed only an assumption that dinosaurs everywhere died out at the end of the Cretaceous, and there is a modest amount of apparently reliable evidence that the group survived in some regions into the early Tertiary.

It has long been recognized that islands are natural laboratories for the study of evolution and extinction. Among the most intriguing of all fossils are those of Neogene age found in several Mediterranean islands, with dwarfed forms of large and giant forms of small mammals. Jaeger and Hartenberger report on the results of an intensive study of diversification and extinction patterns among Neogene rodents in the Mediterranean countries. Island faunas exhibit a rapid turnover and high extinction rate. Most of the evolutionary novelties originated in larger faunal provinces where the driving force was more global. It is sobering that our knowledge of the key environmental change is still tantalisingly limited even in such a well-studied region. Williamson considers neontological data from island faunas in the British area, necessarily confining himself to population disappearances rather than species extinctions, but he maintains that there is continuity between the two processes. Rarity is thought to be the most significant factor in determining extinction vulnerability, which raises the question of why some species are rare.

The book concludes with two essays devoted to our own species, by a polymathic biologist and a distinguished science journalist. Diamond's review of extinctions caused by humans makes sombre, if eloquent, reading, arguing as he does that such extinctions have probably been severely underestimated, especially in the tropics. Though human agency in the extinction of large mammals thousands of years ago is still controversial, there can be no doubt of its importance in the past few hundred years, through such factors as overhunting, introduction of other species and habitat destruction. The current rate of species extinction seems certain to increase as human populations continue to expand, and Diamond, though pessimistic, invites attempts to counter this depressing trend.

Tudge's thought-provoking essay addresses itself to humanity itself, sketching out several possible scenarios about the world we may inherit in the future. Are we ourselves in any danger of extinction? If not, what will happen to us and the rest of the biosphere? Bearing in mind most people's anthropocentric view, those scientists who wish to preserve the biosphere much as it is will have to be prepared to argue their case not on materialistic but on aesthetic, even religious grounds, taking as a starting point their awe of the natural world.

A. HALLAM

Phil. Trans. R. Soc. Lond. B **325**, 241–252 (1989)   241

Printed in Great Britain

The causes of extinction

By J. MAYNARD SMITH, F.R.S.

School of Biological Sciences, University of Sussex, Falmer, Brighton BN1 9QG, U.K.

A species may go extinct either because it is unable to evolve rapidly enough to meet changing circumstances, or because its niche disappears and no capacity for rapid evolution could have saved it. Although recent extinctions can usually be interpreted as resulting from niche disappearance, the taxonomic distribution of parthenogens suggests that inability to evolve may also be important.

A second distinction is between physical and biotic causes of extinction. Fossil evidence for constant taxonomic diversity, combined with species turnover, implies that biotic factors have been important. A similar conclusion emerges from studies of recent introductions of predators, competitors and parasites into new areas.

The term 'species selection' should be confined to cases in which the outcome of selection is determined by properties of the population as a whole, rather than of individuals. The process has been of only trivial importance in producing complex adaptations, but of major importance in determining the distribution of different types of organisms.

An adequate interpretation of the fossil record requires a theory of the coevolution of many interacting species. Such a theory is at present lacking, but various approaches to it are discussed.

1. PALAEONTOLOGY, ECOLOGY AND GENETICS

The relation between population genetics and palaeontology is unsatisfactory. It is not uncommon for palaeontologists to assert that population genetics cannot account for the fossil record, whereas population geneticists hold that there is nothing in the record that they cannot explain. Much misunderstanding would be avoided if it were acknowledged that there is a sense in which both these statements are correct. Consider the matter first from the point of view of a geneticist. Whether the pattern of evolution is punctuational or gradualistic, it is compatible with genetic theory. Stasis can be explained by stabilizing selection, and punctuational change by directional selection; even if the changes are rapid by geological standards, they are slow compared to the changes that occur under artificial selection, and occasionally in the wild. There is no need to invoke Goldschmidtian systemic mutations to account for punctuation, or developmental constraints to account for stasis.

Such a reply, quite properly, would not satisfy a palaeontologist, who wants to know why evolution shows the patterns that it does. Why should there sometimes be normalizing selection for millions of years? What causes punctuational events? The answer to such questions cannot be supplied by population genetics, whose theories typically take the form 'IF the following genotypes have the following fitnesses, THEN the population will change in the following way'. It is in this sense that palaeontologists are correct to claim that population genetics cannot account for evolution.

How can this gap between theory and observation be bridged? Anatomists tend to turn to development. I think that this can provide only partial answers. It is certainly true that

[1]   23-2

developmental constraints may determine the way in which a population responds to selection. For example, Vrba (1984*a*) points out that herbivorous primates respond to hard-vegetation diets by evolving thickened tooth enamel, whereas perissodactyls and artiodactyls, faced with similar environments, have repeatedly evolved extreme hypsodonty. In some cases, developmental constraints may altogether preclude a taxon from evolving in a particular direction. For example, mammals seem to be unable to evolve parthenogenesis, perhaps because of differential gene imprinting from the two parents.

What we most need, however, is a theory that says something about the nature of the selective forces operating. Experimental evidence shows that most populations will respond rapidly to directional selection for most characteristics. Although many experiments in artificial selection lead rather rapidly to a plateau, beyond which further response is difficult, it seems that this happens because experimental populations are usually small, and therefore soon run out of genetic variance. Experience with domesticated plants and animals, and experiments (see, for example, Yoo 1980), show that long-continued responses are possible in larger populations. We therefore need a theory of selective forces, which can only come from ecology. For most animals, the major selective forces come from their competitors, their predators or prey, and their parasites.

Unfortunately, we lack a satisfactory theory of the coevolution of many interacting species. I return to this problem in the last section, but, to illustrate the point, consider two very general questions to which theory cannot at present provide an answer. First, imagine a set of interacting species in a physical environment that does not change in time. Each species may nevertheless evolve; for example, to escape from its parasites or to become better at catching its prey. As each species evolves, this will constitute a change in the biotic environment of at least some other species. Hence there will ensue an evolutionary dance, called by Van Valen (1973) the 'Red Queen', in which each species evolves because the others do. The unanswered question is this: will such a dance continue indefinitely, even in an unchanging physical environment, or will it slow down and stop? More realistically, what features would favour continued evolution, and what stasis? A second unanswered question is this. Does the continued existence of many species of plants, herbivores, carnivores, etc., require a physical environment that varies in space and/or time, or is it possible in a uniform environment? Again, more realistically, what features would favour species diversity, and what features would favour the elimination of all but one species at each trophic level?

Of course, most evolutionary biologists have some idea of the answer to these questions. But our ideas depend more on casual observation and guesswork than they do on theory, and the ideas of different biologists do not coincide. In this paper I shall attack the gap between population genetics theory and palaeontological observation from two sides. First, I shall discuss some qualitative questions that arise from the fossil record, and ask how ecological considerations might help to answer them. Second, I shall discuss the state of ecological theory, and suggest how it might be developed to be of greater relevance to macroevolutionary problems.

2. THE CAUSES OF EXTINCTION

(a) *Lack of evolvability, or running out of niche?*

If a species goes extinct, this is probably caused by a change in its circumstances, biotic or physical. (I say 'probably' because a rare species might go extinct by chance, without any change in its environment; but there remains the question why it was rare in the first place.) One question we can ask is the following: would the species have survived if it had been able to evolve more rapidly? Sometimes the answer is clearly no. If a species died out because its members could not survive the immediate consequences of a meteorite impact, or because the island to which it was confined sank beneath the sea, no capacity for rapid evolution would have saved it. Such unavoidable extinctions can be caused by biotic changes also: an example is G. C. Williams's (imaginary) flea confined to the passenger pigeon. We can say that such a species 'runs out of niche'.

Are all extinctions unavoidable in this sense? One body of evidence suggests that they are not. When discussing the forces responsible for the maintenance of sexual reproduction (Maynard Smith 1978), I pointed out that the taxonomic distribution of parthenogens suggests that parthenogenetic populations (I prefer not to use the term 'species' in the absence of sex) are short-lived in evolutionary time. With few exceptions, there are no high-ranking taxa (genera, families) consisting solely of parthenogens; instead, almost all parthenogens have close sexual relatives. In contrast, the haplodiploid sexual system has arisen rather seldom, but has given rise to large taxonomic groups (e.g. the hymenoptera, the monogonont rotifers). A plausible explanation (although not the only possible one) for the short evolutionary lifespan of parthenogenetic populations is that such populations evolve more slowly to meet changed circumstances than do sexual ones. If so, an inability to evolve rapidly predisposes such populations to extinction.

There is no reason why this method should not be used to identify other characteristics that predispose species to extinction. For example, I have the impression (not checked by serious analysis) that, in flowering plants, dioecy has a taxonomic distribution similar to, but far less extreme than, parthenogenesis.

(b) Physical or biotic causes of extinction?

Are the changes causing extinctions typically biotic or abiotic? I am not competent to discuss the relative importance of meteorite impacts, volcanic activity, climatic change and changes of sea level. Certainly I do not wish to argue that such events have been unimportant. But I think that palaeontologists should bear in mind that biological events, acting alone or in conjunction with physical change, may have been of equal importance. Indeed, I see no reason why extinctions – even mass extinctions – should not take place in the absence of any change in the physical environment, although it would be hard to account for a mass-extinction event affecting both terrestrial and marine organisms in this way.

A number of authors have reported a surprisingly constant diversity (as measured by numbers of genera or families) within particular taxa over many millions of years, despite continued extinction and speciation (see, for example, Webb (1969) for Cainozoic land mammals in North America and Boucot (1978) for Palaeozoic brachiopods). This implies that either speciation rate or extinction rate, or both, are functions of standing diversity: by analogy, constant population size implies density-dependent birth or death rates (Rosenzweig 1975). This suggests that biotic factors are important in extinction: it does not amount to proof, because regulation may be via speciation rate. In some cases, it is possible to follow changes in diversity resulting from identifiable environmental change. Boucot (1978) reports that in the Silurian, when only a single marine province is recognized, there were 90 genera of articulate brachiopods: this number rose to 350 in the lower Devonian, when there were six marine provinces, and fell again to 93 genera in the Mid-Devonian when there was a reduction in provinciality. These changes can be ascribed to physical changes (i.e. in provinciality), but they were presumably mediated by competition. Marshall *et al.* (1982) analyse the changes in mammalian diversity resulting from the union of North and South America 3 Ma BP. Before the join, there was a rather constant extinction rate of 0.3 genera per genus per million years in the north, and of 0.4 genera per genus per million years in the south. At the time of the junction, the proportions of southern mammals moving north and of northern mammals moving south were approximately equal, although the effect on the standing diversity was greater in South America, which had a smaller number of genera before the junction. In fact, the diversity increased by 50 % in the south, and this was followed by a 70 % increase in extinction rate. All these observations are consistent with the hypothesis that there was an equilibrium between extinction and speciation, and hence an important role for biotic factors. The final diversity in South America was greater after the junction, but this was caused by speciation of Northern rodent species after they had moved south.

Figure 1 shows an example of long-term constancy in plant diversity. There is a contrast between the approximate constancy before the origin of the angiosperms, and the steady increase since. Knoll (1986) suggests that this can be explained by the fact that rare animal-pollinated plants can persist, but rare wind-pollinated plants cannot. This is plausible, and confirmed by recent experience in plant conservation, but leaves unexplained a similar change from approximate constancy to steady increase in marine invertebrates in the Cainozoic (Sepkoski 1984).

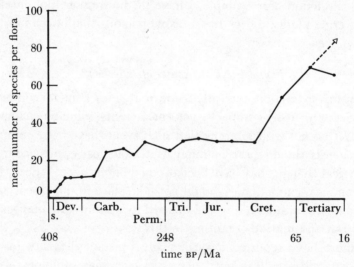

FIGURE 1. Mean number of species per flora during the Phanerozoic. The broken line is an estimate for Neogene floras in moist subtropical to tropical environments, similar to those in which the earlier floras are thought to have grown. (After Knoll 1986.)

What can we learn by looking at recent extinctions? Some, such as that of the passenger pigeon, and perhaps that of the large mammals of North and South America, were the direct result of human overkill. Others, such as the extinction of the large flightless birds of New Zealand, resulted from human overkill combined with destruction of habitat; in that case the burning of forests. In many cases, habitat destruction is probably by itself a sufficient explanation. Typically, habitat destruction results from human actions to which there is no obvious analogue in the past. A change in climate might destroy a particular habitat in one place, but the relevant species have often shifted to higher or lower latitudes or elevations, rather than disappeared. In some cases, however, barriers to migration have prevented such a shift, and climatic change has led to habitat destruction globally, not just locally. Also, climatic change can produce large changes in the total area covered by particular habitat types, and this would result in changes in total diversity. An alternative cause of habitat destruction at the present time is the introduction of foreign species; for example, the native vegetation of New Zealand has been replaced over large areas by the introduced gorse *Ulex*. Comparable habitat destruction may have been caused in the past by plant species that arrived from elsewhere, or which evolved *in situ*. It is illuminating to review the extinctions that have resulted from the introduction, accidental or otherwise, of species into new areas: in particular, onto islands, or, for aquatic organisms, into lakes or rivers. But in comparing these data with those of palaeontology, it is as well to remember that they concern mainly terrestrial and fresh-water vertebrates, whereas the fossil data concern mainly marine invertebrates.

Diamond (1984) has argued convincingly that introduced predators have been one of the commonest causes of extinction on islands. Although human predation was the major cause of the extinction of the Moas, many smaller New Zealand birds succumbed to the rats that the humans brought with them. On islands lacking native rats (e.g. Hawaii, Midway, Lord Howe), introduced rats have quickly eliminated native birds. In contrast, introduced rats have had little effect on the birds of islands (e.g. Solomons, Christmas Island, Galapagos) that had native rats. On some islands (e.g. Tonga, Samoa, Marquesas, Aldabra) with no native rats, but with land crabs, native birds have survived the introduction of rats. Thus a taxonomic group can be driven to extinction by a predator of an ecologically unfamiliar type. Other examples are the extinction of many endemic birds on Guam by the brown tree snake, and (particularly infuriating for population geneticists) the extinction of the native snail, *Partula*, on Moorea by a carnivorous snail, introduced to control a herbivorous snail that had itself been introduced to give the natives something to eat. The introduction as a game fish of the peacock bass, *Cichla ocularis*, into Lake Gatun in Panama led to the elimination of eight of the 11 commonest native fish, and had secondary effects on the abundance of other fish species, of the zooplankton, and of herons, terns and kingfishers (Zaret & Paine 1973).

It is harder to find clear examples of extinction caused by the introduction of a competitor. For freshwater fish, there is ample evidence (Werner 1986) that the introduction of a new species can drastically reduce the numbers and growth rate of competitors, and may restrict them to part of their previous ecological range, but outright extinction seems to be unusual. The work of Moulton & Pimm (1986) on introduced birds on the Hawaiian islands is illuminating in this context. They record 18 cases in which closely related species were introduced onto the same island. In three cases both are extinct, in nine cases one is extinct, and in six cases neither is yet extinct. In those cases in which one species is extinct, the difference in beak size was significantly less than in cases in which both survive (a mean of 9 % in the former, and 22 % in the latter). This suggests that competitive exclusion does occur, but requires very close similarity in resource utilization. Unless two species are very similar, they are likely to share the available resources. This is in line with the conclusion (Lawton 1982) that, in herbivorous insects, many possible ecological niches are vacant in most communities. However, this conclusion should be treated with caution. In particular, it is dangerous to reason as I have just done, jumping from fish to birds, and thence to insects. The role of

[5]

competition may be different in different taxa, and at different trophic levels. In particular, competition may play an important role in plant extinction.

A cause of extinction that would be particularly difficult to detect in the fossil record, and hard enough in recent communities, is disease. Rinderpest was introduced into Africa towards the end of the last century by Indian bullocks used in Kitchener's army to pull guns. It decimated the native ungulates, and caused the extinction of one of them, Swayne's Hartebeest. In Hawaii, the endemic birds are now confined to altitudes above 3000 feet† by avian malaria, brought by resistant introduced bird species. These examples show that extinction can be caused by disease, but it will be difficult to decide how frequent such events have been.

The contemporary evidence, then, shows that extinctions can be caused by single introduced predators, competitors or parasites. Is there evidence from the fossil record for comparable biologically caused extinctions? Perhaps the strongest evidence that competition plays an important role in determining either extinction rates or speciation rates, or both, comes from the observation of constant taxonomic diversity despite species turnover, discussed above. What of the evidence for the competitive displacement of one taxon by another in the fossil record; for example, of brachiopods by bivalves, of cynodonts by thecodonts, or of multituberculates by rodents? Confronted by data of the kind illustrated in figure 2, it is natural to suppose that taxon A has been competitively displaced by taxon B, and to seek some adaptive reason for the displacement. In this case any such attempt would be misguided, because the figure has been drawn to represent data generated by a computer program in which the probabilities of speciation and extinction were constant, subject to the constraint that a constant total diversity be maintained. This illustrates the point, first made by Raup *et al.* (1973), that it is fatally easy to read a pattern into stochastically generated data.

The first task, then, is to provide convincing evidence that a pattern really exists. This is not easy. If the available data consist of numbers of genera (or of species, or families) at different time horizons, the first requirement is to subdivide the data by geographical region and ecological function, and not merely by taxonomic criteria. This point is nicely illustrated by Thayer's (1983) criticism of an argument by Gould & Calloway (1980). The latter had argued that there is no evidence that the rise of the bivalves had caused the decline of the brachiopods: the two groups were 'ships that pass in the night'. Their reason was that there is a positive correlation between the levels of diversity of the two groups at different horizons, whereas if competition caused displacement the correlation should be negative. Thayer argued that the observed correlation could arise from factors increasing, or decreasing, taxonomic richness in general, and that, if functional groups are taken into account, the data are more in accord with the competitive hypothesis. For example, a decline in pedunculate brachiopods has been accompanied by a rise in epibyssate bivalves; a comparison of groups living on hard substrates. Although Thayer's point is a good one, however, I have the impression that a more detailed statistical analysis of such cases is needed before firm conclusions are drawn.

A first and difficult task, then, is to show that a pattern calling for a causal explanation really exists. Even if it does, there remain two possible explanations: competitive displacement or vacancy occupation. According to the former, taxon A went extinct because of competition from taxon B: according to the latter, taxon A went extinct for reasons that had little or nothing to do with taxon B, and the latter subsequently evolved to fill the vacant niches. Note that both

† 1 foot ≈ 30.48 cm.

FIGURE 2. A phylogeny in which the horizontal width of a taxon at any time is approximately equal to the number of genera.

explanations imply competition; vacancy occupation supposes that taxon B was prevented by competition from evolving into niches occupied by A. Thus the conventional explanation of the extinction of the dinosaurs and the subsequent radiation of the mammals assumes that the former event owed nothing to the latter, but that the latter was made possible by the former.

The evidence of recent extinctions, suggesting that habitat destruction, predators, and perhaps parasites, are more effective agents of extinction than competitors, might seem to suggest that competitive displacement of one taxon by another will be rare. I am not sure we should take this argument seriously. There is plenty of evidence that competition can reduce the abundance and ecological range of a species, and in the long run this must make extinction more likely. One palaeontologist willing to argue for competitive displacement is Knoll (1986). During the Silurian and Devonian, a series of vascular spore-bearing taxa replaced one another. Each taxon appeared first as a minor element in the community, expanded to become dominant, and then dwindled as a new taxon appeared. Knoll argues that the morphological characteristics of the successive dominants would ensure increased capacity for gathering light and absorbing water and nutrients, and so could account for competitive displacement. He can find no evidence for mass extinctions followed by rediversification, or for a significant role for climatic change or of herbivory.

Is there evidence that predators have caused the extinction of major taxonomic groups? One example is worth describing in a little detail, although it should perhaps not come under the heading of predation. Thayer (1983) argues that there has been a steady increase in the abundance and variety of animals that disturb marine mud and sand deposits, for example by swallowing it and digesting any organisms it contains; examples are holothurians, malacostracans, tellinid bivalves, and polychaets such as *Arenicola*. A consequence of the activities of these 'bulldozers', not all of which are predators, is to make life difficult for other organisms, and in particular for immobile filter feeders living on soft substrates. For example, bulldozers may make larval settlement impossible, either by swallowing or accidentally burying the larvae. As a consequence, there has been a decline in the number and variety of immobile organisms on soft substrates; for example, free-lying brachiopods, tabulate and rugose corals, and blastoid echinoderms. Modern representatives of these taxa either live on hard substrates or are themselves mobile.

One interesting feature is that these immobile groups suffered particularly severely in the Permo-Triassic mass extinction. This is a bit puzzling. If Thayer is right, they were on their way out anyway. But why should they have been particularly susceptible to a mass-extinction event? It is reasonable that organisms with different ecologies should be differentially susceptible to a major physical disaster. For example, if the Cretaceous extinction was caused by a meteorite impact, it is reasonable that organisms in the water column should have been more susceptible than benthic forms, and that direct herbivores should have been more susceptible than deposit feeders. But why should organisms that were in decline anyway be particularly susceptible? Of course, if the 'mass extinction' was merely a more intense continuation of processes that were already occurring, the coincidence is no longer puzzling. Nor is there any convincing evidence that different mass extinctions had similar causes. Jablonski (1986 a, b) has reviewed the similarities and differences between the patterns of disappearance during mass extinctions and at other times.

3. The role of species selection

Several authors (Maynard Smith 1983; Vrba & Eldredge 1984; Vrba & Gould 1986) have drawn a distinction between two processes that can lead to the extinction of some species and the survival of others. The distinction is best explained by a hypothetical example. Suppose that two species of lizard, A and B, differ in that A individuals are better at catching insects, and that in consequence species A survives and B goes extinct. Contrast this with a second case in which species B is parthenogenetic; as the biotic environment changes, species A evolves to meet the change, but species B does not, and in consequence goes extinct. In the first of these cases, the 'performance' (Emerson & Arnold 1989) that determines survival is the performance of individuals: in the second, it is the performance of the species as a whole. Thus individuals catch flies, but populations evolve. All the authors listed above have argued that only the second case should be referred to as 'species selection'. According to this view, the criterion for species selection is that the performance that determines survival is one that could be measured only on the species as a whole, and not on an individual (although the performance, rate of evolution, depends on a characteristic of individuals, namely sexual reproduction).

Species selection, in this narrow sense, can also affect traits that influence speciation rate. For example, if marine invertebrates with direct development speciate more frequently than those with planktotrophic larvae, this could lead to an increase in the proportion of species with direct development (Jablonski 1986c). In such a case, however, it is well to remember that an increase in number of species does not necessarily imply an increase in number of individuals (Vrba 1984b), and that the increased speciation rate may be balanced by increased extinction.

Some taxa may diversify more extensively than others because they have a developmental system that favours diversification. (For a recent discussion of this idea, see the papers by Dawkins, Liem and Vrba in Wake & Roth (1989).) One cannot think of this as a case of individual selection because individuals do not diversify. However, I would prefer the term 'lineage selection' to 'species selection' for such cases, because the process does not require that the lineages be sexual.

Vrba & Eldredge (1984) used the term 'species sorting' to refer to all cases of differential survival and extinction of species, however caused. How important has species sorting been in evolution? I think that it has been of only trivial importance in giving rise to the complex and integrated adaptations that are characteristic of all living organisms, but that it has been of major importance in determining the abundance and distribution of organisms with different structures and adaptations. The reason why it is of trivial importance in giving rise to adaptations is a quantitative one. It is useful to quantify selection in terms of the bits of information that a given selective process can add to the genome. Thus one bit can be added by the selective removal of half the population, two bits by the removal of three quarters of the population, and so on. On this measure, five mass extinctions, each leading to the loss of 97 % of the extant species, could add at most 25 bits; equivalent to culling half a population selectively for 25 generations. Thus species sorting, whether the target of selection is the individual or the population, is not a relevant process in the generation of individual adaptation. But as a process influencing the abundance and distribution of organismal types it is crucial. I suspect that most species sorting is a consequence of individual rather than species selection. However, the individual characteristics that prevail need not be those that determined survival. To take a hypothetical example, suppose that the characteristics that enabled many mammalian species to survive the end-Cretaceous extinction were small size and the habit of eating seeds or insects. It would be a consequence of this that the characteristic mammalian gait and jaw mechanism replaced those characteristic of the dinosaurs, but their survival would not be evidence of their selective superiority. It think that much of the disagreement between palaeontologists and neontologists on the role and importance of species sorting has arisen because they want to explain different phenomena. A palaeontologist sees dramatic changes in the kinds of organisms that inhabit the Earth, but is often blind to adaptation; a neontologist sees function everywhere.

4. MODELS OF COEVOLUTION

Interpretation of the fossil record would be greatly helped by a theory of the coevolution of many interacting species. Unfortunately, no satisfactory theory exists; indeed, no fully satisfactory theory exists for the stability of ecosystems composed of non-evolving species. It may be that different systems – for example, marine and terrestrial, tropical and temperate – require fundamentally different models. Nevertheless, I now present a brief and contentious account of where we stand. It is convenient to start with the still widespread idea that the stability of an ecosystem (in the sense of approximate constancy of species composition and relative numbers) is a consequence of the large number of species and the complexity of their interactions. Elton (1958) argued for this view, pointing to the relative instability of species-poor subarctic (and to a still greater extent, agricultural) systems. This idea may be correct, but stability is not a necessary mathematical consequence of species richness and complex interactions. Indeed, May (1972) showed that precisely the opposite is the case: in model systems, based on Lotka–Volterra dynamics with randomly chosen parameters, stability becomes less likely as the number of species, and the number of interactions per species, increases. The conclusion from May's work is that complexity itself is not sufficient. One or both of two additional features must be incorporated: the parameters are not random, but are influenced by selection, and there may be structural features common to all ecosystems that favour stability.

A general structural feature of ecosystems is that they are composed of plants, herbivores, carnivores and decomposers (although some species may play more than one role, and a few systems do not depend on plants). Three important suggestions have been made about this structure. Hairston *et al.* (1960) argued that, at least in terrestrial systems, herbivores are usually limited by their predators and parasites, so that they are unable to reduce the level of plant cover substantially below that determined by the resources of light, water and nutrients. If this is correct, it would explain why botanists have been more successful than zoologists in detecting competitive replacement in the fossil record. However, Connell (1970) and Janzen (1970), in a proposed explanation for the diversity of tropical forests, assumed that herbivores are more successful in controlling plants. They proposed that no one tree species can become abundant, because it is kept rare by specialist herbivores and parasites: the resulting tree diversity then supports diversity at higher levels. This model would explain increasing taxonomic diversity by the increasing stenophagy of herbivores, and host specificity and virulence of plant pathogens. A third general point (Pimm 1982) is that food chains will not lengthen indefinitely, because long food chains are dynamically unstable. It follows that any substantial increase in diversity must occur by increasing the number of food chains, and not their length.

A second necessary extension of May's model is to suppose that parameters are selected and not random. There are two ways in which non-randomness can arise in the interactions between species in an ecosystem. First, consider the species on an island not too far distant from the mainland. Those that are actually present are the survivors of the probably much larger number of immigrants. Thus the existence of a stable ecosystem on an island does not prove that any arbitrary set of species will constitute a stable system, but only that some subset of the total set of immigrants will do so. In fact, stability is harder to explain than this might suggest, because, as MacArthur & Wilson (1967) pointed out, the number of species on islands commonly remains rather constant despite continuing turnover of actual composition, owing to immigration and extinction.

What of species on the mainland? The interactions of these species are non-random because they have evolved together. Is there any theoretical reason to expect that coevolution should lead to greater stability? We can answer this question only in part, and only for pairs of interacting species (for reviews see Roughgarden (1983); Slatkin & Maynard Smith (1979)). For competitive interactions, theory does suggest that the typical result will be that competitors will specialize on different resources ('character displacement'), and hence that coexistence will become more stable, although the empirical support for this prediction is by no means overwhelming. For symbiotic interactions, Law & Koptur (1986) have argued that evolution has led to stability, and to low taxonomic diversity of endosymbionts. The real difficulty arises for exploitative interactions: predation, herbivory and parasitism. Here, evolution seems likely to lead to continuing 'arms races', and recurrent extinctions (for a contrary view, see Schaffer & Rosenzweig (1978)).

These models deal only with pairwise interactions. Yet Futuyma (1983) has argued that long-term coevolution of pairs of species is the exception, and that most biotic interactions of evolutionary significance are diffuse and short-lived. His reason is as follows. If, for example, the interaction between a plant species and a herbivorous insect was stable and long-lasting, we would expect to find that the phylogeny of a plant family, and of the herbivorous insects feeding on it, would map closely onto one another. In general this is not what is observed, although it is true that particular insect families often attack one or a few plant families. If,

[10]

as Futuyma thinks, coevolution is usually diffuse rather than pairwise, we need a theory of the coevolution of many species. Is there anything we can usefully say? Van Valen (1973) proposed the 'Red Queen' model, according to which an evolutionary improvement by one species would be experienced as an environmental deterioration by others, thus leading to a continuing evolutionary dance. He went further, arguing for a 'zero-sum' assumption, according to which a unit improvement in one species is exactly balanced by the summed deteriorations for others, and deduced from this a 'law of constant extinction'. Stenseth & Maynard Smith (1984) accepted his general model, but argued that his zero-sum assumption rested on a confusion between ecological and evolutionary advantage, and was mistaken. They concluded that an ecosystem could behave in one of two ways:

(i) 'Stasis': evolution would slow down and cease, until kicked into motion by changes in the physical environment; or

(ii) 'Red Queen' dynamics: evolution, even in a constant physical environment, would continue indefinitely, with recurrent speciation and extinction.

The obvious weakness of this theory is that it predicts nothing testable. We need a model of coevolution that includes more ecological structure. If May's model leaves out evolution, our model leaves out ecology. We need models which would help to answer such questions as the following.

(i) Which ecological factors favour stasis, and which favour 'Red Queen' dynamics?

(ii) In what circumstances will ecological diversity evolve and be stable? Are spatial and/or temporal diversity necessary?

(iii) Can apparently stable ecosystems collapse because of the invasion of novel species? If so, how different must the invader be from any existing species?

REFERENCES

Boucot, A. J. 1978 Community evolution and rates of cladogenesis. *Evol. Biol.* **11**, 545–655.

Connell, J. H. 1970 On the role of natural enemies in preventing competitive exclusion in some marine animals and in rain forest trees. In *Proc. Adv. Study Inst. Dynamics Popul. (Oosterbeck)* (ed. P. J. den Boer & G. R. Gradwell), pp. 298–312.

Diamond, J. M. 1984 Historic extinctions: a Rosetta Stone for understanding prehistoric extinctions. In *Quaternary extinctions: a prehistoric revolution* (ed. P. S. Martin & R. G. Klein), pp. 824–862. Tucson: University of Arizona Press.

Emerson, S. B. & Arnold, S. J. 1989 Intra- and interspecific relationships between morphology, performance and fitness. In *Complex organismal functions: integration and evolution in vertebrates* (ed. D. B. Wake & G. Roth), pp. 293–314.

Elton, C. S. 1958 *The ecology of invasion by animals and plants.* London: Methuen.

Futuyma, D. J. 1983 Evolutionary interactions among herbivorous insects and plants. In *Coevolution* (ed. D. J. Futuyma & M. Slatkin), pp. 207–231. Sunderland, Massachusetts: Sinauer.

Gould, J. J. & Calloway, C. B. 1980 Clams and brachiopods – ships that pass in the night. *Paleobiology* **6**, 383–396.

Hairston, N., Smith, F. & Slobodkin, L. 1960 Community structure, population control and competition. *Am. Nat.* **94**, 421–425.

Jablonski, D. 1986a Evolutionary consequences of mass extinctions. In *Patterns and processes in the evolution of life* (ed. D. M. Raup & D. Jablonski), pp. 313–329. Berlin: Springer-Verlag.

Jablonski, D. 1986b Mass and background extinctions: the alternation of macro evolutionary regimes. *Science, Wash.* **231**, 129–133.

Jablonski, D. 1986c Larval ecology and macroevolution in marine invertebrates. *Bull. marine Sci.* **39**, 565–587.

Janzen, D. H. 1970 Herbivores and the number of tree species in tropical forests. *Am. Nat.* **104**, 501–528.

Knoll, A. H. 1986 Patterns of change in plant communities through geological time. In *Community ecology* (ed. J. Diamond & T. J. Case), pp. 126–141. New York: Harper & Row.

Law, R. & Koptur, S. 1986 On the evolution of non-specific mutualism. *Biol. J. Linn. Soc.* **27**, 251–267.

Lawton, J. H. 1982 Vacant niches and unsaturated communities: a comparison of bracken herbivores at sites on two continents. *J. anim. Ecol.* **51**, 573–595.

MacArthur, R. H. & Wilson, E. O. 1967 *The theory of island biogeography*. Princeton University Press.

Marshall, L. G., Webb, S. D., Sepkoski, J. J. & Raup, D. M. 1982 Mammalian evolution and the great American interchange. *Science, Wash.* **215**, 1351–1357.

May, R. M. 1972 Will a large complex system be stable? *Nature, Lond.* **238**, 413–414.

Maynard Smith, J. 1978 *The evolution of sex*. Cambridge University Press.

Maynard Smith, J. 1983 Current controversies in evolutionary biology. In *Dimensions of Darwinism* (ed. M. Grene), pp. 273–286. Cambridge University Press.

Moulton, M. P. & Pimm, S. L. 1986 The extent of competition in shaping an introduced avifauna. In *Community ecology* (ed. J. Diamond & T. J. Case), pp. 80–97. New York: Harper & Row.

Pimm, S. L. 1982 *Food webs*. London: Chapman & Hall.

Raup, D. M., Gould, S. J., Schopf, T. J. M. & Simberloff, D. J. 1973 Stochastic models of phylogeny and the evolution of diversity. *J. Geol.* **81**, 525–542.

Rosenzweig, M. L. 1975 On continental steady states of species diversity. In *Ecology and evolution of communities* (ed. M. L. Cody & J. M. Diamond), pp. 121–140. Cambridge, Massachusetts: Harvard University Press.

Roughgarden, J. 1983 The theory of coevolution. In *Coevolution* (ed. D. J. Futuyma & M. Slatkin), pp. 33–64. Sunderland, Massachusetts: Sinauer.

Sepkoski, J. J. Jr 1984 A kinetic model of Phanerozoic taxonomic diversity. III. Post-palaeozoic families and mass extinctions. *Paleobiology* **10**, 246–267.

Schaffer, W. M. & Rosenzweig, M. L. 1978 Homage to the red queen. I. Coevolution of predators and their victims. *Theor. Popul. Biol.* **14**, 135–157.

Slatkin, M. & Maynard Smith, J. 1979 Models of coevolution. *Q. Rev. Biol.* **54**, 233–263.

Stenseth, N. C. & Maynard Smith, J. 1984 Coevolution in ecosystems: red queen evolution or stasis? *Evolution* **38**, 870–880.

Thayer, C. W. 1983 Sediment-mediated biological disturbance and the evolution of marine benthos. In *Biological interactions in recent and fossil benthic communities* (ed. M. J. S. Teresj & P. L. McCall), pp. 479–625. New York: Plenum.

Van Valen, L. 1973 A new evolutionary law. *Evol. Theory* **1**, 1–30.

Vrba, E. S. 1984*a* Patterns in the fossil record and evolutionary processes. In *Beyond neo-Darwinism* (ed. M. W. Ho & P. S. Saunders), pp. 115–142. London: Academic Press.

Vrba, E. S. 1984*b* What is species selection? *Syst. Zool.* **33**, 318–328.

Vrba, E. S. & Eldredge, N. 1984 Individuals, hierarchies and processes: towards a more complete evolutionary theory. *Paleobiology* **10**, 146–171.

Vrba, E. S. & Gould, S. J. 1986 The hierarchical expansion of sorting and selection: sorting and selection cannot be equated. *Paleobiology* **12**, 217–228.

Wake, D. B. & Roth, G. 1989 *Complex organismic functions: integration and evolution in vertebrates*. Chichester: John Wiley.

Webb, S. D. 1969 Extinction–origination equilibria in late cenozoic land mammals of North America. *Evolution* **23**, 688–702.

Werner, E. E. 1986 Species interactions in freshwater fish communities. In *Community ecology* (ed. J. Diamond & T. J. Case), pp. 344–358. New York: Harper & Row.

Yoo, B. H. 1980 Long-term selection for a quantitative character in large replicate populations of *Drosophila melanogaster*. I. Response to selection. *Genet. Res.* **35**, 1–17.

Zaret, T. M. & Paine, R. T. 1973 Species introduction in a tropical lake. *Science, Wash.* **182**, 449–455.

Phil. Trans. R. Soc. Lond. B, **325**, 253–261 (1989)

Printed in Great Britain

What, if anything, are mass extinctions?

By A. Hoffman

*Institute of Paleobiology, Polish Academy of Sciences, Al. Zwirki i Wigury 93,
PL-02-089 Warsaw, Poland*

Many phenomena that have traditionally been called 'mass extinctions' are in fact clusters of extinction episodes roughly associated in geological time. This is the case with the latest Ordovician, late Devonian, mid-Cretaceous, latest Cretaceous and Late Eocene–Oligocene extinctions. Several of these clusters are caused, each episode by a different causal factor. Such mass extinctions are then due to the coincidence of various processes in the environment, and they can hardly be considered as individual events. The latest Permian mass extinction, however, is caused by a single process that affected the global ocean–atmosphere system. In the late Permian, the world ocean was full of deposits rich in organic matter, which enhanced nutrient recycling. After oxygen was brought to the sea floor (by whatever process), nutrients began to sink to the sea-bottom, and the resulting nutrient deficiency must have caused mass extinction in the sea. Oxidation of huge amounts of organic matter and associated sediments at the sea bottom must have drawn oxygen from the atmosphere, and the resulting fall in atmospheric oxygen must have contributed to extinctions on land.

1. A historical perspective

'Mass extinctions' is the term traditionally used in geology and palaeontology to denote those relatively short intervals of geological time when rather large and diverse segments of the world's biota underwent extinction. Historically, five mass extinctions have been distinguished: the latest Ordovician, late Devonian, latest Permian, latest Triassic and latest Cretaceous; but several other time intervals often are also interpreted as mass extinction phenomena, especially the Eocene–Oligocene (in the Tertiary) and Cenomanian–Turonian (in the mid-Cretaceous) transitions.

The problem of mass extinctions has only very recently (in the 1980's) become one of the most studied topics in historical geology and biology. The main factors for this are the spectacular and provocative hypotheses that, first, the latest Cretaceous mass extinction was caused by the impact of a huge bolide (Alvarez *et al.* 1980); second, that mass extinctions are periodic (Raup & Sepkoski 1984) and caused by comet showers triggered by an unseen solar companion (Davis *et al.* 1984; Whitmire & Jackson 1984); and third, that the biotic effects of mass extinctions are qualitatively different from all other phenomena in the history of life on Earth (Jablonski 1986a). All these exciting hypotheses have become the subject of intense debates and, perhaps more importantly, have stimulated much new and very productive empirical and theoretical research. The problem of mass extinctions, however, is not at all new, and the current controversies fit very well into a history that extends well back into the 19th century.

Ever since Georges Curvier had, in the early 19th century, observed several faunal breaks in the geological strata exposed in the Paris Basin, more or less dramatic changes of the fossil contents in the stratigraphic column were widely regarded as indications of mass extinctions.

This view was particularly developed by Alcide d'Orbigny, who established a whole time-series of such biotic catastrophes punctuating the history of life on Earth. On the other hand, Charles Lyell professed a more gradualist view of species extinction, which he interpreted as a result of individual species sooner or later encountering the conditions to which they could not adapt. In this perceptive, there was no room for mass-extinction phenomena. Charles Darwin adopted a more pluralist viewpoint, and although he suggested that the latest Permian and latest Cretaceous mass extinctions were in reality artefacts of large gaps in the fossil record, he also allowed for profound environmental changes leading to protracted but roughly simultaneous extinction of a large variety of species.

Lyell's view became untenable in the 20th century because the existence of mass-extinction phenomena in the fossil record could no longer be denied and demanded a causal explanation. The opposition was between the concept of each mass extinction being a real catastrophe brought upon life by whatever extraordinary environmental factor, and the notion of mass-extinction phenomena reflecting periods of considerably accelerated extinction rate, perhaps additionally amplified by some gaps in the record. Both these viewpoints were strongly advocated by several scientists. For example, Marshall (1928), Hennig (1932) and Krasovskiy & Shklovskiy (1957) regarded the latest Cretaceous mass extinction as geologically instantaneous and attributed it to a wave of cosmic radiation, whereas Pavlova (1924), Sobolev (1928) and Newell (1967) interpreted it as a prolonged side-effect of the diastrophic cycle. Similarly, Schindewolf (1954) considered the latest Permian mass extinction as an abrupt catastrophe caused perhaps by the explosion of a supernova, whereas Schopf (1974) viewed this biotic phenomenon as extending over several million years in parallel to major changes in geographical distribution of the continental plates.

Such opposition between the catastrophic and the more gradualist interpretations of mass-extinction phenomena has persisted until today, and many of the current debates on mass extinctions can indeed be presented in these terms (see reviews by Jablonski (1986b); Hoffman (1989a)). This is, however, only one set of contrasting viewpoints on mass extinctions. In a sense, these two interpretations resemble each other in that they both regard mass-extinction phenomena as each being caused by one single factor, most commonly, though not necessarily, all of them by the same factor. The causal agent invoked to explain mass extinctions may be an extraterrestrial impact, or an extraordinary volcanic eruption, or a major palaeo-geographical, oceanographic or climatic change, but it is always just one single driving force that is conceived as being the ultimate cause of all extinctions during the time interval interpreted as a mass extinction. I have recently proposed a contrasting perspective, however (Hoffman 1989a). I suggested that many mass-extinction phenomena in fact represent clusters of extinction episodes, each of them triggered by another casual factor or set of factors and accidentally aggregated within time periods of a few million years. This would imply that they are not natural phenomena at all, but rather mental constructs of the geologist or palaeontologist. In this view, then, there is no ultimate cause, no process of overwhelming physical environmental or biotic change responsible for many apparent mass extinctions, but merely coincidence on the geological timescale that makes them look like single events in the history of the biosphere.

2. MASS EXTINCTIONS AS CLUSTERS OF EVENTS

There are in fact several examples that clearly demonstrate that what has been traditionally considered as mass-extinction events are actually clusters of minor episodes of extinction. For the late Devonian extinctions, the stratigraphic data collected by Buggisch (1972) show that, although the stromatoporoid-coral reef structures disappeared with the onset of an anoxic régime (Kellwasser Limestone) in the late Frasnian, the pelagic biota survived this crisis and underwent extinction only a couple of conodont zones later. House (1985) demonstrated several extinction pulses among ammonoids. Farsan's (1986) detailed studies on continuous sections of strata deposited in the late Devonian indicate a series of pulses of extinction among shallow-water benthic animals that are spread throughout the entire Frasnian rather than limited solely to its terminal portion. Moreover, Stearn (1987) shows that although the late Devonian reef structures largely disappeared toward the end of the Frasnian, the extinction of reef-building stromatoporoids actually took place only during the early Famennian. The view that unrelated causes triggered two separate pulses of extinction in the late Frasnian is also strongly supported by Sandberg *et al.* (1988).

For the latest Ordovician extinctions, Brenchley (1984) observes that they almost certainly include two separate waves of extinction: one at the latest Caradocian – earliest Ashgillian transition when many trilobite and brachiopod taxa disappeared, and another one several million years later, during the very latest Ashgillian. This second wave of extinction, in turn, extended over a period of some two million years and encompassed several pulses that separately affected graptolites, trilobites and cystoids, brachiopods and rugose corals, once again brachiopods and corals, and finally (perhaps as late as in the earliest Silurian) conodonts.

Also well documented is the aggregate nature of extinctions at the Cenomanian–Turonian and Eocene–Oligocene transitions. For the mid-Cretaceous events, much – though perhaps not yet compelling – evidence has been presented by Kauffman (1984) and Elder (1987) based on very detailed studied on strata deposited in a broad epicontinental sea in North America. For the Eocene–Oligocene extinctions, evidence for a series of minor pulses extending over several million years comes from the open ocean (Corliss *et al.* 1984) as well as from shallow-water benthic biota (Hansen 1987) and terrestrial ecosystems (Prothero 1985).

By now, there is also little doubt that the latest Cretaceous extinctions were not confined to a single holocaust at the very end of the Cretaceous. The catastrophist scheme for this mass extinction holds, at the minimum, that the pelagic plankton and much of terrestrial flora and large vertebrate fauna underwent a tremendous and exactly simultaneous extinction. This claim is debatable, to say the least, because there is clearly a temporal structure to the plankton extinctions (Smit & Romein, 1985; Keller 1987; Lindinger & Keller 1987) and there is no compelling evidence for their simultaneity with the dinosaur extinction in North America (Fastovsky 1987; Rigby *et al.* 1987; Smit *et al.* 1987). Even taking this part of the scheme for granted, however, there is unequivocal evidence for an earlier extinction step that exterminated the rudistid reefs, many ammonites and other marine macroinvertebrates; two further pulses of extinction in the marine realm seem to have taken place still before the terminal Cretaceous episode (Kauffman 1986; Mount *et al.* 1986; Ward *et al.* 1986).

That many mass extinctions are series of extinction episodes does not, of course, prove the point that they are sets of separate events accidentally aggregated in time. Each series of episodes might also be caused by one agent, whose action was extended in time. This is indeed the concept advocated by Stanley (1984, 1988) who argues for global climatic cooling as the main cause of all mass extinctions. On the other hand, it is also put forth by Hut *et al.* (1987) who maintain that because (i) the mid-Cretaceous, latest Cretaceous, and Eocene–Oligocene extinctions have composite nature, (ii) evidence exists for several bolide impacts at the Eocene–Oligocene transition, and (iii) impact causation has been demonstrated for the terminal Cretaceous extinction episode, these three mass extinctions, and most probably also the others, are caused by comet showers. In my opinion, however, a more comprehensive analysis of historical geological data on events associated in time with some mass extinctions suggests coincidence of independent processes as a more, or at least equally, likely explanation.

There is indeed evidence for several impacts at the Eocene–Oligocene transition (microtektite fields, iridium anomalies), but their association with extinction episodes is at best tenuous. As clearly visible in the diagrams provided by Hut *et al.* (1987), some impacts are not associated with extinction and some extinction episodes are not associated with impact fingerprints. On the other hand, evidence for global climatic cooling appears fully convincing (Keller 1983; Stanely 1984; Prothero 1985; Hansen 1987). Moreover, extraordinary volcanic activity took place during this time interval (Kennett *et al.* 1985). There is no reason to believe that all these extraterrestrial, climatic and geodynamic events were causally related to each other rather than that they merely coincided within the same several-million-year-long time interval.

The case for the terminal Cretaceous impact is by now very strong. The inference made originally from the iridium anomaly and corroborated later by osmium isotopes, shocked quartz, and possible soot enrichment (see reviews by Jablonski (1986*b*); Hoffman (1989*a*)) is currently supported by new lines of evidence. Rhodium isotope data clearly indicate extraterrestrial, rather than crustal, origin of this element in the Cretaceous–Tertiary boundary clay (Bekov *et al.* 1988), cathodoluminescence analysis of the shocked quartz grains in this clay corroborates the postulate of their non-volcanic nature (Owen & Anders 1988), possible tsunami deposits occur as predicted at the iridium-enriched Cretaceous–Tertiary boundary horizon (Bourgeois *et al.* 1988), and the concentration of apparent soot is much greater than previously estimated and seems to reach a maximum exactly at the iridium peak (Wolbach *et al.* 1988). On the other hand, however, Deccan-trap volcanism – which has also been implicated as the ultimate cause of the terminal Cretaceous extinctions (Officer *et al.* 1987; Hallam 1987) – is now known to have antedated the impact and lasted no more than a million years (see Courtillot *et al.* 1988; Duncan & Pyle 1988). Thus it could not be caused by the impact but its environmental effects could contribute to the extinctions. Neither impact nor volcanic eruptions, however, could have been responsible for destruction of the rudistid reefs, which had begun a couple of million years earlier. The Cretaceous–Tertiary transition, moreover, is the time of a major sea-level fall (Hallam 1984; Haq *et al.* 1987). This process, again, does not seem to be causally related to either impact or volcanic activity.

Historical geological data are much worse for the other clusters of extinction episodes that are traditionally called mass extinctions, but the very existence of empirical evidence to support several different causal explanations in each case suggests that they also might represent coincidence on the geological timescale of a variety of processes. Such coincidence, in fact, is not at all extraordinary or unlikely. Consider the number of various types of physical environmental event that might lead to substantial, though not necessarily mass, extinctions. At least five types of event immediately come to the mind: large bolide impacts, sustained and unusually intense volcanic activity, global climatic change, major sea-level fluctuations, and oceanic anoxic events. Provided that events of each kind occur at random in geological time and with the average frequency of one event per 50 Ma – which seems to be a reasonable estimate – there is a greater than 50% chance that, within a period of 100 Ma, two or more different events will roughly coincide in time (that is, will occur during a 2 Ma time interval); and there is almost certainty that coincidence during a 4 Ma time interval will take place within a period of 250 Ma (Hoffman 1989 *b*).

I submit that many mass extinctions may actually be caused each by a variety of factors that by pure chance happened to roughly coincide within time intervals of 2–4 Ma. Should this hypothesis withstand the scrutiny of empirical testing, these mass extinctions could not be rightly called mass extinctions at all, for they would then represent artificial collections of causally independent phenomena instead of causally coherent events. The latest Cretaceous extinctions, for example, may perhaps include a mass-extinction phenomenon – if the terminal Cretaceous event is indeed attributable to one environmental process (but even this assertion is far from being firmly established) – the preceding steps of extinction, however, may only by chance appear related to this phenomenon. The Eocene–Oligocene extinctions may not include any mass extinction at all, but only a set of minor episodes of extinction.

3. A TRUE MASS EXTINCTION

This is not to say, however, that all the traditionally recognized mass extinctions comprise several independent but accidentally aggregated events. The latest Permian extinctions among marine animals that could be fossilized were undoubtedly the most severe in the Phanerozoic. Because of a huge marine regression and consequent extraordinarily intense erosion, however, the geological record at the Permo-Triassic transition is so poor that the temporal pattern of extinctions cannot be firmly established. This record nevertheless bears strong evidence for a change in the world ocean–atmosphere system, which appears to have been much more dramatic than any other in the Phanerozoic. If our current causal interpretation of this phenomenon is correct, then the latest Permian biotic change fully deserves to be called 'mass extinction' because the record seems to indicate a process which must have caused a true mass extinction.

The evidence comes from studies of stable carbon and oxygen isotopic changes in seawater composition during the latest Permian (Gruszczyński *et al.* 1989; Małkowski *et al.* 1989). Carbon isotopic composition (that is, the ratio between the light and the heavy isotopes of carbon, which is usually shown in 'delta' (δ) notation, as deviations from a certain standard) varies between major carbon reservoirs in the Earth's system: organic carbon in the biosphere and its products, carbonate sediments, the ocean, and the atmosphere. Therefore a change in the oceanic carbon isotopic composition must indicate either a change in fluxes between the reservoirs or a change in juvenile carbon input to the system controlled by degassing of the Earth's mantle. Because of the very rapid cycling of carbon both within and between the ocean and the atmosphere, major changes in stable carbon isotopic composition of seawater must be recorded everywhere in the ocean, unless we deal with a very extreme and unusual environment where some local processes overwhelm the global ones. As indicated by analysis of carbon isotopes in the calcite of productacean brachiopod shells from uppermost Permian outer shelf deposits of west Spitsbergen, which had at the time free and wide connections with the world ocean, seawater was relatively enriched in the heavy isotope of carbon during the late Permian (δ^{13}C of approximately $+4$‰). Subsequently, however, its carbon isotopic composition rapidly shifted towards still heavier values (δ^{13}C of more than $+7$‰) and then declined down to very light values (δ^{13}C of less than -3‰). This dramatic drop in the oceanic carbon isotopic values took place within no more, but very likely less, than a few million years. It is roughly paralleled by a similar decline in the oxygen isotopic composition of seawater. This pattern is generally consistent with observations made in other geological sections (Holser & Magaritz 1987) and represents the most spectacular change observed thus far in the geological record.

Because of the nature of the isotope fractionation processes, which cause differences in carbon isotopic composition between the reservoirs, this change must be explained by shifts of carbon masses enriched in the light isotope (cf. Hoefs 1987). Knowing the initial δ^{13}C values of the late Permian ocean and assuming its total carbon contents to have been comparable to the current value, one can easily calculate that the observed decline in δ^{13}C can only be plausibly explained by rapid input to seawater of huge amounts of oxidized organic carbon; these amounts are approximately equal to the total carbon contents in the ocean and almost two orders of magnitude greater than the present standing crop of the entire biosphere. This interpretation is indeed consistent also with the parallel decline in oxygen isotope values because such a massive oxidation would preferentially involve the light isotope of oxygen and cause its disproportionate enrichment in dissolved carbonates in seawater. The shape of the δ^{13}C curve in the preceding times, in turn, implies removal of great amounts of organic carbon from the ocean–atmosphere system.

The best explanation for the whole pattern is that deposits rich in organic matter accumulated in the Permian at the sea floor, and the rate of this accumulation rapidly increased in the late Permian. This is possible solely under strongly reducing conditions in the absence of such deepwater circulation as it exists today in the ocean, for it would bring oxygen down to the bottom and prevent organic matter from accumulating. Thus the oceanographic situation must have been fundamentally different from the one observed today. In the latest Permian, however, some geodynamic process brought oxygen down to the sea bottom and hence the amassed organic carbon underwent rapid oxidation and uptake by seawater (Gruszczyński et al. 1989).

This process must have had tremendous and detrimental consequences for the marine biosphere (Małkowski et al. 1989). Under reducing conditions, phosphorus and nitrogen recycling is enhanced because they are liberated from decaying organic matter and return to seawater; this is why upwelling waters are so rich in these nutrients. Thus the main limiting factors on the growth of the marine biosphere were relaxed during the late Permian. The subsequent development of oxidizing conditions at the sea bottom dramatically changed the situation because under such conditions both phosphorus and nitrogen sink at the sea floor, as it is observed presently. Thus the massive oxidation of organic matter that accumulated previously at the sea bottom removed large amounts of nutrients from seawater. The resulting nutrient deficiency must have led to a collapse of the marine ecosystem and hence to a mass extinction in the sea. At the same time, massive oxidation of organic carbon and the associated chemical compounds (primarily iron and manganese sulphides) must have drawn oxygen from the atmosphere. The resulting drop in atmospheric oxygen levels most likely led to extinctions among land biotas.

Thus the latest Permian was the time of a change in the Earth's system that must have caused a mass extinction. Given the poor stratigraphic resolution at the Permo-Triassic transition, it is not known whether or not all the extinctions occurred simultaneously with the observed carbon and oxygen isotopic event; nor is it known what happened during the earliest Triassic, in the aftermath of the latest Permian paleoceanographic change. The Permo-Triassic mass extinction, however, differs from the latest Cretaceous one because we know that various processes were at work during the latter time interval. And it dramatically differs from events during the latest Ordovician, late Devonian, mid-Cretaceous and Eocene–Oligocene transition because it is known that all of these phenomena were in fact clusters of episodes.

4. CONCLUSIONS

Many phenomena that have been traditionally called 'mass extinctions' are in fact sets of lesser episodes of extinction. These lesser episodes, moreover, often seem to be each caused by another environmental factor. The causal plexus of such mass extinctions involves, then, coincidence of various processes in the physical environment of life, these processes happening to co-occur roughly in time (on the geological timescale). The Permo-Triassic mass extinction, however, seems to be different from the others in that it may indeed be caused by one, although not truly catastrophic, process of environmental change that affected the global ocean–atmosphere system.

REFERENCES

Alvarez, L. W., Asaro, F., Michel, H. V. & Alvarez, W. 1980 Extraterrestrial cause for the Cretaceous–Tertiary extinction. *Science, Wash.* **208**, 1095–1108.

Bekov, G. I., Letokhov, V. S., Radaev, V. N., Badyukov, D. D. & Nazarov, M. A. 1988 Rhodium distribution at the Cretaceous/Tertiary boundary analysed by ultrasensitive laser photoionization. *Nature, Lond.* **332**, 146–148.

Bourgeois, J., Hansen, T. A., Wiberg, P. L. & Kauffman, E. G. 1988 A tsunami deposit at the Cretaceous–Tertiary boundary in Texas. *Science, Wash.* **241**, 567–570.

Brenchley, P. J. 1984 Late Ordovician extinctions and their relationship to the Gondwana glaciation. In *Fossils and climate* (ed. P. J. Brenchley), pp. 291–316. Chichester: Wiley.

Buggisch, W. 1972 Zur Geologie und Geochemie der Kellwasserkalke und ihrer begleitenden Sedimente (unteres Oberdevon). *Abh. hess. Landesamt Bodenforsch.* **62**, 1–68.

Corliss, B. H., Aubry, M. P., Berggren, W. A., Fenner, J. M., Keigwin, L. D. & Keller, G. 1984 The Eocene/Oligocene boundary event in the deep sea. *Science, Wash.* **226**, 806–810.

Courtillot, V., Feraud, G., Maluski, H., Vandamme, G., Moreau, M. G. & Besse, J. 1988 Deccan flood basalts and the Cretaceous/Tertiary boundary. *Nature, Lond.* **333**, 843–846.

Davis, M., Hut, P. & Muller, R. A. 1984 Extinction of species by periodic comet showers. *Nature, Lond.* **308**, 715–718.

Duncan, R. A. & Pyle, D. G. 1988 Rapid eruption of the Deccan flood basalts at the Cretaceous/Tertiary boundary. *Nature, Lond.* **333**, 841–843.

Elder, W. P. 1987 The paleoecology of the Cenomanian–Turonian (Cretaceous) stage boundary extinctions at Black Mesa, Arizona. *Palaios* **2**, 24–40.

Farsan, N. M. 1986 Faunenwandel oder Faunenkrise? Faunistische Untersuchung der Grenze Frasnium–Famennium im mittleren Südasien. *Newsl. Stratigr.* **16**, 113–131.

Fastovsky, D. 1987 Paleoenvironments of vertebrate-bearing strata during the Cretaceous–Paleogene transition, eastern Montana and western North Dakota. *Palaios* **2**, 282–295.

Gruszczyński, M., Hałas, S., Hoffman, A. & Małkowski, K. 1989 A brachiopod calcite record of the oceanic carbon and oxygen isotopic shifts at the Permo/Triassic transition. *Nature, Lond.* **337**, 64–68.

Hallam, A. 1984 Pre-Quaternary sea-level changes. *A. Rev. Earth planet. Sci.* **12**, 205–243.

Hallam, A. 1987 End-Cretaceous mass extinction event: argument for terrestrial causation. *Science, Wash.* **238**, 1237–1242.

Hansen, T. A. 1987 Extinction of Late Eocene to Oligocene molluscs: relationship to shelf area, temperature changes, and impact events. *Palaios* **2**, 69–75.

Haq, B. U., Hardenbol, J. & Vail, P. R. 1987 Chronology of fluctuating sea levels since the Triassic. *Science, Wash.* **235**, 1156–1167.

Hennig, E. 1932 *Wege und Wesen der Paläontologie*. Berlin: Borntraeger.

Hoefs, J. 1987 *Stable isotope geochemistry*, 3rd edn. Berlin: Springer.

Hoffman, A. 1989a Mass extinctions: the view of a sceptic. *J. geol. Soc. Lond.* **146**, 21–35.

Hoffman, A. 1989b Changing paleontological views on mass extinction phenomena. In *Mass extinctions: processes and evidence* (ed. S. K. Donovan). London: Belhaven Press. (In the press.)

Holser, W. T. & Magaritz, M. 1987 Events near the Permian–Triassic boundary. *Mod. Geol.* **11**, 155–179.

House, M. R. 1985 Correlation of mid-Palaeozoic ammonoid evolutionary events with global evolutionary perturbations. *Nature, Lond.* **313**, 17–22.

Hut, P., Alvarez, W., Elder, W. P., Hansen, T., Kauffman, E. G., Keller, G., Shoemaker, E. M. & Weissman, P. R. 1987 Comet showers as a cause of mass extinctions. *Nature, Lond.* **329**, 118–126.

Jablonski, D. 1986a Background and mass extinctions: the alternation of macroevolutionary regimes. *Science, Wash.* **231**, 129–133.

Jablonski, D. 1986b Causes and consequences of mass extinctions: a comparative approach. In *Dynamics of extinction* (ed. D. K. Elliott), pp. 183–230. New York: Wiley.

Kauffman, E. G. 1984 The fabric of Cretaceous marine extinctions. In *Catastrophes and Earth history* (ed. W. A. Berggren & J. A. Van Couvering), pp. 151–246. Princeton University Press.

Kauffman, E. G. 1986 High-resolution event stratigraphy: regional and global Cretaceous bio-events. In *Global bio-events* (ed. O. H. Walliser), pp. 279–335. Berlin: Springer.

Keller, G. 1983 Biochronology and paleoclimatic implications of middle Eocene to Oligocene planktic foraminiferal faunas. *Mar. Micropaleont.* **7**, 463–486.

Keller, G. 1987 Prolonged biotic stress and species survivorship across the Cretaceous–Tertiary boundary. *Abstr. Progm. geol. Soc. Am.* **19**, 724.

Kennett, J. P., Borch, C. van der, Baker, P. A., Barton, C. E., Boersma, A., Cauler, J. P., Dudley, W. C., Gardner, J. V., Jenkins, D. G., Lohman, W. H., Martini, E., Merrill, R. B., Morin, R., Nelson, C. S., Robert, C., Srinivasan, M. S., Stein, R., Takeuchi, A. & Murphy, M. G. 1985 Paleotectonic implications of increased late Eocene–early Oligocene volcanism from South Pacific DSDP sites. *Nature, Lond.* **316**, 507–511.

Krasovskiy, V. I. & Shklovskiy, I. S. 1957 The possible influence of explosions of supernovae on the evolution of life on Earth. (In Russian.) *Dokl. Akad. Nauk SSSR* **116** (2), 197–199.

Lindinger, M. & Keller, G. 1987 Stable isotope stratigraphy across the Cretaceous/Tertiary boundary in Tunisia: evidence for a multiple extinction mechanism? *Abstr. Progm. geol. Soc. Am.* **19**, 747.

Małkowski, K., Gruszczyński, M., Hoffman, A. & Hałas, S. 1989 Oceanic stable isotope composition and a scenario for the Permo-Triassic crisis. *Hist. Biol.* **2**. (In the press.)

Marshall, H. T. 1928 Ultra-violet and extinction. *Am. Nat.* **62**, 165–187.

Mount, J. F., Margolis, S. V., Showers, W., Ward, P. & Doehne, E. 1986 Carbon and oxygen isotope stratigraphy of the Upper Maastrichtian, Zumaya, Spain: a record of oceanographic and biologic changes at the end of the Cretaceous Period. *Palaios* **1**, 87–91.

Newell, N. D. 1967 Revolutions in the history of life. *Spec. Pap. geol. Soc. Am.* **89**, 63–91.

Officer, C. B., Hallam, A., Drake, C. L. & Devine, J. D. 1987 Late Cretaceous and paroxysmal Cretaceous/ Tertiary extinctions. *Nature, Lond.* **326**, 143–149.

Owen, M. R. & Anders, M. H. 1988 Evidence from cathodoluminescence for non-volcanic origin of shocked quartz at the Cretaceous/Tertiary boundary. *Nature, Lond.* **334**, 145–147.

Pavlova, M. V. 1924 *The causes of extinctions of animals in the last geological period*. (In Russian.) Moscow: Nauka.

Prothero, D. R. 1985 North American mammalian diversity and Eocene–Oligocene extinctions. *Paleobiology* **11**, 363–388.

Raup, D. M. & Sepkoski, J. J. 1984 Periodicity of extinctions in the geological past. *Proc. natn. Acad. Sci. U.S.A.* **81**, 801–805.

Rigby, J. K. Jr., Newman, K. R., Smit, J., Kaars, S. van der, Sloan, R. E. & Rigby, J. K. 1987 Dinosaurs from the Paleocene part of the Hell Creek Formation, McCone County, Montana. *Palaios* **2**, 296–302.

Sandberg, C. A., Ziegler, W., Dreesen, R. & Butler, J. L. 1988 Late Frasnian mass extinction: conodont event stratigraphy, global changes, and possible causes. *Courier Forschungsinst. Senckenberg* **102**, 263–307.

Schindewolf, O. H. 1954 Über die möglichen Ursachen der grossen erdgeschichtlichen Faunenschnitte. *Neues Jb. Geol. Paläont. Mh.* **1954**, 457–465.

Schopf, T. J. M. 1974 Permo-Triassic extinctions: relation to seafloor spreading. *J. Geol.* **82**, 129–143.

Smit, J., Kaars, W. A. van der & Rigby, J. K. Jr. 1987 Stratigraphic aspects of the Cretaceous–Tertiary boundary in the Bug Creek area of eastern Montana, USA. *Mém. Soc. géol. Fr.* (N.S.) **150**, 53–74.

Smit, J. & Romein, A. J. T. 1985 A sequence of events across the Cretaceous–Tertiary boundary. *Earth planet. Sci. Lett.* **74**, 155–170.

Sobolev, D. N. 1928 Earth and life. On the causes of extinction of organisms. (In Russian.) Kiev.

Stanley, S. M. 1984 Marine mass extinctions: a dominant role for temperature. In *Extinctions* (ed. M. H. Nitecki), pp. 69–117. Chicago University Press.

Stanley, S. M. 1988 Paleozoic mass extinctions: shared patterns suggest global cooling as a common cause. *Am. J. Sci.* **288**, 334–352.

Stearn, C. W. 1987 Effect of the Frasnian–Famennian extinction event on the stromatoporoids. *Geology* **15**, 677–679.

Ward, P., Wiedmann, J. & Mount, J. F. 1986 Maastrichtian molluscan biostratigraphy and extinction patterns in a Cretaceous/Tertiary boundary section exposed at Zumaya, Spain. *Geology* **14**, 899–903.

Whitmire, D. P. & Jackson, A. A. 1984 Are periodic mass extinctions driven by a distant solar companion? *Nature, Lond.* **308**, 713–715.

Wolbach, W. S., Gilmour, I., Anders, E., Orth, C. J. & Brooks, R. R. 1988 Global fire at the Cretaceous–Tertiary boundary. *Nature, Lond.* **334**, 665–669.

Discussion

P. W. KING (*Department of Biology, University College London, U.K.*). Would it not be much better in plotting a graph of last appearance against time to use a natural genetic unit, i.e. the species, rather than the family, because the amount of difference between species in a family may differ considerably from one taxonomic group to another, and because different families are not equal to each other as they include very different numbers of species?

A. HOFFMAN. Yes, it would be much better to employ species as the unit of quantitative analysis of extinction patterns. This is often done in studies of local-scale patterns. In global-scale studies, however, the use of species-level data must be very seriously hampered by the notorious vagaries of the fossil record, which make such data extremely biased and unreliable. Family-level data have therefore been customarily accepted as a compromise between biological meaningfulness and palaeontological reliability.

It is with great regret that we note the tragically premature death on 7 November 1992 of Antoni Hoffman. He was a person of exceptional ability who will be greatly missed by the palaeontological community.

Phil. Trans. R. Soc. Lond. B **325**, 263–277 (1989)

Printed in Great Britain

Synchronology, taxonomy and reality

By C. H. Holland

Department of Geology, Trinity College, Dublin 2, Republic of Ireland

Geochronometry is only of limited help in establishing the timing of evolutionary events. Biostratigraphy and the establishment of a global standard stratigraphic scale are essential. These must be handled sensibly. Suggested periodicity of extinctions is dismissed. So called 'mass extinctions' are assessed by reference to the Ordovician–Silurian, Frasnian–Famennian and Cretaceous–Tertiary examples. Too ready use of the term 'mass extinction' tends to over-dramatize the patterns truly obtainable from the fossil record. It is easier to play with secondary data than to collect primary data.

Limitations of geochronometry

The present symposium is concerned largely with extinction rather than with evolution in general. Extinction is, of course, a limiting case of evolutionary rate, before which come all the problems of punctuation versus stasis, of macroevolution compared with microevolution, of tempo and mode. All these questions are time related. One wants to know when an evolutionary event took place; one wants to know the rate of an evolutionary change. It is still not sufficiently recognized that geochronometry – the science of dating rocks in years – is singularly insensitive in providing answers to these questions. The past 40 years have seen the advent of additional and improved methods for the radiometric dating of rocks and an increase in the number of laboratories where this work continues. Yet, as I have previously stated: 'The point is seldom explicitly made that the broadly consistent, but ever changing, geochronometric scale is somewhat loosely attached to the stratigraphic scale within ranges of dates in years' (Holland 1986). The point is illustrated in figure 1, where dates for the beginning and end, and hence the duration, of three Palaeozoic periods (Ordovician, Silurian and Devonian) are given from several well-known and authoritative compilations of the radiometric timescale. It is not perhaps impossible that at some future time a new and magical 'black box' will be invented that can date rocks of many kinds expeditiously and very accurately. For the foreseeable future, however, investment in improved radiometric dating, in the expansion of the database, and in the efficient dissemination of data should be encouraged and will be to our benefit.

There is, of course, an additional snag: the persistent problem of relating radiometric dates to the stratigraphic scale. To take an example relevant to our present purpose, McKerrow *et al.* (1985) have constructed a timescale for the Ordovician, Silurian and Devonian of which the most accurate points are listed as five in number. Stratigraphic interpolation between these points provides their summary scale. Thus the Ordovician–Silurian boundary must be interpolated between an accurate point of 450 ± 7 Ma in the early Cardoc and one at 431 ± 6 Ma at the end of the early Llandovery. Their fully plotted graph crosses only one additional relevant point within this interval, which can be placed only as somewhere between the beginning of the *bicornis* graptolite biozone, through the *wilsoni* biozone, to the middle of the *clingani* biozone, whereas the two accurate points already mentioned are confined respectively

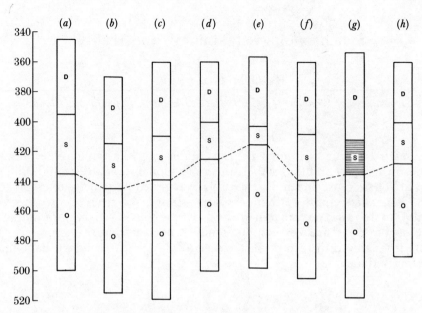

FIGURE 1. Age ranges of the Ordovician, Silurian and Devonian Systems according to (*a*) Geological Society (1964), (*b*) Lambert (1971), (*c*) McKerrow *et al.* (1980), (*d*) Gale *et al.* (1980), (*e*) Odin *et al.* (1982), (*f*) Harland *et al.* (1982), (*g*) McKerrow *et al.* (1985) and (*h*) Gale (1985). Column (*g*) is divided into 20 equal parts representing the standard graptolite biozones of Koren (1984).

to the *clingani* and *cyphus* Biozones. Thus, in terms strictly of radiometric dating, the Ordovician–Silurian boundary lies within an interval of about 20 Ma.

This is not, of course, to detract from the tremendous philosophical value of the sense of scale provided by radiometric dating: the vastness of geological time, the vast length of Precambrian time which is now known to have passed between the origin of life and the Cambrian explosion of invertebrate evolution.

But here I am concerned with the recognition of isochronous evolutionary and evolutionary related events, and with knowledge of the temporal spacing of these, and consequently of evolutionary rates. Things are simpler in the Cainozoic, when so many different strands of evidence may be brought to bear. I confine my comments to the Palaeozoic and Mesozoic, where much of the recent discussion has centred.

Let it be said at once that there are certain situations in which an immensely detailed event stratigraphy is achievable. One thinks immediately of the work of E. G. Kauffman and his colleagues in the Western Interior Cretaceous Basin of North America (Kauffman 1970, 1977), where a sequence containing abundant ammonites and bivalves also includes over 400 bentonites or other layers of volcanic material which provide record of isochronous events within the basin. The sequence has been subjected to a programme of K–Ar dating such that each biozone may represent even as little as 0.25 Ma. But let us accept that this kind of opportunity is very rare and is most likely to remain rare.

What can be more readily done is to subject a stratigraphical sequence simply to a rigorous biostratigraphic subdivision. Thus if one takes the figure of 27 Ma for the length of the Silurian Period (McKerrow *et al.* 1985), this can presently, under the usual scheme, be divided into about 41 graptolite biozones. If we employ the more widely applicable set of standard graptolite biozones which Koren (1984) has been developing, there are still 20 of these, their rocks representing an average figure of only about 1.35 Ma each. I have made this point in a previous figure (Holland 1986, fig. 1) and do so again on one column of figure 1 of the present paper.

BIOSTRATIGRAPHY AND THE GLOBAL STANDARD STRATIGRAPHIC SCALE

This is not the place to discuss the whole topic of methods and resolution in stratigraphy, but there are several points that deserve emphasis. The first is that the work of the Commission on Stratigraphy of the International Union of Geological Sciences, and of its various subsidiary Subcommissions and Boundary Working Groups, should be encouraged, supported and accelerated. This work is not some luxury to be savoured at leisure. It is to provide an agreed international standard of stratigraphic language, such that our research can continue against a stable and understood background. This procedure is not helped by stratigraphic nationalism, which has certainly been in evidence in recent years.

Secondly, it is important that as soon as possible this work should be carried down to the level of the chronozone: the lowest category in the Global Standard Stratigraphical hierarchy, which comes below the stage. In the Lower Palaeozoic, Koren (1984) has shown the way ahead with her first attempt at a set of Silurian Standard Graptolite Biozones, already mentioned above. Eventually such divisions should be related to boundary stratotype sections. In the Mesozoic, the standardization of biozonal schemes is further advanced. T. P. Poulton writes in a recent letter to the Chairman of the International Subcommission on Stratigraphical Classification: '...as correlations become more and more confident, the finest useful worldwide biostratigraphic unit is coming, in Mesozoic circles anyway, to be called the Standard Zone or Chronozone...'. 'Insofar as the zones are subunits of stages, and systems, they require stratotypes, in which the base of each is clearly defined.' I emphasized this coming together of the once very different approaches of Mesozoic and Palaeozoic stratigraphy in a previous paper (Holland 1986, see fig. 9).

Thirdly, in discussing the primary evidence for stasis, Schopf (1981) has very usefully reviewed the limitations associated with the plotting of species durations through stages, series and systems. His first limitation is expressed as follows: 'In sum, the largest amount of direct stratigraphic evidence for stasis over millions of years comes from macroscopic taxa which occur discontinuously in a few localities and to which one has extrapolated over millions of years a polytypic model of species variation designed for the modern world.' A second limitation, which has frequently been mentioned elsewhere, is that when a species is found within a stage it is usually assumed to have the total duration of that stage. I remember the late Professor Lecompte graphically declaiming how such plots showed taxa halting at stratigraphic boundaries like battalions of soldiers. A small part of a range chart of brachiopods compiled by Amsden & Barrick (1988) illustrates a more helpful approach (figure 2). A third limitation worth recording is that short-ranged, rare species are less likely to be preserved, found and described than more abundant forms of longer duration.

To this we may add the advantage of assessing a biozonal scheme based upon one group of fossils against that provided by another. For example, Sevastopulo & Nudds (1987) compared the coral and conodont biozones of the Courceyan Stage (lower Dinantian) of the British Isles. These rocks have been extensively investigated. A modification of the coral biozonal scheme of Ramsbottom & Mitchell (1980) was plotted for several well-known sections against an

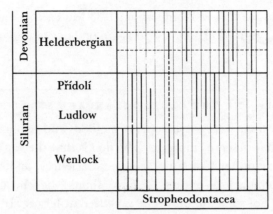

FIGURE 2. A small part of a range chart of brachiopod genera from plate 4 of Amsden & Barrick (1988). The grey area in the Wenlock and the divisions within the Devonian separated by broken lines are individual members.

integrated set of conodont biozones and sub-biozones. As shown in figure 3, the bases of the coral biozones do not appear consistently at the same level relative to the conodont biozonation. Most of the conodont biozones are based upon lineages, whereas the coral biozones are based upon cryptogenic taxa, although this does not automatically imply that the former are isochronous. Analysis suggested that the conodont biozones are not facies controlled; they may occur in different lithologies. Sevastopulo & Nudds conclude from their analysis that the corals, on the other hand, represent ecologically controlled assemblages.

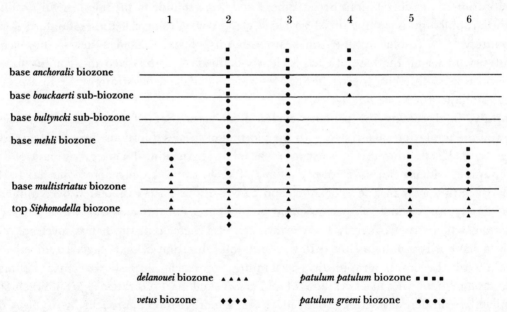

FIGURE 3. Courceyan coral biozones plotted against conodont biozones and sub-biozones for six localities in south Wales, southwest England and Ireland. After Sevastopulo & Nudds (1987), who give names and positions of the localities.

Horowitz *et al.* (1979) compared a Lower Carboniferous subsurface section in Tennessee with one in the Illinois Basin and with the classic section in Belgium, using algae, conodonts, foraminiferans and spores. The last of their conclusions is that 'The combined efforts of several paleontologists working on different groups of fossils is both necessary and important if chronological correlations are to transcend problems associated with different rock types (facies).'

May I suggest, then, in this first part of the paper in which I am simply looking in general ways at the acquisition of primary data, that one always try to say what is meant by the various kinds of biozones that are used; that one is as rigorous as possible in taxonomy (and I can return to this point later); that, in plotting ranges of fossils, one tries to indicate the beginnings and endings of these as closely as possible; that if one is connecting two occurrences through a significant stretch of barren strata one indicate this to be the case; and that if records are thought to be doubtful it is made clear that this is so. It is immensely important that there should be continuing support for taxonomic and biostratigraphic research and that the notion should be dispelled that it is more significant to theorize about palaeobiology by playing with secondary data.

There is increasing preoccupation with quantitative stratigraphy, which is obviously very useful in borehole and offshore work. It has been thoroughly reviewed by Agterberg & Gradstein (1988) and Thomas et al. (1988) have provided a useful bibliography of 637 items. The methods may be too laborious for widespread and worldwide use, though carefully designed experiments at particular levels and in chosen places may prove to be of interest. We have one such project in hand, Transhemisphere Telychian (Holland 1988), which is designed to test the limits of precision in correlation between rocks of the uppermost stage of the Llandovery Series in the British Isles and in China, and to investigate such matters as palaeobiogeography and coevolution at this level.

In what follows, I can myself, of course, be accused of dwelling on secondary data. The literature on cases of extinction (and particularly the so called mass extinctions) and their explanation is now so vast that I can also be accused (as one so often is) of providing information that is simply anecdotal. But in this paper I make no apology for being selective about selected cases.

SUPPOSED PERIODICITY OF EXTINCTIONS

It seems to me that periodicity of extinctions, rather then their irregular occurrence at irregular intervals is unlikely. Suggested causes of an astronomical nature are highly speculative. Worse still, as Teichert (1988a) notes, the discussion 'is increasingly being dominated by geophysicists, astronomers, statisticians, and others (even psychologists have joined the fun) who have little or no first hand knowledge of the fossil record which provides the only data base for all speculations and statistics'. Of course, this does not mean that a scientific investigation should not be undertaken. The question involves taxonomy as well as timing. We find that the approximately 26 Ma periodicity suggested by Raup & Sepkosi (1982, 1984) has, I believe, been disposed of by the work of Patterson & Smith (1987). These authors, experienced respectively in working with fish and echinoderms, checked the families within these two groups, which formed about one fifth of Raup & Sepkosi's original data. All the extinctions recorded by these last two authors were treated as occurring at the end of the stage in which each taxon was recorded. Patterson & Smith note that with acceptance of the cladistic approach there is understanding of monophyletic groups (which can become extinct) and non-monophyletic groups (which cannot become extinct). The latter are artificially circumscribed and so their ending does not 'directly reflect extinction of a lineage'. Patterson & Smith recognize seven categories of taxa in Raup & Sepkosi's data:

(1) the acceptable cases of monophyletic families with two or more species whose extinction is attributed to the correct stage;

(2) monophyletic families whose disappearance is attributed to the wrong stage;

(3) paraphyletic 'ancestral groups' that do not become extinct in a biological sense;

(4) polyphyletic families united by convergent features, which, being artificial, again cannot become extinct;

(5) families which may be either paraphyletic or polyphyletic;

(6) monotypic families whose real distribution in time and space is poorly known;

(7) non-marine families.

In terms of their own records, the one valid category expressed as (1) above provides components of only 27 % in the echinoderms and 23 % in the fishes. The remaining 'noise' component in Raup & Sepkosi's data comes mainly from paraphyletic groups in the echinoderms and from monotypic families in the fishes. Mistaken dating ((2) above) makes a significant contribution in both cases. The families do show five of the eight extinction peaks that Raup & Sepkosi had recognized, but it is the 'noise' that provides these peaks. Patterson & Smith speculate that the cause of the peaks in 'noise' may be taphonomic or monographic.

In a recent consideration of arthropod extinctions, Briggs *et al.* (1988) have also made the distinction 'between the true extinction of major clades and "taxonomic pseudoextinction", where the "termination" of a group is a reflection of taxonomic practice, such as the acceptance of paraphyletic groups.'

Teichert (1988a) reminds us that families are of very different sizes. 'Obviously, the extinction of a monogeneric family is an event of vastly different importance from that of the extinction of a family with 50 genera.'

There have been other criticisms of proposed periodicity in the extinction of marine families, notably that of Quinn's (1987) statistical analysis.

On the other hand, it is surely likely that there were times of conspicuous extinction in some groups (to avoid for now use of the expression 'mass extinction') related to times of major change in the tectonic evolution of the earth.

There is a possible explanation of major cycles in plate tectonic evolution as one convection system switches over to another, with considerable consequences for continental distribution, climatic change and above all for sea level. Such events would necessarily have their effects upon living things, variously of course, according to the type of organism in question, and over an extended period.

So-called mass extinctions

The so-called mass extinctions, in which large proportions, or even all, of the members of many different groups of living things disappeared, presumably rapidly in terms of geological time, are now the subject of a vast literature. Teichert (1988a) referred to them as 'still fashionably described as global catastrophies that wiped out up to 90 percent of the biomass on earth'. As Benton (1985a) put it: 'It is becoming clear that the history of life has been punctuated by a series of events during which, typically, an apparently random selection of organisms died out. These mass extinction events can be seen to have "re-set" the evolutionary clock....'

Benton (1985b) attempted a survey of non-marine tetrapods on the lines of those developed for marine invertebrates. He found several mass extinctions but none of these associated with a statistically high extinction rate. 'The extinction events, including the famous terminal Cretaceous extinction, were the result of a slightly elevated extinction rate combined with a depressed origination rate, and the present evidence does not support the view that mass extinctions are statistically distinguishable from background extinctions.' Benton (1985 b) does warn us (and this cannot be said too often) that the time resolution in his studies is very coarse which 'could minimize the true size of a sudden event by averaging it out over several million years'.

But each case should really be taken on its merits and the literature now covers various individual cases of supposed mass extinction and, naturally, with attempts to relate these to particular events. Three examples, taken in turn, will illustrate different aspects of the problem.

THE ORDOVICIAN–SILURIAN BOUNDARY EVENT

The boundary between the Ordovician and Silurian systems was the subject of investigation by a Working Group of the Commission on Stratigraphy of the International Union of Geological Sciences between 1974 and 1985. Its achievements have been summarized by Cocks (1985, 1988) and by Cocks & Rickards (1988). The recent volume edited by Cocks & Rickards (1988) provides abundant data on sections about the world and on the behaviour of selected fossil groups across the boundary. In this case a significant event, namely the glaciation centred upon Gondwana, was already recognized and had been well described (Beuff *et al.* 1971; Allen 1975; Spjeldnaes 1981; Brenchley 1984, 1988). Spjeldnaes (1981, plate 2) provided a striking illustration of glacial striations from the Upper Ordovician of southern Algeria, but there are many other convincing features of the glaciation in the central Sahara and its extension to South Africa and North Africa (figure 4). Evidence from South America may indicate a separate ice sheet and there are glaciofluvial deposits in Europe. In the present context, the dating of this event must be of concern. Spjeldnaes (1981) referred to studies by Destombes in Morocco as giving a precise date of Upper Asgill for the glacial beds. He concluded, however,

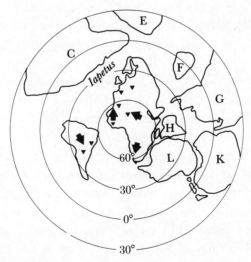

FIGURE 4. After a reconstruction of palaeogeography and climatic zonation in the Ordovician by Spjeldnaes (1981), but with data here confined to evidence for the late Ordovician glaciation. Triangles show glacial beds, arrows glacial striations. C, North America; E, northeast Asia; F, Kazakhstan; G, Southeast Asia; K, Australia; L, Antartica.

that the dating of the Ordovician glaciation as a whole is 'still somewhat doubtful'. He noted that time resolution in the Ordovician is normally not better than ± 1 Ma and that within this margin there may have been more than one glacial episode. In summary, he concluded that the evidence suggests a polar ice cap present from the Arenig onwards and that this 'expanded drastically' in the Upper Ordovician.

One of the most recent assessments of the age of the glaciation is by Brenchley (1988), who concerns himself with the regression–transgression phase in the late Ordovician and early Silurian, which is of the same age on different Lower Palaeozoic plates, thus indicating eustatic change. The fall in sea level started at the beginning of the Hirnatian stage and the subsequent rise in level was largely completed by the end of Hirnatian time, that is within the Ordovician period, though, of course, close to its end. Such radiometric dates as are available can be used in different ways to suggest a duration for the Hirnatian. Brenchley suggests that a figure between 1 and 2 Ma is probable.

Barnes's (1986) review of the Ordovician–Silurian boundary event mentions 'three or four major glacial phases' in North Africa, dated imprecisely from late Caradoc through Ashgill. He refers to the maximum glaciation as in the Hirnantian 'but with significant fluctuations even within this interval'. The Hirnantian is estimated as covering 2 Ma and the subsequent earliest Silurian deglaciation is thought to have been rapid. He considers the influence of plate motions through this brief time to have been minimal. Indeed one can agree in suggesting that over an interval of 2 Ma a figure of only 40 km of lateral movement is reasonable and for the Ashgill as a whole a figure of 200 km. Barnes (1986) argues that draining of warm seas on low-latitude platforms, combined with widespread cooling, would destroy endemic faunas. The more cosmopolitan faunas of the slope and oceanic environments were modified during the Ashgill. The main turnover of both was in the Hirnatian.

Cocks & Rickards (1988) suggest that the maximum glacio-eustatic drop in sea level was probably about half-way through the (top Ordovician) *persculptus* Biozone, on the basis of well-dated *persculptus* bearing post-glacial transgressive beds in North Africa. Incidentally, the international agreement to move the Ordovician–Silurian boundary from the base of the *persculptus* biozone to the base of the succeeding *acuminatus* biozone is obviously something to beware of in assessing older and modern literature together. They consider that the glacial episode varied from place to place, commencing even in late Caradoc times in parts of Gondwana.

It is to be expected that such changes in environment would have affected different groups of organisms in different ways in different parts of the world. Analysis of faunal distributions suggests that this is indeed the case. Particularly striking in this context is the so-called *Hirnantia* fauna, which is diachronous through three pre-*persculptus* biozones in China (Rong 1984) but extends into the *persculptus* biozone in Kazakhstan (Apollonov et al. 1988; Koren et al. 1988), in Quebec (Lespérance 1985), and in the English Lake District (Cocks 1988).

Some major groups of fossils are now briefly considered. Scrutton (1988), in his general review of patterns of extinction and survival in Palaeozoic corals finds a 'significant end Ordovician extinction event' followed by an increased percentage of new genera. This applies to both rugose corals and tabulates, but in neither case is the extinction shown as nearly complete.

According to Taylor & Larwood (1988), the Bryozoa at neither family nor generic level reach a high level of extinction at the time of the end-Ordovician 'mass extinction'.

In his review of brachiopods across the boundary, Cocks (1988) considers that 'in general, however, the degree of extinction across the boundary appears to have been less than previously reported...'. The extinctions at the end of the Hirnantian do not appear to be greater than at the end Caradoc or end Rawtheyan.'

Briggs *et al.* (1988), on the other hand, see the end-Ordovician event as significant in the extinction of major clades of trilobites. The associated reduction in trilobite diversity may be attributed to the extension of cold-water regions. The low diversity cool-water faunas themselves are typified by the widespread occurrence of the *Mucronaspis* fauna. 'With such an important glacial event recognised from independent evidence there seems to be no good reason to look further for the cause of the Hirnantian decline. However, it is not obvious that the same cause was necessarily responsible for the extinction of families at the *end* of the Hirnantian.' Possibly widespread anoxia in the oceans at the transgression may have caused the selective extinctions there.

Lespérance (1988) refers to the distinctive and impoverished trilobite fauna of the Hirnantian Stage, but notes that the degree and nature of the impoverishment varied from region to region. He considers that the major extinction of the trilobites came near the boundary between the Rawtheyan and Hirnatian Stages, rather than at the base of the Silurian System.

Dr A. W. Owen's view of patterns of extinction at this level (expressed in an abstract prepared for the Annual Conference of the Palaeontological Association at the University of Aston in December 1988) suggests that they 'were disproportionately concentrated in deeper water trilobite assemblages. This indicates that although draining of the shelves and migration of the thermocline in response to the end Ordovician glaciation may have been contributory factors, many extinctions may have been caused by the chemical effects of oceanic overturn.'

The history of the nautiloid cephalopods as a whole has been reviewed by Teichert (1988b) and Holland (1987). There are some losses, such as the almost complete disappearance of the Endoceratida; but most groups decline from their middle Ordovician acme on through the Ordovician–Silurian boundary.

Eckhert's (1988) analysis of the crinoids reveals a diversity peak in the Caradoc, a descent to a minimum in the early Ashgill, and a major extinction in the late Ashgill. Glacio-eustatic lowering of sea level is blamed for this. The Silurian radiation of crinoids originated from assemblages of Hirnantian crinoids.

Aldridge (1988) sees conodont evolution as follows: 'The Caradoc was punctuated by an interval in which extinction rates exceeded origination rates,.... Following a brief late Caradoc – early Ashgill recovery, the latest Ordovician saw severe faunal changes, with extinctions reaching a rate of seven genera per Ma, the highest in conodont history.' The survivors led to little innovation in the earliest Silurian. Barnes & Bergstrom (1988), in their elaborate consideration of conodonts at this level, consider this perhaps the most striking faunal turnover in the whole history of the group.

The graptolites (Rickards 1988), so important themselves in dating these horizons, show a gradual change between the *persculptus* and *atavus* biozones. The widespread fauna of the *acuminatus* biozone, between them, 'represents a distinctive stage in the evolution of Silurian graptoloids reflecting a very advanced stage of post-glacial marine transgression and the development of widespread anaerobic black shales and the re-establishment of a rich, marine, tropical plankton'. There is a pronounced increase in diversity at this level, though this had begun even before the *persculptus* biozone.

To conclude, there is a clear evidence of a significant event in the latest Ordovician, the timing of the associated changes, although imprecise, does appear to involve a period of only a few million years. In terms of extinction, one sees, as might be expected, different effects in different groups. To call this a mass extinction is to dramatize unduly.

THE FRASNIAN–FAMENNIAN EXTINCTIONS

According to McLaren (1982), at the Frasnian–Famennian boundary there was a 'huge disappearance of biomass in tropical and sub-tropical shallow seas with an almost total change in fauna which occurred within one conodont sub-zone of less; that is in 0.5–1 Ma. Pedder (1982) presented detailed documentation on the rugose corals at this level. Only 4% of the 148 shallow water species survived, compared with 3 or 4 out of the 10 deeper-water forms.

An attempt was made to locate an iridium anomaly at the boundary in the now well-known and beautifully displayed sections of Upper Devonian rocks in the Canning Basin, Western Australia (Playford et al. 1984). One such was found in the Virgin Hills formation on the McWhae Ridge. In a verbal discussion at the International Geological Congress in Moscow in 1984, I suggested, as Ager (1988) has since done, that perhaps iridium anomalies should be looked for in places where it would be regarded as not so useful to find them. In the event, McLaren's (1985) well-balanced account of the matter suggested that sedimentation here was very slow. The iridium anomaly occurs in a 12 cm stromatolitic bed, containing the cyanobacterium *Frutexites* and showing enrichment in iron oxide minerals. One of the possibilities here is therefore that the *Frutexites* not only fixed iron but also concentrated other elements from seawater.

Sandberg et al. (1988) have provided a long discussion of the 'Late Frasnian Mass Extinction', with particular reference to conodont stratigraphy. Documentation is provided (with reference to sea-level curves) that a rapid eustatic rise followed by an abrupt fall in sea level immediately preceded the extinction and that this continued into the early Famennian, though the extinction event was clearly over. The mass extinction is deduced to be very short because of the thinness of sediments involved and the lack of significant evolutionary change in the surviving conodonts. A duration of some 12500 to 21500 years is estimated. These figures are decided by the quantification of conodont biozones within a known radiometric time span. The extinction mechanism, it is suggested, was caused by a variety of interrelated factors including sea-level changes and changes in ocean currents. The authors 'theorize' that a large bolide impact was ultimately responsible.

The Systematics Association's Special Volume on *Extinction and survival in the fossil record* (Larwood 1988) is a useful source on the behaviour of different groups at this (and other) levels. Brasier (1988) here reminds us of the extinction of reef faunas at the end of the Frasnian, associated with major transgressive pulses and anoxic black shales, especially of the 'Kelwasser Event' (House 1975, 1985). This event affected the foraminiferans, eliminating 'faunas of relatively advanced architecture...while those of primitive to intermediate architecture survived. Even these suffered a setback in the number of species.'

On the Bryozoa, Taylor & Larwood (1988) note that 'The mid–late Devonian falls within a general trough of low bryozoan family diversity and it is not possible to distinguish any times of outstanding reduction in diversity.'

In their survey of the arthropods, Briggs et al. (1988) give particular attention to patterns of trilobite diversity across this boundary. They conclude that 'Extinctions of major clades coinciding exactly with the end of the Frasnian are relatively few...'. The clades terminating at the boundary had already become reduced. 'The Frasnian–Famennian drop is much less than that between the Eifelian and Givetian.'

Stratigraphic control at this level is already good. There were certainly changes of sea level and climate (the drop-off in well-developed reefs of Frasnian times alone would suggest this); but it does not seem very satisfactory to speak of mass extinction when the behaviour of different groups is so variable.

HAPPENINGS AT THE CRETACEOUS–TERTIARY BOUNDARY

One hesitates to penetrate the jungle of this matter. The literature is now enormous and in some cases emotive. In any case, Hallam (1987) has provided an excellent review of the case for a terrestrial explanation of the supposed mass extinction at the close of the Cretaceous. For my part, I have already referred to Benton's (1985b) views on the quantitative aspect of the extinction of terrestrial vertebrates. My own interest in the cephalopods, though largely older ones, has led to concern with the Upper Cretaceous record of the ammonites.

Professor J. M. Hancock has kindly provided fuel, both from his own wide experience of the Cretaceous ammonites and by drawing attention to references. He writes that the ammonites did not fade away quite so gradually as he used to think, in particular because of more accurate stratigraphical assignment of rocks from the Southern Hemisphere. Nevertheless, the decrease in the number of families still shows a pattern similar to that he recorded in *The fossil record* (Harland *et al.* 1967). Hancock provides a more recent graph (figure 5), in which the stages are given a more nearly correct proportion of time. He notes that even this diagram is now some seven years old. The numbers of families are derived from the new edition of the relevant volume of the *Treatise on invertebrate paleontology*, not yet published.

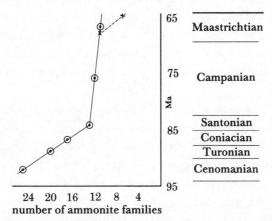

FIGURE 5. Numbers of Upper Cretaceous ammonite families. Numbers for each stage placed at the middle point of the stage. The broken lines shows the reduction in ammonite families during the Maastrichtian. (Original kindly provided by Professor J. M. Hancock).

Professor Hancock notes that there are only two places presently available where the ammonite succession close to the top of the Cretaceous is even potentially known, respectively in Denmark and northern Spain. Birkelund (1979) demonstrated the reduction of ammonite genera within the Maastrichtian of Denmark from 34 to about half that number. He saw this as part of a general decline through the Cretaceous. Seven genera were recorded in the 'topmost layer of the Upper Maastrichtian'. There are large numbers of juveniles at the top and regression of the sea is suggested as the cause.

In an abstract by Ward & Wiedman (1983), reference was made to poorly preserved ammonites from northern Spain. The highest ammonites here came from strata 12 m below the boundary clay with its iridium enrichment. Abundance and diversity had dropped off long before the close of the Cretaceous. Wiedmann (1986), in an extended review, places emphasis on sea-level changes. He observes that the Bryozoa and ostracods, both sensitive to water quality, actually reached their peak density at the time of the 'mass extinction'. The point was made verbally at this Discussion Meeting that Ward has now collected additional ammonites from other sections in northern Spain, taking the record to very close to the boundary, but not altering the picture of an earlier established decline.

Kennedy (1984), too, noted that the ammonites declined throughout the late Cretaceous and that most of the decline was pre-Santonian. He pointed out that the diversity of ammonites during the Maastrichtian was no lower than during the Berriasian.

In consideration of a very different group of animals, Whalley (1988), writing about evolutionary changes in the insects over the Cretaceous–Tertiary boundary, notes that they 'provide no evidence of abrupt or catastrophic changes'. I am indebted to Dr D. C. G. Briggs for this reference.

Once again, at this level, stratigraphic control is possible, but there is always room for improvement; and, once again, different groups may have been affected in different ways, as one would expect if changes in sea level, continental distribution and climate were involved. Even here it seems that the term 'mass extinction' may be over dramatic. One must continue to work at biostratigraphy, at radiometric dating and at taxonomy seen to some extent at least with a cladistic eye. From such acquisition of primary data better models can perhaps be built.

The sociologists of science will make much of the story of theorizing about the Cretaceous–Tertiary boundary event. One of them (Elizabeth S. Clemens) has already suggested that the Alvarez hypothesis owes much of its support to its relative simplicity and a favourable press. 'It offered an elegant and parsimonious solution to a question firmly embedded in popular culture' (quoted by Malcolm W. Browne in the *New York Times* for 19 January 1988).

I have not mentioned as such the end-Cretaceous bolide. I simply do not believe in it. Unseen bolides dropping into an unseen sea are not for me. It must be said, though, that things reach a low level when palaeontologists, who after all alone can provide the necessary data, are described by a proponent as 'not very good scientists', 'more like stamp collectors' (quoted by M. W. Browne, reference cited above). I leave some of the last words to two exceedingly experienced palaeontologists who are certainly not 'stamp collectors'.

Teichert (1988b) contrived to say many important things in a paper only one page long. It includes the following statement: 'There has been since early days, a tendency to search for a unified theory of the causes of extinction, preferably the same cause for all extinctions of all ages. These attempts are, in my opinion, simplistic and misguided.' He continues later: '…I believe on the contrary that no single cause is sufficient to explain mass extinctions and "partial extinctions", and that the explanation must be sought, in each case, in random combinations of a number of factors that ordinarily and continuously influence the course of organic evolution and its record in the rocks.'

Ager (1988) has this to say: 'The most publicized mass extinction in the whole fossil record is that at the end of the Cretaceous. All that I can say about it here is that no-one told the brachiopods.'

REFERENCES

Ager, D. V. 1988 Extinctions and survivals in the Brachiopoda and the dangers of data bases. In *Extinction and survival in the fossil record* (ed. G. P. Larwood), pp. 89–97. Oxford: Clarendon Press.

Agterberg, F. P. & Gradstein, F. M. 1988 Recent developments in quantitative stratigraphy. *Earth Sci. Rev.* **25**, 1–73.

Aldridge, R. J. 1988 Extinction and survival in the Conodonta. In *Extinction and survival in the fossil record* (ed. G. P. Larwood), pp. 231–256. Oxford: Clarendon Press.

Allen, P. 1975 Ordovician glacials of the central Sahara. In *Ice ages: ancient and modern* (ed. A. E. Wright & F. Mosley), pp. 275–285. *Geol. J.* special issue no. 6.

Amsden, T. W. & Barrick, J. E. 1988 Late Ordovician through early Devonian annotated correlation chart and brachiopod range charts for the southern midcontinent region, U.S.A., with a discussion of Silurian and Devonian conodont faunas. *Oklahoma Geol. Surv. Bull.* **143**, 1–66.

Apollonov, M. K., Koren, T. N., Nikitin, I. F., Paletz, L. M & Tsai, D. T. 1988 Nature of the Ordovician–Silurian boundary in south Kazakhstan, USSR. *Bull. Br. Mus. nat. Hist.* A **43**, 145–154.

Barnes, C. R. 1986 The faunal extinction event near the Ordovician–Silurian boundary: a climatically induced crisis. In *Lecture notes in Earth Sciences*, (*Global bio-events*) (ed. O. Walliser), pp. 121–126. Berlin: Springer-Verlag.

Barnes, C. R. & Bergstrom, S. M. 1988 Conodont biostratigraphy of the Uppermost Ordovician and Lowermost Silurian. *Bull. Br. Mus. nat. Hist.* A **43**, 325–343.

Benton, M. J. 1985*a* Interpretations of mass extinction. *Nature, Lond.* **314**, 496–497.

Benton, M. J. 1985*b* Mass extinction among non-marine tetrapods. *Nature, Lond.* **316**, 811–814.

Beuff, S., Biju-Duval, B., Chaperal, O. de, Rognon, R., Gariel, O. & Bennacef, A. 1971 Les Grès du Paléozoique inférieur au Sahara – sédimentation et discontinuités, evolution structurale d'un craton. *Sci. Téch. Petrol* **18**, 1–464.

Birkelund, T. 1979 The last Maastrichtian ammonites. In *Cretaceous/Tertiary boundary events: I. The Maastrichtian and Danian of Denmark* (ed. T. Birkelund & R. G. Bromley), pp. 51–57. University of Copenhagen.

Brasier, M. D. 1988 Foraminiferid extinction and ecological collapse during global events. In *Extinction and survival in the fossil record* (ed. G. P. Larwood), pp. 37–64. Oxford: Clarendon Press.

Brenchley, P. J. 1984 Late Ordovician extinctions and their relationship to the Gondwana glaciation. In *Fossils and climate* (ed. P. J. Brenchley), pp. 291–315. London: J. Wiley.

Brenchley, P. J. 1988 Environmental changes close to the Ordovician–Silurian boundary. *Bull. Br. Mus. nat. Hist.* A **43**, 377–385.

Briggs, D. E. G., Fortey, R. A. & Clarkson, E. N. K. 1988 Extinction and the fossil record of the arthropods. In *Extinction and survival in the fossil record* (ed. G. P. Larwood), pp. 171–209. Oxford: Clarendon Press.

Cocks, L. R. M. 1985 The Ordovician–Silurian Boundary. *Episodes, Ottawa* **8**, 98–100.

Cocks, L. R. M. 1988 The Ordovician-Silurian Boundary and its working group. *Bull. Br. Mus. nat. Hist.* A **43**, 9–15.

Cocks, L. R. M. & Rickards, R. B. (eds) 1988 A global analysis of the Ordovician–Silurian boundary. *Bull. Br. Mus. nat. Hist.* A **43**, 1–394.

Eckhert, J. D. 1988 Late Ordovician extinction of North American and British crinoids. *Lethaia* **21**, 147–167.

Gale, N. H., Beckinsale, R. D. & Wadge, A. J. 1980 Discussion of a paper by McKerrow, Lambert and Chamberlain on the Ordovician, Silurian and Devonian time-scales. *Earth planet. Sci. Lett.* **51**, 9–17. 1980

Gale, N. H. 1985 Numerical calibration of the Palaeozoic time-scale; Ordovician, Silurian and Devonian periods. In *The chronology of the geological record* (ed. N. J. Snelling), pp. 81–88. London: Geological Society.

Geological Society 1964 Phanerozoic timescale. *Q. Jl geol. Soc. Lond.* **120** S, 260–262.

Hallam, A. 1987 End-Cretaceous mass extinction event: argument for terrestrial causation. *Science, Wash.* **238**, 1237–1242.

Harland, W. B., Cox, A. V., Llewellyn, P. G., Pickton, C. A. G., Smith, A. G. & Walters, R. 1982 *A geologic timescale*. Cambridge University Press.

Harland, W. B. et al. (eds) 1967 *The fossil record*. (828 pages.) London: Geological Society.

Holland, C. H. 1986 Does the golden spike still glitter. *J. geol. Soc. Lond.* **143**, 3–21.

Holland, C. H. 1987 The nautiloid cephalopods: a strange success. *J. geol. Soc. Lond.* **144**, 1–15.

Holland, C. H. 1988 Transhemisphere Telychian: a biostratigraphical experiment. *Lethaia* **21**, 188.

Horowitz, A. S., Mamet, B. L., Neves, R., Potter, P. E. & Rexroad, C. B. 1979 Carboniferous paleontological zonation and intercontinental correlation of the Fowler No. 1 Traders Core, Scott County, Tennessee, U.S.A. *Southeastern Geol.* **20**, 205–228.

House, M. R. 1975 Facies and time in the marine Devonian. *Proc. Yorks. geol Soc.* **40**, 233–288.

House, M. R. 1985 Correlation of mid-Palaeozoic ammonoid evolutionary events with global sedimentary perturbations. *Nature, Lond.* **313**, 17–22.

Kaufmann, E. G. 1970 Population systematics, radiometrics, and zonation – a new biostratigraphy. *Proceedings of the North American Paleontological Convention* **1F**, 612–666.

Kaufmann, E. G. 1977 Geological and biological overview: western interior Cretaceous basin. *Mountain Geologist* **14**, 75–99.

Kennedy, W. J. 1984 The extinction of the ammonites. *Bull. Section Sci.* **6**, 107–108.

Koren, T. N. 1984 Graptolite zones and standard stratigraphic scale of Silurian. *Proceedings of the 27th International Geological Congress (Stratigraphy)* **1**, 47–76. Utrecht: VNU Science Press.

Koren, T. N., Oradovskaya, M. M. & Sobolevskaya, R. F. 1988 The Ordovician–Silurian boundary beds of the north-east USSR. *Bull. Br. Mus. nat. Hist.* A **43**, 133–138.

Lambert, R. St. J. 1971 The pre-Pleistocene Phanerozoic timescale – a review. In *The Phanerozoic timescale – a supplement* (ed. W. B. Harland & E. H. Francis). *Spec. pub. geol. Soc. Lond.* **5**, 9–31.

Larwood, G. P. (ed.) 1988 *Extinction and survival in the fossil record*, Systematics Association special volume no. **34**. (365 pages.) Oxford: Clarendon Press.

Lespérance, P. J. 1985 Faunal distributions across the Ordovician–Silurian boundary, Anticosti Island and Perce, Quebec, Canada. *Can. J. Earth Sci.* **22**, 838–849.

Lespérance, P. J. 1988 Trilobites. *Bull. Br. Mus. nat. Hist.* **43**, 359–376.

McKerrow, W. S., Lambert, R. St. J. & Chamberlain, V. E. 1980 The Ordovician, Silurian and Devonian timescales. *Earth planet. Sci. Lett.* **51**, 1–8.

McKerrow, W. S., Lambert, R. St. J. & Cocks, L. R. M. 1985 The Ordovician, Silurian and Devonian periods. In *The chronology of the geological record* (ed. N. J. Snelling), pp. 73–80. London: Geological Society.

McLaren, D. J. 1982 Large Body impacts and terrestrial evolution: geological, climatological, and biological implications. *Geosci. Can.* **9**, 74–76.

McLaren, D. J. 1985 Mass extinction and iridium anomaly in the Upper Devonian of Western Australia: a commentary. *Geology* **13**, 170–172.

Odin, S. G. (ed.) 1982 *Numerical dating in stratigraphy.* Chichester: John Wiley.

Patterson, C. & Smith, A. B. 1987 Is the periodicity of extinctions a taxonomic artefact? *Nature, Lond.* **330**, 248–251.

Pedder, A. E. H. 1982 *The rugose coral record across the Frasnian/Famennian boundary.* Geological Society of America Special Paper, pp. 485–489.

Playford, P. E., McLaren, D. J., Orth, C. J., Gilmore, J. S. & Goodfellow, W. D. 1984 Iridium anomaly in the Upper Devonian of the Canning Basin, Western Australia. *Science Wash.* **226**, 437–439.

Quinn, J. F. 1987 On the statistical detection of cycles in the marine fossil record. *Paleobiology* **13**, 465–478.

Ramsbottom, W. H. C. & Mitchell, M. 1980 The recognition and division of the Tournaisian Series in Britain. *J. geol. Soc. Lond.* **137**, 61–63.

Raup, D. M. & Sepkosi, J. 1982 Mass extinctions in the marine fossil record. *Science, Wash.* **215**, 1501–1503.

Raup, D. M. & Sepkosi, J. 1984 Periodicity of extinctions in the geologic past. *Proc. natn. Acad. Sci. U.S.A.* **81**, 801–805.

Rickards, R. B. 1988 Graptolite faunas at the base of the Silurian. *Bull. Br. Mus. nat. Hist.* A **43**, 345–349.

Rong, Jia-Yu 1984 Distribution of the *Hirnantia* fauna and its meaning. In *Aspects of the Ordovician System, Palaeontological Contributions from the University of Oslo* (ed. D. L. Bruton), no. 295. pp. 101–112, Oslo: Universitetsforlaget.

Sandberg, C. A., Ziegler, W., Dreesen, R. & Butler, J. L. 1988 Late Frasnian mass extinction: conodont event stratigraphy, global changes, and possible causes. *Cour. Forsch.-Inst. Senckenberg* **102**, 263–303.

Schopf, T. J. M. 1981 Punctuated equilibrium and evolutionary stasis. *Paleobiology* **7**, 156–166.

Scrutton, C. T. 1988 Patterns of extinction and survival in Palaeozoic corals. In *Extinction and survival in the fossil record*, Systematics Association special volume no. 34 (ed. G. P. Larwood), pp. 65–88. Oxford: Clarendon Press.

Sevastopulo, G. D. & Nudds, J. R. 1987 Courceyan (early Dinantian) biostratigraphy of Britain and Ireland: coral and conodont zones compared. *Cour. Forsch.-Inst. Senckenberg* **98**, 39–46.

Spjeldnaes, N. 1981 Lower Palaeozoic palaeoclimatology. In *Lower Palaeozoic of the Middle East, Eastern and Southern Africa, and Antarctica* (ed. C. H. Holland), pp. 199–256. Chichester: John Wiley.

Taylor, P. D. & Larwood, G. P. 1988 Mass extinctions and the pattern of bryozoan evolution. In *Extinction and survival in the fossil record* (ed. G. P. Larwood), pp. 99–119. Oxford: Clarendon Press.

Teichert, C. 1988a Extinctions and extinctions. *Palaios* **2**, 411.

Teichert, C. 1988b Main features of cephalopod evolution. In *The Mollusca*, vol. 12 (*Paleontology and neontology of* Committee on Quantitative Stratigraphy special publication no. 1, pp. 1–58. International Commission on

Thomas, F. C., Gradstein, F. M. & Griffiths, C. M. 1988 *Bibliography and index of quantitative biostratigraphy*, Committee on Quantitative Stratigraphy special publication no. 1, pp. 1–58. International Commission on Stratigraphy.

Ward, P. D. & Wiedmann, J. 1983 The Maastrichtian ammonite succession at Zumaya, Spain. *Abstracts, Symposium on Cretaceous Stage Boundaries, Subcommission on Cretaceous Stratigraphy*, pp. 205–208. Copenhagen.

Wiedmann, J. 1986 Macro-invertebrates and the Cretaceous–Tertiary boundary. In *Lecture notes in Earth Sciences*, vol. 8, (*Global bio-events*) (ed. O. Walliser), pp. 397–409. Berlin: Springer-Verlag.

Whalley, P. 1988 Insect evolution during the extinction of the Dinosauria. *Entomol. gen., Stuttgart* **13**, 119–124.

Discussion

L. B. Halstead (*Department of Geology, University of Reading, U.K.*). Over two decades ago during the compilation *The Fossil Record* (Geological Society of London), which listed first and last occurrences or most major taxa of the animal and plant kingdoms, the arbitrary decision was made to mark each last record as terminating at the end of the stage in which it was last deemed to have been present. This convention was for ease of drafting.

It is important to recognize the limitations of such ranges of fossils and their apparent extinction at exact geological boundaries. I was responsible for the chapter on the Agnatha (the jawless vertebrates). All the major groups finally became extinct at the end of the Frasnian stage of the Upper Devonian; one of the five major mass extinctions recognized in the fossil record. It is certainly the case that the last examples of, say, the cephalaspids and anaspids are from the Frasnian of Nova Scotia. But this is quite misleading. The mass of cephalaspids died out at the end of the Lower Devonian, one genus is known from the Middle Devonian of the Orcadian Basin of Scotland and two isolated forms are known from the Upper Devonian of Nova Scotia. These were essentially 'living fossils' just hanging on in isolated regions, they were not part of the mainstream and their final demise is hardly significant.

To all intents and purposes the more streamlined anaspids also died out at the end of the Lower Devonian, but once again a couple of genera managed to hang on in Nova Scotia until the Frasnian.

With the thelodonts the story is somewhat different. The main extinction was at the end of the Lower Devonian but a few forms continued into the lower part of the Middle Devonian of Spitsbergen. Recently further Middle Devonian thelodonts have been found in Iran, Thailand and Australia and in Australia (and perhaps Antarctica) they flourished into the Upper Devonian, finally disappearing during the Frasnian. In this case the thelodonts lived on in Australia long after they had died out throughout the rest of the world.

It is important to determine whether the extinctions being discussed are regional or global; and whether they involve the main faunas or merely just the last few stragglers.

Phil. Trans. R. Soc. Lond. B **325**, 279–290 (1989)
Printed in Great Britain

Evolution and extinction in the marine realm: some constraints imposed by phytoplankton

By A. H. Knoll

Botanical Museum, Harvard University, Cambridge, Massachusetts 02138, U.S.A.

The organic and mineralized remains of planktonic algae provide a rich record of microplankton evolution extending over nearly half of the preserved geological record. In general, Phanerozoic patterns of phytoplankton radiation and extinction parallel those documented for skeletonized marine invertebrates, both augmenting and constraining thought about evolution in the oceans. Rapidly increasing knowledge of Proterozoic plankton is making possible the recognition of additional episodes of diversification and extinction that antedate the Ediacaran radiation of macroscopic animals. In contrast to earlier phytoplankton history, the late Mesozoic and Cainozoic record is documented in sufficient detail to constrain theories of mass extinction in more than a general way. Broad patterns of diversity change in planktonic algae show similarities across the Cretaceous–Tertiary and Eocene–Oligocene boundaries, but detailed comparisons of origination and extinction rates in calcareous nannoplankton, as well as other algae and skeletonized protozoans, suggest that the two episodes were quite distinct. Common causation appears unlikely, casting doubt on monolithic theories of mass extinction, whether periodic or not. Studies of mass extinction highlight a broader class of insights that palaeontologists can contribute to evolutionary biology: the evaluation of evolutionary change in the context of evolving Earth-surface environments.

1. Introduction

Palaeontologists seeking to understand evolution from the evidence of the fossil record are a bit like astronomers who rely on Earth-based telescopes to study the heavens. The atmosphere is transparent or nearly so to radiation in some spectral regions, but opaque for others. Thus there are some astrophysical phenomena that the astronomer can observe clearly and others about which he or she can infer nothing at all. Because of the selective information loss attendant on fossilization, the palaeontological record is similarly 'transparent' to some evolutionary phenomena, permitting one to document palaeobiological pattern and draw reasonable inferences about process, whereas it is 'opaque' for others. Complicating interpretation and the search for evolutionary generality is the fact that the several major fossil records – those of terrestrial animals, land plants, marine invertebrate animals, benthic protists and microplankton – have different regions of evolutionary 'opacity' and 'transparency'. Thus palaeobotanical patterns permit inferences about the importance of resource competition in tracheophyte evolution but shed relatively little light on the evolutionary role of herbivory (see, for example, Knoll 1986; Knoll & Niklas 1987; Lidgard & Crane 1988; but see Wing & Tiffney 1987), whereas for the marine invertebrate record, much the opposite appears to be true (see, for example, Benton 1987; but see Jackson 1988; Vermeij 1987). This, of course, does not necessarily indicate that plant and marine invertebrate evolution are driven primarily by competition and predation, respectively. It means that the nature of plants and animals and the differing properties of their fossil records make such comparisons extremely difficult.

[39]

Studies of evolutionary responses to environmental change are attractive, in part, because they provide a focus for the direct palaeobiological comparison of disparate taxonomic and ecological groups of organisms. In this context, the fossil record of phytoplankton assumes particular significance. The distinctive ecology, life cycles and metabolism of planktonic algae provide useful counterpoint to often zoocentric discussions of evolution and extinction in the marine realm. Further, phytoplankton are widely distributed in sedimentary rocks and have a long evolutionary record; plants and animals may be the most conspicuous fossils in Phanerozoic rocks, but planktonic protists are the most abundant. Particularly important, the calcareous tests of at least some photosynthetic and heterotrophic microplankton species preserve isotopic indications of the relative depth, temperature and productivity of the water masses in which they formed (Douglas & Savin 1978; Boersma *et al.* 1987). Thus planktonic microfossils can provide sensitive and stratigraphically precise indications of change in physical environments. In so far as mass extinctions are widely viewed as the consequences of environmental change or perturbation, phytoplankton, then, can do much to constrain thinking about these signal evolutionary events.

2. The earliest record: Proterozoic microplankton

Prokaryotes undoubtedly colonized marine water columns early in Earth history, but the oldest fossils reasonably interpreted as plankters are spheroidal Problematica found in *ca.* 2000 Ma old rocks from Canada (Hofmann 1976; Knoll *et al.* 1978). The oldest probable eukaryotic microplankton are assemblages of acritarchs from 1400–1800 Ma old successions in China (Zhang 1986), the Soviet Union (Keller & Jankauskas 1982), Australia (Peat *et al.* 1978), and the United States (Horodyski 1980). All consist of morphologically simple, spheroidal cysts of limited apparent diversity; at least in Australia, this morphological record is complemented by abundant steranes, biomarker molecules that independently confirm the role of eukaryotes in Middle Proterozoic ecosystems (Jackson *et al.* 1986).

The patchy stratigraphic coverage and limited morphological complexity of these early remains preclude statements about radiation and extinction, but it is clear that by 850–950 Ma ago, a significant morphological radiation of phytoplankton had begun. The plankton record of this era has long been known to include spheroidal acritarchs with modestly sculptured walls (see, for example, Vidal 1976; Jankauskas 1982), but discoveries over the past decade indicate that a wide variety of acanthomorphic and otherwise complexly ornamented forms also evolved, many of them unusually large relative to Palaeozoic taxa (see, for example, Timofeev *et al.* 1976; Knoll 1984; Yin 1985, 1987; Butterfield *et al.* 1988; Zang 1988; Zang & Walter 1989). Critical data on stratigraphic distribution remain limited, but on present evidence it appears that the Proterozoic acritarch biota reached its zenith in both diversity and morphological complexity during the Vendian interval (*ca.* 600 Ma ago) (Zang 1988; Zang & Walter 1989). The youngest assemblages that contain distinctively Proterozoic, large, process-bearing cysts occur in rocks stratigraphically above Varangian glacial deposits but below Ediacaran metazoans (see Awramik *et al.* 1985; Yin 1985, 1987; Zang 1988; Knoll & Ohta 1988). Latest Proterozoic biotas are dominated by simple leiosphaerid acritachs and small, spiny micrhystrids (Volkova 1972; Vidal & Knoll 1983). Thus by the time large animals first diversified, planktonic protists had already undergone at least one protracted radiation and one or more episodes of major extinction (Vidal & Knoll 1983; Zang 1988; Zang & Walter 1989). The early phytoplankton record effectively doubles, and with work may treble, the span of geological history through which palaeontological patterns of extinction can be studied.

3. Increasing diversity and turnover: the Palaeozoic record

Prasinophyte green algae and other cyst-forming phytoplankton diversified markedly during the early Cambrian, and an even more pronounced radiation followed during the Ordovician (Volkova *et al.* 1979; Tappan 1980; Vidal & Knoll 1983). Skeletonized microplankton – the siliceous radiolaria and problematic scale-forming protists – also appeared during this interval (Nazarov & Ormiston 1985; Allison & Hilgert 1986). Recently, it has been established that phytoplankton experienced a brief but pronounced extinction episode at the end of the Ordovician period (Colbath 1986; Martin 1988). Recovery was rapid, as previously subordinate as well as newly evolved genera diversified to establish a diverse but distinctively Silurian phytoplankton biota (Downie 1973; Dorning 1981). Widespread extinctions again reduced acritarch diversity markedly near the end of the Devonian Period, and, curiously, the diversity of organic-walled phytoplankton remained low throughout the remainder of the Palaeozoic era (Downie 1973; Tappan 1980). The precise timing of this diversity collapse is not clear, but judging from recent publications (see, for example, Martin 1984; Wicander & Playford 1985), many late Devonian extinctions may have occurred within the Famennian or close to the Devonian–Carboniferous boundary, rather than being concentrated at the Frasnian–Fammenian boundary, the point most commonly inferred to have been a time of abrupt ecosystem collapse (see, for example, McGhee 1988). Because some extant microalgae do not produce preservable cysts, and some form them only under particular environmental conditions, the absence of an obvious post-Devonian rebound in acritarch diversity does not necessarily mean that planktonic algae remained depauperate for the remainder of the era. It is equally possible that late Palaeozoic oceans teemed with phytoplankton, but that the dominant taxa did not produce fossilizable cysts. (Although this confounding feature of organic-walled microplankton fossils is particularly pertinent to discussions of late Palaeozoic phytoplankton, it must, of course, be kept in mind in all considerations of acritarch and dinoflagellate diversity.) According to Tappan (1982), acritarchs suffered further extinctions at the close of the Permain.

The major intervals of radiation (early Cambrian, Ordovician) and extinction (end Ordovician, late Devonian, late Permian) seen in Palaeozoic phytoplankton clearly parallel those documented for skeletonized marine invertebrates (Sepkoski 1982) (figure 1). Benthic calcareous algae also approximate this pattern: lightly calcified Problematica diversified early in the Cambrian and more-heavily skeletonized red and green algae radiated during the Ordovician, whereas major extinctions mark the ends of the Devonian and Permian (Chuvashov & Riding 1984; Flügel 1985). At the very least, these parallels constrain thinking about Palaeozoic evolution by indicating that the major episodes of diversification, extinction, and innovation in biomineralization documented for marine invertebrates were not limited to animals, but were ocean-wide in extent. Hypotheses advanced to explain any one of these events must apply to photoautotrophs as well as heterotrophs, single-celled protists as well as developmentally complex animals, and plankton as well as benthos.

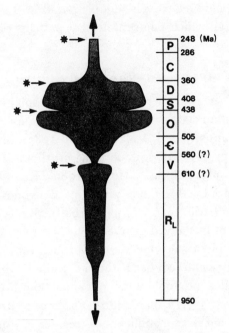

FIGURE 1. Diagram showing major patterns of change in phytoplankton cyst diversity during the late Proterozoic and Palaeozoic eras. Much remains to be accomplished in the taxonomic description of early phytoplankton; thus the figure is drawn to convey only general patterns of diversification and extinction, and a diversity scale for the diagram has been omitted intentionally. Asterisks with arrows indicate times of major extinction. Stratigraphic intervals are indicated by initials as follows: R_L, Late Riphean; V, Vendian; €, Cambrian; O, Ordovician; S, Silurian; D, Devonian; C, Carboniferous; P, Permian. The timescale is that of Harland *et al.* (1982), except for the beginning and end of the Vendian interval. Question marks denote uncertainties in those dates.

4. MESOZOIC AND CAINOZOIC RADIATIONS AND EXTINCTIONS

The Mesozoic era was a time of major morphological and physiological innovation among pelagic phytoplankton. At least some Palaeozoic acritarchs may have dinoflagellate affinities, but dinoflagellates that produce cysts unambiguously attributable to that group first became diverse and abundant during the late Triassic, with notable Jurassic and, especially, mid- to late Cretaceous radiations. Calcareous nannoplankton (largey attributable to the Coccolithophyceae, but also including the extinct discoasters and nannoconids, as well as – in some treatments – a few calcareous dinoflagellates) also have a scattered and sometimes questioned Palaeozoic record, but they, too, rose to ecological abundance in the Jurassic and diversified markedly during the second half of the Cretaceous. Diatoms and silicoflagellates similarly show Mesozoic origins and notable late Cretaceous increases in abundance and diversity.

Modern ocean basins date from this same time, providing the geological advantage of thick, often relatively complete and little-altered sedimentary sequences, as well as a much improved record of palaeogeography and palaeoceanographic circulation. (Undoubtedly, this circumstance contributes to the *perception* of Mesozoic phytoplankton radiations. The minute calcite plates of calcareous nannoplankton and delicate opaline silica frustrules produced by diatoms might well be destroyed in older, more highly altered sediments, and if early dinoflagellates lacked archaeopyles, their affinities would not be recognizable. None the less, the clear diversity changes recorded within Mesozoic sedimentary successions containing well-preserved fossils indicate that during the late Mesozoic, a genuine radiation of microplankton took place, one that included heterotrophic as well as photoautotrophic protists.) Carbon, oxygen and strontium isotopic ratios in microplankton tests themselves contribute to a refined understanding of climatic, oceanographic, and biogeochemical change over the last 100 Ma. Thus, compared with those of earlier eras, late Mesozoic and Cainozoic sediments not only contain a much expanded and improved record of phytoplankton evolution; they also provide a far better documentation of the environmental context in which that evolution took place.

Arguments for cyclicity in mass extinction (Raup & Sepkoski 1984), with their implied unity of causation, rest largely on the timing and nature of events during this interval. At first glance, the Mesozoic–Cainozoic phytoplankton record, like that of Palaeozoic acritarchs, appears to corroborate palaeozoological data on the timing and severity of mass extinctions. For example, dinoflagellates show significant decreases in diversity during the Norian, Tithonian, Cenomanian, Campanian–Maastrichtian, late Eocene, and Middle Miocene ages (Bujak & Williams 1979). Calcareous nannoplankton also show diversity drops during the late Jurassic, Cenomanian, Maastrichtian, and late Eocene (Haq 1973; Roth 1987). However, closer examination of these diversity changes, especially the relatively well documented events at the Cretaceous–Tertiary and Eocene–Oligocene boundaries, reveals a more complex picture that places sharp constraints on our general thinking about these events. In the following paragraphs, I briefly review the much scrutinized record of phytoplankton extinctions across the Cretaceous–Tertiary (K–T) and Eocene–Oligocene (E–O) boundaries and then record a few observations about the less frequently discussed patterns of phytoplankton originations associated with the same events.

Pelagic ecosystems experienced a profound disruption at the end of the Cretaceous period. Among calcareous nannoplankton and planktonic foraminifera, rates of extinction increased sharply at the K–T boundary, resulting in nearly total devastation of the diverse biotas that had populated late Cretaceous oceans (Bramlette & Martini 1964; Percival & Fischer 1977; Perch-Nielsen *et al.* 1982; Smit & Romein 1985; Jiang & Gartner 1986; Brinkhuis & Zachariasse 1988). Arguments persist as to whether the extinctions occurred instantly, in rapid pulses, or gradually through 10000 to 100000 years, but there is no question that the terminal Cretaceous event must be considered abrupt on any save perhaps the finest of geological timescales.

This is not to say that the terminal Cretaceous event was visited upon an evolutionarily static world. Several marine invertebrate groups declined in diversity though part or all of the Maastrichtian, and a similar pattern can be seen in the calcareous nannoplankton (Roth 1987) (figure 2).

The dinoflagellate record also shows a major late Cretaceous diversity reduction, but extinctions do not cluster at the K–T boundary. After a Santonian peak in species diversity, dinoflagellate extinctions reduced diversity more or less continuously throughout the Campanian and Maastrichtian (Bujak & Williams 1979). Detailed analyses of dinoflagellate assemblages in K–T boundary sections show changes in relative abundance but few if any extinctions at or near the boundary (Hultberg 1986; Brinkhuis & Leereveldt 1988; Brinkhuis & Zachariasse 1988). Diatoms and prasinophytes also appear to have weathered the K–T event relatively well (Tappen 1982; Fenner 1985; Kitchell *et al.* 1986). A characteristic common to many species within these three groups is the capacity to form a non-planktonic resting stage during periods of nutrient deprivation or other ecological stress (see, for example, Tappan 1980; Kitchell *et al.* 1986).

Extinction patterns across the Eocene–Oligocene boundary contrast with those seen at the end of the Cretaceous. As in the Maastrichtian, late Eocene to Early Oligocene phytoplankton extinctions took place throughout an interval of several million years. However, the younger episode is not terminated by a sudden dramatic rise in extinction rates. Phytoplankton do not show uniquely high or, in some groups, even notable increases extinction rates at the E–O boundary (Corliss *et al.* (1984); Perch-Nielsen (1986), and other papers in Pomerol & Premoli-Silva (1986)) (see figure 2). Last appearances among species are not synchronous (see Beckman

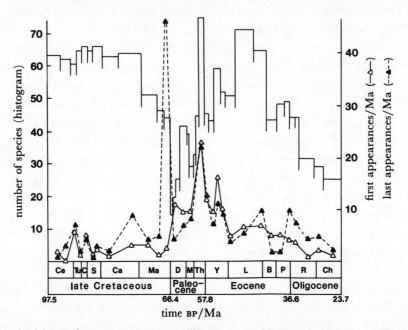

FIGURE 2. Taxonomic richness, first appearances per million years, and last appearances per million years for Late Cretaceous and Palaeogene calcareous nannoplankton species. Not all living species leave readily preservable coccoliths in sea-floor sediments. Therefore the fossil record probably provides an incomplete sampling of coccolithophorid diversity at any one time, a problem well known to invertebrate palaeontologists and palaeobotanists. In terms of the present discussion, which simply seeks to ascertain whether or not first and last appearances are clustered at or near stratigraphic horizons perviously identified as times of mass extinction, the data are considered adequate to the task. Species are defined palaeontologically as discrete morphological entities; palaeospecies do not necessarily correspond to biological species recognized in the living biota (see, for example, Aubry 1988). Data are plotted by nannoplankton zone, with the exception of origination and extinction data for zones shorter than one million years. In those cases, first and last appearances have been combined with those of the shorter adjacent zone. Cretaceous data are replotted from Roth (1987); Palaeogene data were compiled from Perch-Nielsen (1985a). The timescale follows Berggren *et al.* (1985) and Kent & Gradstein (1985). Ages indicated by initials are as follows: Ce, Cenomanian; Tu, Turonian; C, Coniacian; Ca, Campanian; Ma, Maastrichtian; D, Danian, M, Montian; Th, Thanetian; Y, Ypresian; L, Lutetian; B, Bartonian; P, Priabonian; R, Rupelian; Ch, Chattian. See Roth (1987) for a discussion of how such data are assembled, their advantages and disadvantages.

et al. 1981; Corliss *et al.* 1984; Saunders *et al.* 1984), and even within species last appearances at different latitudes may be diachronous by as much as 500000 years (see, for example, Beckman *et al.* 1981). With some justification, Corliss & Keigwin (1986) have questioned whether the E–O event should be called a mass extinction at all. Calcareous nannoplankton, among other groups, experienced equally high or higher rates of extinction several times earlier in the Palaeogene (figure 2).

Inspection of origination rates highlights additional distinctions between the K–T and E–O boundaries. Immediately after the K–T extinction episode, so-called 'disaster plankton', especially the calcareous dinoflagellate *Thoracosphaera*, bloomed throughout depopulated oceans (Perch-Nielsen 1985 a). Subsequent early Paleocene origination rates were quite high in calcareous nannoplankton (Perch-Nielsen 1985 a) (figure 2), as well as in foraminifera. In contrast, the most interesting feature of turnover at the E–O boundary are the unusually low rates of origination characterizing early Oligocene phytoplankton. This can be seen in figure 2, and additionally characterizes silicoflagellates, ebridians and dinoflagellates (Bujak & Williams 1979; Tappan 1980). In many phytoplankton groups, the relative abundances of species change at several points during late Eocene and early Oligocene time, including at or near the epoch boundary (see, for example, Fenner 1985); however, unlike the aftermath of the K–T extinction, there is no spike of 'disaster' forms.

Thus, to judge from the marine phytoplankton record, the K–T and E–O events appear to be quite different. The K–T boundary is characterized by truly high rates extinction through a very short interval followed by expansion of 'disaster' species and, shortly thereafter, high rates of diversification. This pattern is consistent with hypotheses favouring ecological catastrophy, in particular the physical, chemical, climatic and/or productivity effects of a bolide impact (Alvarez et al. 1980). Despite the presence of microtektites and an iridium anomaly near the E–O boundary (Alverez et al. 1982; Ganapathy 1982), this transition is characterized by moderate rates of extinction spread through several million years followed by unusually low rates of origination, a pattern consistent with other data indicating a complex series of climatic and oceanographic changes that reduced both productivity and water-mass heterogeneity, eliminating the habitats of narrowly adapted species and simultaneously limiting the oceanographic opportunities for continued diversification (Corliss et al. 1984; Lipps 1986). My intent here is not so much to advocate particular causal explanations as to emphasize that the two best-known episodes widely characterized as mass extinctions were in fact very different events. This casts doubt on any monolithic theory of mass extinction, whether periodic or not.

5. Discussion

The phytoplankton record highlights several general and unresolved issues in the biology of diversification and extinction. First, there is the question of how mass extinctions relate to biological innovation. It has been proposed that mass extinction is a major force in the generation of evolutionary novelty, breaking the 'stagnation [of long running macro-evolutionary régimes] by clearing ecospace for the radiation of new lineages' (Sepkoski 1985, p. 230). This appears to be the case for many animal groups, but phytoplankton present a more complicated picture. Morphological and physiological innovation are not obviously tied to mass extinction events. Microplanktonic species that proliferate after major extinctions look tolerably like those that disappeared; indeed the zooplanktonic foraminifera are famous for their patterns of iterative evolution (Lipps (1986) and papers cited therein). Figure 3 illustrates patterns of origination and extinction for late Cretaceous and Palaeogene calcareous nannoplankton genera. Plots of generic turnover should provide a better indication of morphological innovation than those of species, because genera are erected according to major features of morphology. High rates of origination follow the K–T boundary, but they also characterize other intervals of late Cretaceous and Palaeogene time. These latter bursts of diversification precede rather than follow extinctions and appear to reflect palaeoceanographic changes that promoted opportunities for diversification (see, for example, Roth 1987). Indeed, the most dramatic phytoplankton radiation of the Cainozoic era, the Miocene diversification of

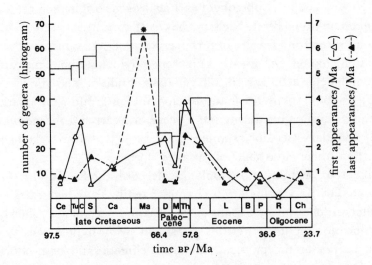

FIGURE 3. Taxonomic richness, first appearances per million years, and last appearances per million years for calcareous nannoplankton genera, plotted by age. The asterisk draws attention to the fact that the severity of the terminal Cretaceous event is underestimated because the Maastrichtian extinction rate was determined by dividing total extinctions by the 8 Ma duration of the age, whereas many last appearances are actually concentrated at or near the boundary. Data are compiled from Perch-Nielsen (1985a, b). Timescale as indicated in figure 2.

diatoms (Tappan 1980), cannot be related to the clearing of ecospace by mass extinction. In that there are long-term evolutionary trends in phytoplankton, these probably reflect continuing selection pressures of organism–organism interactions and, possibly, directional changes in the marine environment.

Another observation whose generality is challenged by the phytoplankton record is Jablonski's (1986a, b) hypothesis that characters selected for during 'normal' macroevolutionary régimes are distinct from those that promote survival during catastrophic mass extinctions. In theory, one might expect that an exception to this would be adaptations that facilitate survival during periods of ecological stress. Such features may be rare in marine invertebrates, but they are characteristic features of the life cycles of many bacteria, protists and plants. As Kitchell *et al.* (1986) have suggested for diatoms, patterns of differential phytoplankton survival across the K–T boundary may reflect just such adaptations.

Seeking another type of generalization, one can, perhaps optimistically, ask whether the details of phytoplankton origination and extinction that shed such useful light on terminal Cretaceous and Tertiary extinctions might prove equally valuable in constraining hypotheses about earlier episodes of mass extinction. Colbath (1986), working on the terminal Ordovician extinction, has made important strides in this direction, but much remains to be learned. Also, can the palaeoceanographic insights that isotopic geochemistry has provided for late Mesozoic and Cainozoic oceans be applied more fruitfully to older systems? Surely, the dearth of earlier calcareous plankton and deep-ocean sedimentary records limits direct extension of Cainozoic techniques, but can we recognize physical, chemical or sedimentological correlates of oceanographic change in shelf or epicratonic sediments that will improve our understanding of the palaeoenvironmental context of Proterozoic or Palaeozoic radiations and extinctions?

Finally, let me suggest that the study of mass extinctions represents a special case of a broader class of inquiry through which palaeontologists can make important contributions to

[46]

evolutionary biology. The fundamental question is, how do populations and communities respond to environmental change? Earth history is marked by repeated changes in climate, oceanographic circulation, productivity, tectonic activity, geography and atmospheric composition of an order that lies outside of the range of ecological observation or experiment. If Cainozoic palaeoceanography provides any indication, such change is a continuing feature of the Earth's surface (see, for example, Boersma *et al.* 1987). On the geological timescales on which evolution is played out, the physical development of our planet may be a major engine of evolutionary change. In making this statement, I do not wish to minimize the importance of the biological component of the environments of the organisms (for microplankton, see Kitchell (1983) and other papers discussed in Knoll (1987)); nor do I deny that evolutionary change is itself a significant contributor to the development of Earth-surface environments. The context of biological history is the interplay between non-directional environmental perturbations of varying magnitude and directionally changing features of the biological and physical environment. Depending on the timescale, spatial dimensions and magnitude of environmental change, population responses may include migration, the evolution of broad tolerances or plastic responses or both, new adaptations, or extinction. The important point is that we can no longer assume in theory or practice that evolution continues for long periods of time under conditions of environmental constancy. Rapidly growing sophistication in our understanding of the historical development of Earth-surface environments provides a rich new framework for the interpretation of palaeobiological pattern, including extinctions.

I thank R. Norris and S. Damassa for helpful discussions and M. R. Walter and Zang Wenlong for permission to cite their as yet unpublished ideas on Proterozoic extinctions. Research leading to this paper was made possible by a fellowship from the John Simon Guggenheim Foundation and grants from NSF and NASA.

REFERENCES

Allison, C. W. & Hilgert, J. W. 1986 Scales microfossils from the early Cambrian of northwest Canada. *J. Paleont.* **60**, 973–1015.

Alvarez, L. W., Alvarez, W., Asaro, F. & Michel, H. V. 1980 Extraterrestrial cause for the Cretaceous–Tertiary extinction. *Science, Wash.* **208**, 1095–1108.

Alvarez, W., Asaro, F., Michel, H. V. & Alvarez, L. 1982 Iridium anomaly approximately synchronous with terminal Eocene extinctions. *Science, Wash.* **216**, 886–888.

Aubry, M.-P. 1988 Phylogeny of the Cenozoic calcareous nannoplankton genus *Helicosphaera*. *Paleobiology* **14**, 64–80.

Awramik, S. M., McMennamin, D. S., Yin, C., Zhao, Z., Ding, Q. & Zhang, S. 1985 Prokaryotic and eukaryotic microfossils from a Proterozoic/Phanerozoic transition in China. *Nature, Lond.* **315**, 655–658.

Beckman, J. P., Bolli, H. M., Perch-Nielsen, K., Decima, F. P., Saunders, J. B. & Toumarkine, M. 1981 Major calcareous nannofossil and foraminiferal events between the middle Eocene and early Miocene. *Palaeogeogr. Palaeoclimat. Palaeoecol.* **36**, 155–190.

Benton, M. J. 1987 Progress and competition in macroevolution. *Biol. Rev.* **62**, 305–338.

Berggren, W. A., Kent, D. V., Flynn, J. J. & van Couvering, J. A. 1985 Cenozoic geochronology. *Bull. geol. Soc. Am.* **96**, 1407–1418.

Boersma, A., Premoli-Silva, I. & Shackleton, N. J. 1987 Atlantic Eocene planktonic foraminiferal paleohydrographic indicators and stable isotope paleoceanography. *Paleoceanography* **2**, 287–331.

Bramlette, M. N. & Martini, E. 1964 The great change in calcareous nannoplankton fossils between the Maastrichtian and the Danian. *Micropaleontology* **10**, 291–322.

Brinkhuis, H. & Leereveldt, H. 1988 Dinoflagellate cysts from the Cretaceous/Tertiary boundary sequence of El Kef, northwest Tunisia. *Rev. Palaeobot. Palynol.* **56**, 5–20.

Brinkhuis, H. & Zachariasse, W. J. 1988 Dinoflagellate cysts, sea level changes, and planktonic foraminifers across the Cretaceous–Tertiary boundary at El Haria, northwest Tunisia. *Mar. Micropaleont.* **13**, 151–191.

Bujak, J. P. & Williams, G. L. 1979 Dinoflagellate diversity through time. *Mar. Micropaleont.* **4**, 1–12.

Butterfield, N. J., Knoll, A. H. & Swett, K. 1988 Exceptional preservation of fossils in an Upper Proterozoic shale. *Nature, Lond.* **334**, 424–427.

Chuvashov, B. & Riding, R. 1984 Principal floras of Palaeozoic marine calcareous algae. *Palaeontology* **27**, 487–500.

Colbath, G. K. 1986 Abrupt terminal Ordovician extinction in phytoplankton associations, southern Appalachians. *Geology* **14**, 943–946.

Corliss, B. H., Aubry, M.-P., Berggren, W. A., Fenner, J. M., Keigwin, L. D. Jr & Keller, G. 1984 The Eocene/Oligocene boundary event in the deep sea. *Science, Wash.* **226**, 806–810.

Corliss, B. H. & Keigwin, L. D. Jr 1986 Eocene–Oligocene paleoceanography. In *Mesozoic and Cenozoic oceans* (ed. K. J. Hsü), pp. 101–118. Washington: American Geophysical Union.

Dorning, K. J. 1981 Silurian acritarchs from the type Wenlock and Ludlow of Shropshire, England. *Rev. Palaeobot. Palynol.* **34**, 175–203.

Douglas, R. G. & Savin, S. M. 1978 Oxygen isotope evidence for the depth stratification of tertiary and Cretaceous planktonic foraminifera. *Mar. Micropaleont.* **3**, 175–196.

Downie, C. 1972 Observations on the nature of acritarchs. *Palaeontology* **16**, 239–259.

Fenner, J. 1985 Late Cretaceous to Oligocene planktic diatoms. In *Plankton stratigraphy* (ed. H. M. Bolli, J. B. Saunders & K. Perch-Nielsen), pp. 713–762. Cambridge University Press.

Flügel, E. 1985 Diversity and environments of Permian and Triassic dasycladacean algae. In *Paleoalgology* (ed. D. M. Toomey & M. Nitecki), pp. 344–351. Berlin: Springer-Verlag.

Ganapathy, R. 1982 Evidence for a major meteorite impact on Earth 34 million years ago. *Science, Wash.* **216**, 885–886.

Harland, W. B., Cox, A. V., Llewellyn, P. G., Pickton, C. A. G., Smith, A. G. & Walters, R. 1982 *A geologic time scale.* Cambridge University Press.

Haq, B. W. 1973 Transgressions, climatic change and the diversity of calcareous nannoplankton. *Mar. Geol.* **15**, M25–M30.

Hofmann, H. J. 1976 Precambrian microflora, Belcher Islands, Canada: significance and systematics. *J. Paleont.* **50**, 1040–1073.

Horodyski, R. J. 1980 Middle Proterozoic shale-facies microbiota from the lower Belt Supergroup, Little Belt Mountains, Montana. *J. Paleont.* **54**, 649–663.

Hultberg, S. U. 1986 Danian dinoflagellate zonation, the C–T boundary and the stratigraphical position of the Fish Clay in southern Scandinavia. *J. Micropalaeont.* **5**, 37–47.

Jablonski, D. 1986a Causes and consequences of mass extinctions: a comparative approach. In *Dynamics of extinction* (ed. D. K. Elliott), pp. 183–229. New York: Wiley.

Jablonski, D. 1986b Background and mass extinctions: the alternation of macroevolutionary regimes. *Science, Wash.* **231**, 129–133.

Jackson, J. B. C. 1988 Does ecology matter? *Paleobiology* **14**, 307–312.

Jackson, M. J., Powell, T. G., Summons, R. E. & Sweet, I. P. 1986 Hydrocarbon shows and petroleum source rocks in sediments as old as 1.7×10^9 years. *Nature, Lond.* **322**, 727–729.

Jankauskas, T. V. 1982 Microfossils from the Riphean of the southern Urals. In *Stratotype of Riphean: paleontology and paleomagnetics* (ed. B. M. Keller), pp. 84–120. Moscow: Nauka Acad. Sci. U.S.S.R. Geol. Inst. (In Russian).

Jiang, M. J. & Gartner, S. 1986 Cretaceous nannofossil succession across the Cretaceous–Tertiary boundary in east-central Texas. *Micropaleont.* **32**, 232–255.

Keller, B. M. & Jankauskas, R. V. 1982 Microfossils in the Riphean stratotype section in the Southern Urals. *Int. Geol. Rev.* **24**, 925–933.

Kent, D. V. & Gradstein, F. M. 1985 A Cretaceous and Jurassic geochronology. *Bull. geol. Soc. Am.* **96**, 1419–1427.

Kitchell, J. A. 1983 Biotic interactions and siliceous marine phytoplankton: an ecological and evolutionary perspective. In *Biotic interactions in Recent and fossil benthic communities* (ed. M. J. S. Tevesz & P. L. McCall), pp. 285–329. New York: Plenum.

Kitchell, J. A., Clark, D. L. & Gombos, A. M. Jr 1986 Biological selectivity of extinction: a link between background and mass extinctions. *Palaios* **1**, 504–511.

Knoll, A. H. 1984 Microbiotas of the late Precambrian Hunnberg Formation. *J. Paleont.* **58**, 131–162.

Knoll, A. H. 1986 Patterns of change in land plant communities through geological time. In *Community ecology* (ed. J. Diamond & T. J. Case), pp. 126–141. New York: Harper & Row.

Knoll, A. H. 1987 Protists and Phanerozoic evolution in the oceans. In *Fossil prokaryotes and protists: notes for a short course* (ed. J. H. Lipps), pp. 248–264. Knoxville, Tennessee: University of Tennessee Studies in Geology No. 18.

Knoll, A. H., Barghoorn, E. S. & Awramik, S. M. 1978 New microorganisms from the Aphebian Gunflint Iron Formation, Ontario. *J. Paleont.* **52**, 976–992.

Knoll, A. H. & Niklas, K. J. 1987 Adaptation, plant evolution, and the fossil record. *Rev. Palaeobot. Palynol.* **50**, 127–149.

Knoll, A. H. & Ohta, Y. 1988 Microfossils in metasediments from Prins Karls Forland, western Svalbard. *Polar Res.* **6**, 59–67.

Lidgard, S. & Crane, P. R. 1988 Quantitative analyses of the early angiosperm radiation. *Nature, Lond.* **331**, 344–346.

Lipps, J. H. 1986 Extinction dynamics in pelagic ecosystems. In *Dynamics of extinction* (ed. D. K. Elliott), pp. 87–104. New York: Wiley.

Martin, F. 1983 (1984) Acritarches du Frasnien superieur et du Fammenien inferieur du bord méridional du Bassin de Dinant (Ardenne belge). *Bull. Inst. r. Sci. nat. Belg.* (Sci. Terre) **55**(7), 1–57.

Martin, F. 1988 Late Ordovician and early Silurian acritarchs. *Bull. Br. Mus. nat. Hist.* A **43**, 299–309.

McGhee, G. R. Jr 1988 The late Devonian extinction event: evidence for abrupt ecosystem collapse. *Paleobiology* **14**, 250–257.

Nazarov, B. & Ormiston, A. R. 1985 Evolution of radiolaria in the Paleozoic and its correlation with the development of other marine fossil groups. *Senckenberg. leth.* **66**, 203–235.

Peat, C. J., Muir, M. D., Plumb, K. A., McKirdy, D. M. & Norvick, M. S. 1978 Proterozoic microfossils from the Roper Group, Northern Territory, Australia. *Bur. Mineral Res. J. Aust. Geol. Geophys.* **3**, 1–17.

Perch-Nielsen, K. 1985*a* Cenozoic calcareous nannofossils. In *Plankton stratigraphy* (ed. H. M. Bolli, J. B. Saunders & K. Perch-Nielsen), pp. 427–554. Cambridge University Press.

Perch-Nielsen, K. 1985*b* Mesozoic calcareous nannofossils. In *Plankton stratigraphy* (eds. H. M. Bolli, J. B. Saunders & K. Perch-Nielsen), pp. 329–426. Cambridge University Press.

Perch-Nielsen, K. 1986 Calcareous nannofossil events near the Eocene/Oligocene boundary. In *Terminal Eocene events* (ed. C. Pomerol & I. Premoli-Silva), pp. 274–287. Amsterdam: Elsevier.

Perch-Nielsen, K., McKenzie, J. & He, Q. 1982 Biostratigraphy and isotopic geochemistry and the "catastrophic" extinctions of calcareous nannoplankton at the Cretaceous/Tertiary boundary. *Spec. Pap. geol. Soc. Am.* **190**, 353–371.

Percival, S. F. & Fischer, A. G. 1977 Changes in calcareous nannoplankton in the Cretaceous–Tertiary biotic crisis at Zumaya, Spain. *Evol. Theor.* **2**, 1–35.

Pomerol, C. & Premoli-Silva, I. (eds) 1986 *Terminal Eocene events.* Amsterdam: Elsevier.

Raup, D. M. & Sepkoski, J. J. Jr 1984 Periodicity of extinctions in the geologic past. *Proc. natn. Acad. Sci. U.S.A.* **81**, 801–805.

Roth, P. H. 1987 Mesozoic calcareous nannofossil evolution: relation to paleoceanographic events. *Paleoceanography* **2**, 601–611.

Saunders, J. B., Bernoulli, D., Müller-Merz, E., Oberhänsli, H., Perch-Nielsen, K., Riedel, W., Sanfilippo. A. & Torrini, R. Jr 1984 Stratigraphy of the late Middle Eocene to Oligocene in the Bath Cliff section, Barbados, West Indies. *Micropaleontology* **30**, 390–425.

Sepkoski, J. J. Jr 1982 Mass extinctions in the Phanerozoic oceans: a review. *Spec. pap. geol. Soc. Am.* **190**, 283–289.

Sepkoski, J. J. Jr 1985 Some implications of mass extinctions for the evolution of complex life. *Int. Astronom. Union Symp.* **112**, 223–232.

Smit, J. & Romein, A. J. T. 1985 A sequence of events across the Cretaceous–Tertiary boundary. *Earth planet. Sci. Lett.* **74**, 155–170.

Tappan, H. 1980 *The paleobiology of plant protists.* San Francisco: Freeman.

Tappan, H. 1982 Extinction or survival: selectivity and causes of Phanerozoic crises. *Spec. Pap. geol. Soc. Am.* **190**, 265–276.

Timofeev, B. V., German, T. N. & Mikhailova, N. 1976 *Plant microfossils of the Precambrian, Cambrian, and Ordovician.* Leningrad: Scientific Institute of Precambrian Geology and Geochronology.

Vermeij, G. J. 1987 *Evolution and escalation: an ecological history of life.* Princeton University Press.

Vidal, G. 1976 Late Precambrian microfossils from the Visngsö Beds in southern Sweden. *Fossils Strata* **9**, 1–54.

Vidal, G. & Knoll, A. H. 1983 Proterozoic plankton. *Mem. geol. Soc. Am.* **161**, 265–277.

Volkova, N. 1973 Acritarchs and the correlation of Vendian and Cambrian strata of the Russian Platform. (In Russian.) *Sov. Geol.* **4**, 48–62.

Volkova, N., Kirjanov, V. V., Piskun, L. V., Paskeviciene, L. T. & Jankauskas, T. V. 1979 (English translation 1983) Plant microfossils. In *Upper Precambrian and Cambrian paleontology of the east European platform* (ed. A. Urbanek & A. Yu. Rozanov), pp. 7–46. Warsaw: Publishing House Wyndawnnictwa.

Wicander, R. & Playford, G. 1985 Acritarchs and spores from the Upper Devonian Lime Creek Formation, Iowa, U.S.A. *Micropaleontology* **31**, 97–138.

Wing, S. L. & Tiffney, B. H. 1987 The reciprocal interaction of angiosperm evolution and tetrapod herbivory. *Rev. Palaeobot. Palynol.* **50**, 179–210.

Yin, L. 1985 Microfossils of the Doushantou Formation in the Yangtze Gorge district, Western Hubei. *Palaeontologia Cathay.* **2**, 229–249.

Yin, L. 1987 Microbiotas of latest Precambrian sequences in China. *Stratigr. Palaeont. Systemic Bound. China, Precambr.-Cambr.* **1**, 415–494.

Zang, W. 1988 Late Proterozoic – early Cambrian microfossils and biostratigraphy in China and Australia. Ph.D. thesis, Australian National University, Canberra, Australia.

Zang, W. & Walter, M. R. 1989 Latest Proterozoic plankton from the Amadeus Basin in central Australia. *Nature, Lond.* **337**, 642–645.

Zhang, Z. 1986 Clastic facies microfossils from the Chuanlinggou Formation (1800 Ma) near Jixian, north China. *J. Micropalaeont.* **5**, 9–16.

Discussion

J.-J. JAEGER (*Department of Vertebrate Palaeontology, Pierre and Marie Curie University, Paris, France*). Would it be possible to relate the increase of the diversity of the phytoplankton through time with an increase in its biomass?

A. H. KNOLL. The short answer to Dr Jaeger's question is probably 'no', but given the importance of the question, let me explain briefly why I hold this view. The first problem is that we have no direct way of estimating phytoplankton biomass for ancient oceans. In that chlorophyll concentration (a parameter related to phytoplankton biomass) provides a useful indication of primary production for water masses in the present-day oceans, estimates of palaeoproductivity drawn from vertical oceanic $\delta^{13}C$ gradients recorded in calcareous microfossil tests might provide indirect insights into secular variations in biomass. For the Cretaceous and Tertiary periods, palaeoproductivity estimates do show a positive correlation with diversity in phytoplankton fossils, as suggested years ago by Tappan, Lipps, Fischer and Arthur, and others. However, the relation is not simple and is complicated by the effects of radiations and mass extinctions: it is doubtful, for example, whether the observed doubling of phytoplankton diversity from the Cambrian to the Ordovician reflects a comparable change in algal biomass. I have no confidence that phytoplankton diversity itself can be used as an index of biomass in pre-Cretaceous rocks. This leads me to the second problem, which is that many algae that contribute in important measure to marine productivity and biomass are not represented in the geological record by recognizable fossils. This further confounds efforts to relate ancient biomass and diversity in any predictive way. I hope, by the way, that my pessimism is misplaced.

Phil. Trans. R. Soc. Lond. B **325**, 291–305 (1989)

Printed in Great Britain

Plants at the Cretaceous–Tertiary boundary

By R. A. Spicer†

Life Sciences Department, Goldsmiths' College, Creek Road, London SE8 3BU, U.K.

Environmental selection determines to a large extent the morphology and anatomy of individual plants, and the composition and structure of vegetation. The intimate relation between vascular land plants, climate and substrate produces an abundant fossil record with a strong inherent signal reflecting, in particular, air temperatures, precipitation and evaporation, light régime and seasonality. Studies of palynomorphs, cuticles and plant megafossils in detailed sedimentological and stratigraphic context across the Cretaceous–Tertiary (K–T) boundary in North America, suggest sudden and traumatic vegetational disturbance, profound and long-lasting climatic change, and survivorship patterns that are palaeogeographically heterogeneous and possibly related to the ability of taxa to enter dormancy. Some of these changes are reflected in palaeosols. Major vegetational changes are also apparent in Europe, and Asia, although the precise timing of these events is less clear. The record, as presently interpreted, is one of ecological catastrophe, some selective extinction of broad-leaved evergreen species, and long-term vegetational restructing, expressed most strongly at middle and lower latitudes in the Northern Hemisphere.

1. Introduction

The terrestrial vascular plant fossil record exhibits several characteristics that allow it to be used to elucidate past environmental change at fine scales of geographical and stratigraphic resolution. The spatial fixity of the sporophyte plant body, its intimate association and modification of substrates (palaeosols), interaction and direct exposure to the atmosphere, interception and utilization of light, and its morphological/anatomical record of growth patterns mediated by temperature and water availability result in a rich legacy of potential plant fossils that give a record of environmental conditions through time. Furthermore, many adaptations to maximize vegetative productivity, and in part therefore fitness, have evolved several times and are not taxonomically restricted. Thus the palaeobotanist interested in assessing environmental change is largely freed from the constraints of precise phylogenetic or taxonomic determinations (Spicer 1989a; Knoll & Niklas 1987). This is a particularly significant consideration because of the fact that much of the historically published palaeobotanical database on systematics contains high levels of investigator bias (Boulter *et al.* 1988).

As with any palaeontological record the primary database provided by the living organisms is modified considerably by the processes of fossilization. Understanding taphonomic biases is vital to any accurate account of ecological trauma, disruption of the physical environment, or extinctions. Many plant parts have a high preservation potential either because of their inherent resistance to decay (e.g. lignin and sporopollenin), or their production in large numbers (e.g. leaves), or both (e.g. pollen). Under certain circumstances even the most ephemeral elements of the life cycle may be preserved (Brack-Hanes 1978; Rothwell 1977).

† Present address: Department of Earth Sciences, Oxford University, Parks Road, Oxford OX1 3PR, U.K.

Although the palaeobotanical record is a rich one it is also bedevilled by two factors that together serve to frustrate assessments of extinction rates. The first is that vascular land plants are very rarely preserved whole. Throughout the lifetime of most plants an indeterminate number of individual organs are shed, some of which enter depositional environments and become mixed with organs from other individuals and taxa. The difficulties of whole plant reconstruction that arise from such fragmentary preservation are compounded by mozaic evolution (see, for example, Sporne 1980; Knoll & Niklas 1987; Boulter *et al.* 1988) in which plant organs evolve to some extent independent of each other. For example, leaves are exposed to different selective pressures than reproductive organs, wood or rooting structures. Thus in angiosperms, for example, one sees intricate evolution of floral form in relation to reproductive pressures (including co-evolution with biotic pollinators), whereas leaf or wood evolutionary rates tend to reflect competition for growth requirements or changes in the physical environment. As a result, extinction rates, or rates of evolutionary radiation, based on one or more separate plant organs are a very poor reflection of what is happening to the coherent biological entities we call whole plants (cf. Lidgard & Crane 1988; Cleal 1988).

The above factors determine to a large extent the contextual framework within which any discussion of vegetational change at the Cretaceous–Tertiary (K–T) boundary must take place. A requirement not unique to the plant record is for accurate high resolution stratigraphy within complete sections using non-biological markers (to avoid circularity) spanning the K–T boundary. Unfortunately, sites where these requirements are met are mostly in the western United States and Canada, which introduces a strong geographical bias. However, less well-constrained evidence from other areas suggests this may indeed be where terrestrial K–T boundary phenomena are expressed most strongly.

2. Late Cretaceous vegetation and climate

Studies of vegetational changes at the K–T boundary have to be viewed in the context of the vegetational structure and dynamics, and climate, of the late Cretaceous. The late Cretaceous was a time of some climatic change and gradual vegetational restructuring, as the angiosperms underwent geographical and ecological diversification. It is now well established that from a late early Cretaceous low latitude cradle the angiosperms spread polewards (Axelrod 1959; Hickey & Doyle 1977) reaching 75° N by the end of the Albian (Scott & Smiley 1979; Spicer & Parrish 1986) and high southern latitudes about the same time (Douglas & Williams, 1982). Ecological diversification apparently proceeded at a slower pace. Concomitant with the rise of angiosperms (whose origin, as far as is known, was unconnected with extinctions or ecological disruption) there was a gradual decline (apparently because of competition with the angiosperms) in major Mesozoic groups such as the Bennetitales, Cycadales and conifers such as the Cheirolepidiaceae.

Using North American data, late Cretaceous vegetational and climatic change may be divided into two phases: (1) Aptian – early Cenomanian, (2) middle Cenomanian – Maastrichtian (Upchurch & Wolfe 1987). The Aptian – early Cenomanian saw the early diversification in which most foliage physiognomic types first appear, multistratal forests have a restricted distribution in time and space, and although angiosperms may dominate megafossil assemblages in certain (notably riparian) facies the palynoflora is always dominated (more than 60%) by spores and pollen from ferns, sphenophytes and gymnosperms (Upchurch &

[52]

Wolfe 1987). Early Cenomanian megafloras are diverse and widespread and the discrepancy between pollen and leaf abundances may be due to high palynomorph productivity by pteridosperms and gymnosperms, the 'preference' of gymnosperms for well-drained soils where preservation potential of megafossils is low, and the poor preservation potential of the pollen of Laurales (Muller 1981), a group that, based on megafloral evidence (Upchurch & Dilcher 1984) formed a significant component of the angiosperm flora (Upchurch & Wolfe 1987).

From the middle Cenomanian to the late Maastrichtian leaf and vegetational physiognomic types stabilize in relation to environmental parameters. At low (less than 30°) and middle (30–65°) palaeolatitudes angiosperms become the most abundant component of regional palynofloras.

According to Upchurch & Wolfe (1987) low–middle northern latitude (30–45°) vegetation was predominantly evergreen megathermal, and the resulting leaf assemblages are characteristically thick-textured and predominantly entire-margined (greater than 70%). Low leaf size in both fine and coarse-grained facies, few species with either drip tips or probable vine habit, and the prevalence of leaves with emarginate apices suggest sub-humid conditions, whereas small diaspore size (Tiffney 1984) is both consistent with sub-humid conditions and indicative of open-canopy structure.

In the Santonian (Coniacian?) palms first appear in megathermal environments, but by the Campanian they were common in mesothermal situations (Bell 1957; Daghlian 1981; Hickey 1984).

The absence of growth rings in fossil wood from low–middle latitudes (Creber & Chaloner 1985; Wolfe & Upchurch 1987) suggests minimal seasonality in the climate. Angiosperm woods from trunks up to 1 m in diameter exhibit large quantities of parenchyma with vessels of large diameter, often solitary, occurring at low frequency (Wheeler *et al.* 1987; Wolfe & Upchurch 1987); features that today are typical of tropical woods. Abundant parenchyma and wide solitary vessels are unusual in present-day sub-humid environments that tend to be highly seasonal. The lack of good modern analogues for this kind of environment makes interpretation of the wood anatomy problematical, particularly because of the physiognomic signal of the leaves. Wolfe & Upchurch (1987) suggest the closest living analogue might be tropical evergreen vegetation on sandy soils, where low nutrient status limits leaf size, and they suggest that abundant parenchyma and solitary vessels may be adaptive under conditions of low seasonal variation in precipitation, with no period of drought.

At high–middle latitudes (45–65° N) angiosperm dominance in the palynofloras is reached between the Turonian and Campanian (D. J. Nichols, unpublished data cited by Upchurch & Wolfe (1987)). By the Santonian the megafossil record is also dominated by angiosperms in terms of species and specimen abundance, at least locally. Physiognomic analysis of leaves suggests mesothermal evergreen vegetation in which 45–65% of species are entire-margined (= 13–20 °C mean annual temperature), but the rarity of drip tips suggests that rainforest was absent. Leaf size is greater than in the lower latitude assemblages, suggesting more humid conditions.

At 45° N early Maastrichtian angiosperm and gymnosperm woods have virtually no growth rings whereas those from the Maastrichtian of central Alberta (65° N) exhibit well-defined rings with a high ratio of late wood to early wood (Ramanujam & Stewart 1969), suggesting growth was limited later in the growing season because of shortened day length, drier conditions or lower temperatures.

At high northern palaeolatitudes there is a rich fossil record (Hollick 1930; Smiley 1966, 1969a, b; Spicer & Parrish 1986; Spicer 1987; Spicer et al. 1987) documenting the existence of a vegetational physiognomic type now extinct: the Polar Broadleaved Deciduous Forest (Wolfe 1985). This forest type is characterized by deciduous conifers plus numerous platanoid, hammamelid and trochodendroid, thin-textured deciduous angiosperm leaves exhibiting a large range of leaf sizes. Entire-margined forms rarely constitute more than 30% of the assemblages. In spite of moderately high angiosperm leaf diversity (Spicer & Parrish 1986; Spicer 1987) in the Cenomanian, the regional vegetation was dominated by large deciduous conifers, together with a variety of ginkgophytes. Ferns, sphenophytes (e.g. *Equisetites*), and deciduous cycads also occurred in some abundance. Leaf and vegetational physiognomic analysis, together with studies of comparative diversity and wood anatomy from northern Alaskan floras, suggest mean annual sea-level temperatures at 75° N were in the order of 10 °C, possibly rising to 13 °C in the Coniacian, before declining by the Maastrichtian to 5–6 °C at 85° N (Parrish & Spicer 1988). By the crude application of the present-day average adiabatic lapse rate of 6 °C per 1000 m these temperatures suggest mean annual temperatures of 0 °C (and therefore the possibility of permanent ice) above 1700 m at 75° N in the Cenomanian and above 1000 m at 85° N in the Maastrichtian (Spicer 1989b).

Sedimentological evidence, together with an abundance of coals, leaf characteristics, and wood relatively free of false rings, suggest water was not limiting to plant growth at high latitudes in the Cenomanian. However, an increase in fossil charcoal may indicate a degree of drying in the Campanian and Maastrichtian (Spicer & Parrish 1987).

3. VEGETATIONAL CHANGES AT THE K–T BOUNDARY
(a) The palynological record

It is now well established that in many parts of the world the boundary between the latest Cretaceous and earliest Tertiary beds is marked by a boundary clay associated with anomalously high iridium concentrations (see, for example, Alvarez et al. 1984), shocked quartz grains (see Izett & Bohor 1986) and, occasionally, mineral spherules (see Bohor et al. (1987)).

The most widely recognized plant–fossil boundary indicator is a sudden and severe attenuation in abundance and diversity of pollen grains typical of the Cretaceous at or near a boundary clay. This is followed by a sudden rise in spore abundance in the immediate post boundary sediments. Subsequent to this so called 'fern spike' the proportion of spores gradually declines as pollen increases once more in abundance and diversity, but this time with characteristic Palaeocene forms (see Tschudy & Tschudy (1986) and references therein). So pronounced is this pattern that it was recognized in the Western Interior of the United States long before the bolide impact scenario (Alvarez et al. 1980) was proposed (see, for example, Leffingwell 1971; Tschudy 1971). So consistent is the palynological signal that it is sometimes used to define the boundary in the absence of corroborating data such as anomalously high iridium levels and shocked quartz (Tschudy 1973): a situation that introduces a certain circularity.

Detailed studies of complete non-marine boundary sections ranging from New Mexico to Saskatchewan show that although the general pattern of palynological change is broadly consistent, the pollen taxa that are affected and the timing of 'originations' and 'extinctions' with respect to the boundary varies geographically (Tschudy & Tschudy 1986; Sweet & Braman 1988). The likelihood of palynomorph reworking across the boundary is minimized in complete coal-bearing boundary sections and thus temporal resolution of changes in the spore assemblages is generally high.

In the northern region of the Western Interior key pollen species such as *Aquilapollenites* spp. (except *A. spinulosus*) and *Proteacidites* are confined to the Cretaceous, with other taxa (*Gunnera microreticulata*, *Liliacidites altimurus*, *Liliacidites complexus* and *Cranwellia striata*) also disappearing at the boundary. Some species (*Wodehouseia spinata*, *Tricolpites parvistriatus* and *Arecipites columellus*) suffered a dramatic decline at the boundary, whereas others (*Kurtzipites trispissatus*, *Triporopollenites plektosus*, *Ulmipollenites* spp. and *Alnipollenites* spp.) appear to pass relatively unaffected into the Tertiary (Leffingwell 1971). In the southern region, the Raton Basin, the only northern region pollen type to occur in Cretaceous rocks and then disappear at the boundary is *Proteacidites* spp. Typical southern region taxa to disappear are '*Tilia*' *wodehousei*, *Trisectoris* and *Trichopeltinites* (probably a fungal thallus). *Trichopeltinites* reappears in the Eocene of the Mississippi Embayment (G. R. Upchurch, Jr, personal communication 1988) and may be an indicator of warm conditions. Taxa common to both the uppermost Cretaceous and continuing into the basal Tertiary include: *Gunnera microreticulata*, *Fraxinopollenites variabilis*, *Liliacidites complexus*, *Thomsonipollis magnificus*, *Tricolpites anguloluminosus*, *Pandaniidites radicus*, *Arecipites columellus* and *Salixipollenites* sp.

Above the boundary clay there are usually a few millimetres of organic-rich rock generally devoid of palynomorphs. Sapropel and fusain are often, but not always, found in this layer suggesting rotting or burnt vegetation or both. This 'barren zone' is (usually) overlain by several centimetres of carbonaceous mudstone or coal which yields abundant fern spores and very few pollen grains. This spore-rich horizon is interpreted as representing the first Tertiary vegetation and indicates an abrupt and profound post-boundary vegetational change in that region. Angiosperm pollen typically undergoes a slow recovery in abundance until about 10–15 cm above the boundary it again predominates (Tschudy & Tschudy 1986). In some instances aberrant grains occur several metres above the boundary suggesting, perhaps, prolonged periods of stress (Fleming 1988).

Careful work by Hotton (1988) at six boundary sites in Montana reveals palynological changes consistent with traumatic ecological disturbance and at least local extinctions. Before the boundary within the Hell Creek Formation (Cretaceous) there is little change in overall diversity, or in the diversity and abundance of 'doomed' species. However, within any given section, major palynological changes occur at a single horizon just above an iridium anomaly. These changes consist of 21–46% of Hell Creek species disappearing or becoming extremely rare, and 18% undergo significant decline. Overall there are three to six times as many rare forms in the Hell Creek than the subsequent Tullock (Palaeocene) Formation. Immediately above the boundary there is typically a coal yielding unusually high percentages of *Cyathidites* fern spores and few other species. Spores, however, tend to be abundant in most Tullock coals. Hotton (1988) considers that 30–40% of the species losses might be attributable to post-boundary facies changes. Although partial recovery takes place within 10–30 cm of the boundary Hell Creek levels of diversity are never achieved within the Tullock.

The range of 21–46% 'extinctions' noted by Hotton in a restricted geographical area exceeds that exhibited across the latitudinal range from the Raton Basin (Tschudy & Tschudy (1986) imply greater than 25%) to Saskatchewan (30%) (Nichols *et al.* 1986) and, although some differences may be due to Hotton's figure including all taxa whereas the other reflects only angiosperm losses, no clear latitudinal trend is demonstrated.

Sweet & Braman (1988) note a different sequence of events in western Canada between palaeolatitudes 60 and 75° N. Differences in assemblage composition can be related to differences in depositional environment and latitudinal shifts in the angiosperm flora in response to climatic differences. Despite this background heterogeneity Sweet & Braman identified four sequential palynofloral changes across the boundary (marked by a geochemical anomaly and shocked quartz). These include an extinction event 5–10 m below the boundary extinctions, a miospore–gymnosperm assemblage beginning 5–10 cm above the boundary within which newly introduced species occur together with Maastrichtian relicts, and a last reduction of diversity approximately coincident with the first occurrences of *Wodehousia fimbriata* approximately 3–15 m above the boundary. Sweet & Braman (1988) consider these phenomena difficult to explain as the result of a single bolide impact and suggest the causes for biotic change are, as yet, undetermined.

At very high northern palaeolatitudes Frederiksen *et al.* (1988) observe a variety of angiosperm pollen forms representing understorey or disturbed-site herbs and shrubs that were probably insect pollinated. Although the latest Maastrichtian and the boundary itself is not apparent early Palaeocene palynofloras have a distinctly different angiosperm flora, reduced in diversity and predominantly representing anemophilous taxa. Megafossil evidence (Spicer & Parrish 1989) supports the interpretation of Frederiksen *et al.* (1988) that the Cretaceous–Tertiary transition on the North Slope of Alaska saw a change from a predominantly conifer-dominated vegetation to one rich in wind-pollinated angiosperm trees. The lack of an intact section raises the possibility that these changes could be the result of long-term climatic fluctuations.

Tschudy & Tschudy (1986) note that all the extinctions they observed were only regional, in that elsewhere in North America all the Cretaceous pollen types continue through into the Tertiary. What Tschudy & Tschudy did not determine, however, was the extent of reworking present in the other North American records. In the case of *Aquilapollenites*, however, there is a genuine continuation into the Tertiary in Japan (Saito *et al.* 1986).

Whether or not the extinctions were regional or global, the North American plant microfossil record demonstrates four types of biotic change at the K–T boundary. In addition to the abrupt loss of certain taxa, as typified by *Proteacidites* spp., some taxa, such as *Wodehouseia*, seem to undergo a pseudoextinction at the 'species' level. In the northern Western Interior the Cretaceous form *W. spinata* is replaced above the boundary by its presumed descendant *W. fimbriata*. *Kurtzipites trispissatus* typifies a third type of boundary change in which a pollen type appears in the Maastrichtian, reaches its zenith at the boundary and subsequently becomes a significant element in the 'recovery' pollen flora. Other species of *Kurtzipites* exhibit this pattern but all become extinct by the middle of the Palaeocene. The fourth type of biotic change is represented by a large group of pollen types that was little affected by the K–T boundary event. Many Cretaceous forms pass into the Palaeocene and some survived to the Eocene or younger. Where ecological trauma is well documented new Palaeocene forms are slow to appear and previously rare Cretaceous forms adopt a new ecologically dominant role in early Palaeocene vegetation (Nichols & Fleming 1988).

Evidence that the terminal Cretaceous non-marine ecological trauma was experienced outside North America comes from a marine boundary sequence at Hokkaido, Japan (Saito *et al.* 1986). Here, pollen and spores of land plants exhibit an immediate post-boundary increase in the proportion of fern spores to pollen, but this is followed by an increase in pine pollen. Saito *et al.* (1986) interpret this as indicating that pine was an early colonizer of the devastated landscape. Although pine can behave as a 'weed' (e.g. *Pinus pinaster* in South Africa (Carlquist 1975, p. 96)) conifers with long life cycles are not well suited to this role unless there is a major deterioration of the climate. Bisaccate pine pollen grains are produced in large numbers and are particularly suited to long distance transport into marine environments. It is likely then that the abundance of pine in the recovery vegetation, relative to other taxa, is exaggerated in this instance.

The El Kef section, Tunisia, preserves Maastrichtian pollen from the palm-rich province of Africa – South America together with pollen of the *Normapolles* type (see Batten (1981) for a review of these forms) characteristic of Europe and eastern North America. In the transition to the Tertiary, palynological diversity declined and only pollen typical of the European source to the north occurs in the Palaeocene sediments (Méon 1988).

Rawat *et al.* (1988) report on pollen zonations that encompass the K–T boundary in the Krishna–Godavari basin, India, and although independent stratigraphic resolution is coarse, some palynofloral turnover has been used to position the boundary. This suggests some vegetational change may be present. Even if, however, independent evidence subsequently shows the boundary is correctly positioned, the degree of vegetational disruption would appear to be milder than that experienced in North America.

No unweathered K–T boundary outcrops are known from Australia (Hannah & Partridge 1988) but well data may indicate some palynological changes (Baird 1988). There appears to be an early Palaeocene diversity decreased within the genus *Proteacidites*, but coarse sampling renders any conclusions regarding actual vegetational change highly speculative.

In New Zealand, K–T palynological changes are considerably less dramatic than those of the mid Cretaceous, the Palaeocene–Eocene boundary, or the end of the Eocene (Raine 1988). Sequences in coal measures indicate little modification of the podocarp-dominated vegetation across the K–T boundary which is marked palynologically by the extinction and appearance of mostly angiosperm taxa.

The K–T boundary record on Seymour Island, Antarctic Peninsula, is similar to that of New Zealand in that there is no evidence for an abrupt change in the podocarp vegetation, despite evidence for reworking of Cretaceous pollen into Tertiary sediments (Askin 1988*a, b*). Podocarp pollen is prevalent throughout succession, together with diversifying angiosperm pollen assemblages. Cryptogam spores are low in diversity and abundance. Gradual long-term turnover in the palynofloras is consistent with coeval changes in regional palaeogeography and climates. Latest Cretaceous angiosperm pollen is diverse but highly provincial; a pattern also seen at high northern latitudes (Frederiksen *et al.* 1988; T. A. Ager, personal communication 1986).

(b) Leaf megafossil studies

Because of their limited potential for transport and reworking, a more spatially and temporally constrained picture of terminal Cretaceous plant disturbance is emerging from the study of plant megafossils and dispersed cuticles. Hickey (1981), working on assemblages from Wyoming and Montana, reported extinctions comparable in intensity to those of the Palaeocene–Eocene boundary, although in northeastern Asia extinctions may have been slightly more severe (Krassilov 1983). These studies suffered, however, by being done at a somewhat coarse stratigraphic resolution.

Wolfe & Upchurch (1986, 1987) have studied in detail leaf and cuticle assemblages at a large number of Western Interior (U.S.A.) boundary sites that display a continuous sedimentary sequence from the Cretaceous to the Tertiary. Instead of concerning themselves with the taxonomic problems inherent in studies of leaves of this age, Wolfe & Upchurch studied environmental and vegetational changes using foliar physiognomy.

Based on 15 sections in the Raton Basin that can be related to the iridium anomaly Wolfe & Upchurch (1987) recognize five vegetational–floristic phases that serve as a general comparison model for sections elsewhere.

Phase 1 (Lancian age – latest Cretaceous) is characterized by broadleaved evergreen vegetation with high diversity. Leaves tend to be small in size with thick hairy cuticles and very few have drip tips. This type of foliar physiognomy indicates a sub-humid (dry) vegetation which persists up to the boundary clay. Evergreen conifers were also present.

Phase 2 (immediately above the boundary) consists entirely of leaves and rhizomes of a fern (morphologically similar to extant *Stenochleana*) and cuticles typical of herbs.

Phase 3 (extending up to 2 m above the boundary) is typified by a depauperate flora of large leaves with drip tips and thick smooth cuticles. Overall, the leaves suggest early successional vegetation in an environment of high precipitation.

Phase 4 (lasted throughout the next 200 m of section) is characterized by an increasing (but still low) leaf diversity. Physiognomically a warm humid environment is indicated.

Phase 5 (throughout the next 150 m of section) indicates low diversity megathermal rainforest.

Wolfe & Upchurch (1986, 1987) suggest that the megafossil pattern of floristic change at the K–T boundary is indicative of an ecological trauma followed by a steady recovery, over perhaps 1.5 Ma, that mimics normal seral succession. The immediate post-boundary vegetation is fern-dominated like modern rainforest vegetation that is devastated by volcanic eruption (Spicer *et al.* 1985), but that is not to say that the K–T boundary trauma was volcanic in origin. It is also evident that although the long-term thermal régime appears to have been little affected, there was a significant increase in humidity (probably precipitation) that lasted well into the Palaeocene. This pattern of increased precipitation is not confined to the Raton Basin and is seen elsewhere in North America.

Leaf data from 66 collections at 8 geographical locations ranging from the Mississippi Embayment to Alberta also show a major shift in vegetational patterns at the boundary. The latest Cretaceous vegetation at palaeolatitudes less than 65° N appears to have been evergreen-rich. However, the post-boundary recovery vegetation was essentially deciduous, even at low middle palaeolatitudes. In the Raton Basin 75 % of leaf taxa become at least locally extinct at the boundary, and extinction rates both in the Raton and Denver Basins are highest in the evergreen taxa. In central Alberta only 24 % of leaf taxa became extinct, although the gymnosperms were strongly affected. In general, extinction was most pronounced in megathermal and mesothermal broadleaved evergreen vegetation and lowest in broadleaved deciduous vegetation. Deciduous elements had the lowest extinction rates in all types of vegetation, whereas in mesothermal vegetation (between palaeolatitudes 45 and 65° N) evergreen elements were particularly hard hit. Similarly, among conifers, evergreen species became extinct whereas deciduous species survived. The latest Cretaceous mesothermal broadleaved evergreen forests of high–middle latitudes of North America were succeeded by dominantly broadleaved deciduous forests in the earliest Palaeocene. Maastrichtian evergreen elements were replaced by taxa from the more northerly polar broadleaved deciduous forests.

Johnson & Hickey (1988) and Johnson (1988), working in Montana and North Dakota, have developed a biostratigraphy using 798 leaf species at 62 localities spanning 110 m of the latest Cretaceous Hell Creek Formation and the overlying 100 m thick earliest Palaeocene Ludlow member of Fort Union. At Mamarth (southwest North Dakota) 7498 specimens have been used to define three biozones in the 'Lancian Stage' of the Hell Creek and one in the Ludlow Member. The boundary, above the basal lignite of Fort Union Formation, is marked by a weak iridium anomaly and constrained by palynology, magnetostratigraphy, vertebrate biostratigraphy, sedimentary facies and lithology (Johnson 1988). The boundary is preceded by considerable floral turnover in latest Cretaceous, including immigration from the south of typically Raton forms near the base of the magnetostratigraphic unit Chron 29R (approximately 40000 years before the end of the Cretaceous), followed by a diverse basal Palaeocene megaflora. The warming implied by the northward migration of taxa is matched by an increase in the proportion of entire-margined leaves, and an increase in leaf size suggests an increased precipitation: evaporation ratio (K. R. Johnson, personal communication 1988). However there is no evidence for a fern spike. Johnson & Hickey suggest that floristic change was due to long-term environmental processes with only a portion due to traumatic disturbance.

The general pattern of increased precipitation is supported by sedimentological evidence. Coal is more abundant in the Palaeocene than the Cretaceous and leaching of calcareous palaeosols in Montana (see Retallack *et al.* 1987) has been attributed to a change from a sub-humid to wet climate, and possibly a period of 'acid rain', following the boundary event.

The expansion of range of broadleaved deciduous forests that occurred in the Palaeocene gave rise to an increase in genetic diversity. This is seen most strongly in the Juglandaceae but overall a threefold increase in dicot families represented in the polar broadleaved deciduous forest occurs during the Palaeocene (Wolfe 1987). Even today the mesic northern hemisphere deciduous forests are diverse compared with Southern Hemisphere forests growing under similar conditions, but even these are much reduced in comparison with those of the late Palaeocene and Eocene (Wolfe 1987). The increase in diversity occurred after the inoculation of mid-latitude vegetation by northern taxa that had a lower latitude origin earlier in the Cretaceous. At present, there is no evidence of major clades (for example families or orders) originating at high latitudes (Spicer *et al.* 1987).

The ability of some plants to enter dormancy appears to have conferred a major advantage at the K–T boundary and this would be consistent with a short duration cold–dark excursion, during which evergreen taxa were adversely affected, before a new stable 'wet' climate régime was established. Such an interpretation clearly favours the possible existence of a 'boundary event winter'.

The reduced extinction rates at low latitudes (less than 45° N) probably reflects the limited magnitude of the low temperature excursion: in tropical régimes freezing sufficient for mass kill of the megathermal vegetation may not have been experienced. Southern Hemisphere vegetation was much less severely affected at the K–T boundary and even today is evergreen-rich compared with the Northern Hemisphere (Wolfe 1987). Evergreen refugia in mesothermal vegetation must have been few because of the prolonged regional effect and the successful influx and persistence of deciduous elements. A viable deciduous-rich ecosystem was established before megathermal evergreens could evolve into mesothermal vegetation. However, increased Palaeocene precipitation in megathermal régimes may have been a critical factor in the origin of angiosperm-dominated paratropical and tropical rainforest. The increase in precipitation may also have played a role in the extinction of some taxa but it would not have brought about the overall pattern of an increase in deciduousness.

Despite the apparent devastation of the northern mid-latitude vegetation 'standing crop' at the time of the boundary event, and attendant ecological and environmental trauma, most plant lineages, including many evergreens, were able to pass through the boundary and evolve in new directions in the changed post-boundary conditions. The key to this success of plants, even under severe environmental stress, undoubtedly lies in their ability to enter dormancy, either in a mature state (deciduousness or dying back to a perennating organ such as a rhizome) or in seed form. The exposure of buried seeds during increased post-boundary erosion of a denuded landscape would have quickly re-established most lineages. Most vertebrate animals, on the other hand, are poorly equipped to survive prolonged adverse conditions and at the very least populations would have been decimated.

If we assume that the events at the K–T boundary were the result of a 'boundary-event winter' the effects on the vegetation might be expected to be strongly dependent on the duration, intensity and timing of the event, and, in the case of the bolide scenario, the location of the impact site. Tinus & Roddy (1988) suggest that if the impact occurred during spring growth and the effects produced no fatal chill, defoliation might affect deciduous plants more strongly than evergreens because not only would an entire year's crop of leaves be lost but so would the stored food reserves invested in the lost leaf crop. All plants would be relatively safe so long as they remained dormant but subsequent bud break would render all plants susceptible to darkness, temperature fluctuations and/or atmospheric pollutants. If the effects were prolonged slow starvation might kill the trees. If the event occurred late in late summer evergreens might be expected to suffer more because deciduous plants would already be prepared for their normal period of dormancy. In view of the fact that respiration increases with temperature the effects of prolonged darkness might be expected to be more severe at lower latitudes, notwithstanding a cooling induced by loss of insolation. A consequence of these predictions is that any global-scale event would be expected to display a degree of symmetry between Northern and Southern Hemispheres.

Unfortunately, these predictions are difficult to reconcile with the observed global patterns of vegetational change and local extinctions. Although clearly at mid-northern latitudes broadleaved evergreens are selectively eliminated, lower latitude evergreens and those in the Southern Hemisphere are not. Although very incomplete, data from the Southern Hemisphere reveal very minor vegetational changes. This suggests the putative dust cloud must have had a relatively localized effect in the Northern Hemisphere, was probably not 'global', and thus was likely to have been of short duration. The selective evergreen losses are a little easier to explain if a cooling of the Northern Hemisphere took place that was of sufficient intensity to chill broadleaved evergreens fatally only in the northern parts of their range. However, such a freeze would be unlikely to kill soil-buried seeds (from which the evergreens could recover) unless it were very severe. Such an implied severity might be expected to have adversely affected local and more northerly deciduous plants to a far greater extent than is evident from the fossil record.

Many boundary sections yield an abundance of fusain or other evidence of post-boundary fire (Tschudy & Tschudy 1986; Saito et al. 1986; Wolbach et al. 1985) and the suggestion has been made that wildfires were started as the result of the presumed bolide impact (Wolbach et al. 1988). Such a scenario is difficult to envisage, however, because at the time of the impact most vegetation would have been living and with a high moisture content. It is more likely that numerous wildfires started in post-event dead vegetation as the result of frequent lightning strikes produced in a destabilized atmosphere. In a post-event world the recently killed and relatively desiccated forests would have provided ample fuel and would have been easily ignited.

These fires probably had little effect on land-plant extinction because the distribution of fusain in the late Cretaceous fossil record suggests that wildfires were common and probably were an integral element in shaping late Cretaceous vegetation. The lack of fusain (which has a very high preservation potential) at all the intact terrestrial boundary sites, and the apparently minor vegetational disturbance in the Southern Hemisphere, argues strongly against a simultaneous global conflagration (cf. Wolbach *et al.* 1988).

4. CONCLUSIONS

Even under the most traumatic of environmental changes plant life shows a remarkable capacity for survival and, as far as can be ascertained, true (that is permanent global loss of the gene line) extinctions of major groups due to catastrophic events are few. When true plant extinctions have occurred they appear to be the result of interplant competition or the loss of specific environments to which a particular group or groups have become irreversibly specialized. Extinctions tend to follow innovation (Knoll 1984; Boulter *et al.* 1988).

If true extinctions at the K–T boundary are few, there is clear evidence for local extinctions and 'lazarus taxa' (taxa which appear to become extinct but which later reappear). Clearly there were numerous refugia. Although this was a time of global warmth, Cretaceous high-latitude coastal vegetation was well adapted to freezing conditions and long periods without sunlight. Mid-latitude coastal vegetation rarely, if ever, experienced frosts. The higher extinction rates among thermophilous broadleaved evergreen taxa at middle northern latitudes and the survival and expansion of deciduous northern taxa suggests strongly that darkness or cold, or both, filtered Northern Hemisphere plants at the end of the Cretaceous. Responses of presumed herbaceous plants at very high northern latitudes, and trees at middle northern latitudes, are consistent with a 'boundary-event winter' in which low temperatures killed mid-latitude evergreens. Limited data from very high northern latitudes do show some vegetational change, which may be used to argue for a northern spring onset for the boundary event in that even deciduous trees and herbaceous plants would be most vulnerable at that time of year. However, until more is known of these high latitude floras our understanding of the timing of the event must remain highly speculative.

Light attenuation with perhaps no evergreen-lethal chill (standing-crop evergreens could have been killed by extended darkness alone) is perhaps less likely because this would have had no effect whatsoever on the evergreen seed bank unless the period of darkness was prolonged (much greater than 10 years). Obviously, survivorship among the deciduous plants would then also have been limited. However, the deleterious effect on the standing crop of low light levels would have been enhanced by relatively warm conditions because respiration rates would have not been depressed by cold.

Southern Hemisphere responses at mid to high latitudes, in so far as they are known, suggest little difference in susceptibility between evergreens and deciduous plants and no wholesale ecological disruption. Clearly, more data are required from Europe, Asia and low palaeolatitudes in both hemispheres, but at present the boundary event seems to have had most effect in northern mid-latitudes and western North America in particular. It is unlikely that Deccan Traps volcanism was responsible for the ecological disruption because floristic changes, although not well defined, are clearly not pronounced in India.

Johnson's (1988) observations of climatic warming, coupled with evidence of at least local increase in water availability in the very latest Cretaceous of the western high plains of the U.S.A., imply the boundary event occurred against a backdrop of changing climates. By way of comparison and contrast, Stott & Kennett (1988) note that oxygen isotope studies using foraminifera suggest that surface waters of high latitude southern oceans experienced a latest Cretaceous cooling of about 5 °C. This cooler temperature was maintained at high latitudes well into the Palaeocene, but at low latitudes temperatures recover after the boundary. Although the significance of these changes is difficult to evaluate at present, the very latest Cretaceous was a time of complex climatic variation that was apparently interrupted by the boundary event. The abrupt climatic change that accompanied the boundary event ushered in a new stable climatic régime that continued well into the Palaeocene. Under the early Palaeocene climate in northern mid-latitudes forest ecosystems evolved a new stable structure and composition. Previously rare species assumed new dominant roles and underwent evolutionary change as a consequence. It is likely that these changes, both climatic and vegetational, induced profound ecological stress in the terrestrial faunal realm, giving rise to enhanced extinction rates.

Existing scenarios are clearly inadequate to explain all the patterns of regional extinction, survivorship and ecological restructuring that took place at the end of the Cretaceous. The difference in seed versus standing crop survivorship is a particularly interesting problem with respect to the 'event winter' scenario. Detailed understanding of latest Cretaceous vegetation and climatic heterogeneity is needed before the K–T boundary events can be properly understood. Even more obscure is the evolution of early Palaeocene forests and yet it is here that we see for the first time the forest types that characterize much of the Northern Hemisphere today. A study of the post K–T boundary recovery, involving as it does evolutionary, ecological and climatic change, offers exciting future challenges.

I am extremely grateful to Jack Wolfe, Gary Upchurch, Kirk Johnson, Rosemary Askin and Doug Nichols for discussing freely their published and unpublished data with me and for sharing their considerable expertise on this subject. I am equally endebted to Professor W. G. Chaloner, F.R.S., for his helpful criticisms of the manuscript.

References

Alvarez, L. W., Alvarez, W., Asaro, F. & Michel, H. V. 1980 Extraterrestrial cause for the Cretaceous–Tertiary extinction. *Science, Wash.* **208**, 1095–1108.

Alvarez, W., Alvarez, L. W., Asaro, F. & Michel, H. V. 1984 The end of the Cretaceous: sharp boundary or gradual transition? *Science, Wash.* **223**, 1183–1186.

Askin, R. A. 1988a Campanian to Paleocene spore and pollen assemblages of Seymour Island, Antarctica. In *Abstracts of the 7th International Palynological Congress, Brisbane*, p. 7.

Askin, R. A. 1988b The palynological record across the Cretaceous/Tertiary transition on Seymour Island, Antarctica. In *Geology and paleontology of Seymour Island, Antarctic peninsula* (*Mem. geol. Soc. Am.* **169**) (ed. R. M. Feldmann & M. O. Woodburne), pp. 155–162.

Axelrod, D. I. 1959 Poleward migation of early angiosperm flora. *Science, Wash.* **130**, 203–207.

Baird, J. G. 1988 Palynostratigraphy of the Late Cretaceous – Early Tertiary of the eastern portion of the Gippsland Basin. In *Abstracts of the 7th International Palynological Congress, Brisbane*, p. 9.

Batten, D. J. 1981 Stratigraphic, palaeogeography and evolutionary significance of late Cretaceous and early Tertiary Normapolles pollen. *Rev. Palaeobot. Palynol.* **35**, 125–137.

Bell, W. E. 1957 Flora of the Upper Cretaceous Nanaimo Group of Vancouver Island, British Columbia. *Mem. geol. Surv. Can.* **293**, 1–84.

Bohor, B. F., Triplehorn, D. M., Nichols, D. J. & Millard, H. T. Jr 1987 Dinosaurs, spherules, and the "magic" layer: a new K–T boundary clay site in Wyoming. *Geology* **15**, 896–899.

Boulter, M. C., Spicer, R. A. & Thomas, B. A. 1988 Patterns of plant extinctions from some palaeobotanical evidence. In *Extinction and survival in the fossil record (Spec. Vol. Syst. Ass.* **34**), pp. 1–36.

Brack-Hanes, S. D. 1978 On the megagametophytes of two lepidodendracean cones. *Bot. Gaz.* **139**, 140–146.

Carlquist, S. 1975 *Ecological strategies of xylem evolution.* (259 pages.) Berkeley: University of California Press.

Cleal, C. J. 1988 Questions of flower power. *Nature, Lond.* **331**, 344–346.

Creber, G. T. & Chaloner, W. G. 1985 Tree growth in the Mesozoic and Early Tertiary and the reconstruction of palaeoclimates. *Palaeogeogr. Palaeoclimatol. Palaeoecol.* **52**, 35–60.

Daghlian, C. P. 1981 A review of the fossil record of monocotyledons. *Bot. Rev.* **47**, 517–555.

Douglas, J. G. & Williams, G. E. 1982 Southern polar forests: the early Cretaceous floras of Victoria and their palaeoclimatic significance. *Palaeogeogr. Palaeoclimatol. Palaeoecol.* **39**, 171–185.

Fleming, R. F. 1988 Palynology of the Cretaceous–Tertiary boundary in the Raton Basin: implications for development of the Tertiary flora. In *Abstracts of the 7th International Palynological Congress, Brisbane*, p. 50.

Frederiksen, N. O., Ager, T. A. & Edwards, L. E. 1988 Palynology of Maastrichtian and Paleocene rocks, lower Colville River region, North Slope of Alaska. *Can. J. Earth Sci.* **25**, 512–527.

Hannah, M. J. & Partridge, A. D. 1988 Cretaceous–Tertiary boundary localities in Australia. In *Abstracts of the 7th International Palynological Congress, Brisbane*, p. 65.

Hickey, L. J. 1981 Land plant evidence compatible with gradual not catastrophic change at the end of the Cretaceous. *Nature, Lond.* **292**, 529–531.

Hickey, L. J. 1984 Changes in the angiosperm flora across the Cretaceous–Tertiary boundary. In *Catastrophies in Earth history: the new uniformitarianism* (ed. W. A. Bergren & J. A. Van Couvering), pp. 279–314. Princeton University Press.

Hickey, L. J. & Doyle, J. A. 1977 Early Cretaceous fossil evidence for angiosperm evolution. *Bot. Rev.* **43**, 3–104.

Hollick, A. 1930 The Upper Cretaceous Floras of Alaska. *U.S. geol. Surv. prof. Pap.* **159**, 1–123.

Hotton, C. 1988 Cretaceous–Tertiary palynostratigraphy in east central Montana, U.S.A. In *Abstracts of the 7th International Palynological Congress, Brisbane*, p. 76.

Izett, G. A. & Bohor, B. F. 1986 Microstratigraphy of continental sedimentary rocks in the Cretaceous–Tertiary boundary interval in the Western Interior of North America. *Geol. Soc. Am. Abstr. Progr.* **18**, 644.

Johnson, K. R. 1988 High resolution megafloral biostratigraphy for the late Cretaceous and early Paleocene of N. Dakota and Montana. *Geol. Soc. Am. Abstr. Prog.* **20**, 379.

Johnson, K. R. & Hickey, L. J. 1988 Patterns of megafloral change across the Cretaceous/Tertiary boundary in the northern Great Plains and Rocky Mountains. In *Abstracts – Global Catastrophies in Earth History, Snowbird, Utah, October 20–23 1988*, p. 87.

Knoll, A. H. 1984 Patterns of extinction in the fossil record of vascular plants. In *Extinctions* (ed. M. Nitecki), pp. 21–68. Chicago: University of Chicago Press.

Knoll, A. H. & Niklas, K. J. 1987 Adaptation, plant evolution, and the fossil record. *Rev. Palaeobotan. Palynol.* **50**, 127–149.

Krassilov, V. A. 1983 Evolution of the flora of the Cretaceous period: is a Cenophytic era necessary? (In Russian.) *Paleontol. Zh.* 93–95.

Leffingwell, H. A. 1971 Palynology of the Lance (Late Cretaceous) and Fort Union (Paleocene) Formations of the Lance area, Wyoming. In *Symposium on palynology of the late Cretaceous and early Tertiary (Spec. Pap. geol. Soc. Am.* no. 127) (ed. R. M. Kosanke & A. T. Cross), pp. 1–64.

Lidgard, S. & Crane, P. R. 1988 Quantitative analyses of the early angiosperm radiation. *Nature, Lond.* **331**, 344–346.

Méon, H. 1988 Sporo-pollenic studies of El Kef outcrop (N.W. Tunisia), palaeogeographic implications. In *Abstracts of the 7th International Palynological Congress, Brisbane*, p. 109.

Muller, J. 1981 Fossil pollen records of extant angiosperms. *Bot. Rev.* **47**, 1–142.

Nichols, D. J., Jarzen, D. M., Orth, C. J. & Oliver, P. Q. 1986 Palynological and iridium anomalies at Cretaceous–Tertiary boundary, south–central Saskatchewan. *Science, Wash.* **231**, 714–717.

Nichols, D. J. & Fleming, R. F. 1988 Plant microfossil record of the terminal Cretaceous event in western United States and Canada. In *Abstracts – Global Catastrophies in Earth History, Snowbird, Utah, October 20–23 1988*, p. 130.

Parrish, J. T. & Spicer, R. A. 1988 North Polar Late Cretaceous temperature curve: evidence from plant fossils. *Geology* **16**, 22–25.

Raine, J. I. 1988 The Cretaceous/Cainozoic boundary in New Zealand terrestrial sequences. In *Abstracts of the 7th International Palynological Congress, Brisbane*, p. 137.

Ramanujam, C. G. K. & Stewart, W. N. 1969 Fossil woods of Taxodiaceae from the Edmonton formation (Upper Cretaceous) of Alberta. *Can. J. Bot.* **47**, 115–124.

Rawat, M. S., Swamy, S. N. & Juyal, N. P. 1988 Cretaceous/Tertiary boundary changes in the palynofossil assemblage of the Krishna–Godavari Basin, India. In *Abstracts of the 7th International Palynological Congress, Brisbane*, p. 138.

Retallack, G. J., Leahy, G. D. & Spoon, M. D. 1987 Evidence from paleosols for ecosystem changes across the Cretaceous/Tertiary boundary in eastern Montana. *Geology* 15, 1090–1093.

Rothwell, G. W. 1977 Evidence for a pollen-drop mechanism in Paleozoic pteridosperms. *Science, Wash.* 198, 1251–1252.

Saito, T., Yamanoi, T. & Kaiho, K. 1986 Devastation of the terrestrial flora at the end of the Cretaceous in the Boreal Far East. *Nature, Lond.* 323, 253–256.

Scott, R. A. & Smiley, C. J. 1979 Some Cretaceous plant megafossils and microfossils from the Nanushuk Group, Northern Alaska, a preliminary report. *Circ. U.S. geol. Surv.* 794, 89–112.

Smiley, C. J. 1966 Cretaceous floras of the Kuk River area, Alaska – stratigraphic and climatic interpretations. *Bull. geol. Soc. Am.* 77, 1–14.

Smiley, C. J. 1969a Cretaceous floras of the Chandler–Colville region, Alaska – stratigraphy and preliminary floristics. *Bull. geol. Soc. Am.* 53, 482–502.

Smiley, C. J. 1969b Floral zones and correlations of Cretaceous Kukpowruk and Corwin formations, northwestern Alaska. *Bull. Am. Ass. Petrol. Geol.* 53, 2079–2093.

Spicer, R. A. 1989a Physiological characteristics of land plants in relation to environment through time. *Proc. R. Soc. Edinb.* (In the press.)

Spicer, R. A. 1989b Reconstructing high latitude Cretaceous vegetation and climate: Arctic and Antarctic compared. In *Antarctic paleobiology and its role in the reconstruction of Gondwana* (ed. T. N. Taylor & E. L. Taylor). Berlin: Springer-Verlag.

Spicer, R. A. 1987 The significance of the Cretaceous flora of northern Alaska for the reconstruction of the Cretaceous climate. In *Das Klima der Kreide-Zeit (Geol. Jb. A 96)* (ed. E. Kemper), pp. 265–291.

Spicer, R. A., Burnham, R. J., Grant, P. R. & Glicken, H. 1985 *Pityrogramma calomelanos*, the primary, post eruption colonizer of Volcán Chichonal, Chiapas, Mexico. *Am. Fern J.* 53, 1–5.

Spicer, R. A. & Parrish, J. T. 1986 Paleobotanical evidence for cool North Polar climates in the mid-Cretaceous (Albian–Cenomanian). *Geology* 14, 703–706.

Spicer, R. A. & Parrish, J. T. 1987 Plant megafossils, vertebrate remains, and paleoclimate of the Kogosukruk Tongue (late Cretaceous), North Slope, Alaska. In *Geologic studies in Alaska by the United States Geological Survey during 1986 (Circ. U.S. geol. Surv. 933)* (ed. T. D. Hamilton & J. P. Galloway), pp. 47–48.

Spicer, R. A. & Parrish, J. T. 1989 Late Cretaceous–early Tertiary palaeoclimates of northern high latitudes: a quantitative view. *J. geol. Soc.* (In the press.)

Spicer, R. A., Wolfe, J. A. & Nichols, D. J. 1987 Alaskan Cretaceous–Tertiary floras and Arctic origins. *Paleobiology* 13, 73–83.

Sporne, K. R. 1980 A re-investigation of character correlations among dicotyledons. *New Phytol.* 85, 419–449.

Stott, L. D. & Kennet, J. P. 1988 Cretaceous/Tertiary boundary in the Antarctic: climatic cooling precedes biotic crisis. *Geol. Soc. Am. Abstr. Prog.* 20, 251.

Sweet, A. R. & Braman, D. R. 1988 Floral Changes within the interval containing the Cretaceous–Tertiary boundary in western Canada: a stratigraphic, palaeogeographic and palaeoenvironmental perspective. In *Abstracts of the 7th International Palynological Congress, Brisbane*, p. 161.

Tiffney, B. H. 1984 Seed size, dispersal syndromes, and the rise of the angiosperms: evidence and hypothesis. *Ann. Mo. bot. Gdn* 71, 551–576.

Tinus, R. W. & Roddy, D. J. 1988 Effects of global atmospheric perturbations on forest ecosystems: predictions of seasonal and cumulative effects. In *Abstracts – Global Catastrophies in Earth History, Snowbird, Utah, October 20–23 1988*, p. 196.

Tschudy, R. H. 1971 Palynology of the Cretaceous–Tertiary boundary in the northern Rocky Mountains and Mississippi Embayment regions. *Spec. Pap. geol. Soc. Am.* no. 127, 65–111.

Tschudy, R. H. 1973 The Gasbuggy Core – a palynological appraisal in Cretaceous and Tertiary rocks of the southern Colorado Plateau. In *Memoir: Durango, Colorado* (ed. J. E. Fasset), pp. 131–143. Colorado: Four Corners Geological Society.

Tschudy, R. H. & Tschudy, B. D. 1986 Extinction and survival of plant life following the Cretaceous–Tertiary boundary event, Western Interior, North America. *Geology* 14, 667–670.

Upchurch, G. R. Jr & Dilcher, D. L. 1984 A magnoliid leaf flora from the mid Cretaceous Dakota Formation of Nebraska. *Am. J. Bot.* 71, 119.

Upchurch G. R. Jr & Wolfe, J. A. 1987 Mid Cretaceous to early Tertiary vegetation and climate: evidence from fossil leaves and woods. In *The origin of angiosperms and their biological consequences* (ed. E. M. Friis, W. G. Chaloner & P. R. Crane). Cambridge University Press.

Wheeler, E., Lee, M. R. & Matten, L. C. 1987 Dicotyledonous woods from the Upper Cretaceous of southern Illinois. *Bot. J. Linn. Soc.* 95, 77–100.

Wolbach, W. S., Lewis, R. S. & Anders, E. 1985 Cretaceous extinctions: evidence for wildfires and search for meteoric material. *Science, Wash.* 230, 167–170.

Wolbach, W. S., Gilmour, I., Anders, E., Orth, C. J. & Brooks, R. R. 1988 Global fire at the Cretaceous–Tertiary boundary. *Nature, Lond.* 334, 665–669.

Wolfe, J. A. 1985 Distribution of major vegetational types during the Tertiary. *Monogr. Am. geophys. Un. Geophys.* **32**, 357–375.

Wolfe, J. A. 1987 Late Cretaceous – Cenozoic history of deciduousness and the terminal Cretaceous event. *Paleobiology* **13**, 215–226.

Wolfe, J. A. & Upchurch, G. R. Jr 1986 Vegetation, climatic and floral changes at the Cretaceous–Tertiary boundary. *Nature, Lond.* **324**, 148–152.

Wolfe, J. A. & Upchurch, G. R. Jr 1987 Leaf assemblages across the Cretaceous–Tertiary boundary in the Raton Basin, New Mexico and Colorado. *Proc. natn. Acad. Sci. U.S.A.* **84**, 5096–5100.

Discussion

N. J. SHACKLETON, F.R.S. (*Sub-Department of Quaternary Research, University of Cambridge, U.K.*). If current understanding of oceanic events at the K–T boundary is correct, atmospheric CO_2 must have risen rapidly by about a factor of two or three as a consequence of the drastic reduction in the biological pumping of carbon from surface to deep ocean. Obviously this would have affected global climate, but I wonder whether the direct effect of this rise in CO_2 on plant water use efficiency might bias Dr Spicer's inference from plant leaf shapes of changing precipitation across the boundary?

R. A. SPICER. Changes in atmospheric CO_2 levels are known to affect stomatal action, and thereby water relations, in several species of modern plants but the extent to which this occurs varies from species to species (see, for example, Bazzaz 1980; Moore 1983). This, in turn, alters competitiveness and, by extrapolation, the composition of vegetation (Jarvis & McNaughton 1986). Unfortunately, far too little is known about this phenomenon and its retrodictive value for the Cretaceous or Palaeocene world for us to quantify its possible effect in biasing estimates of precipitation:evaporation ratios. Intuitively, sudden (say less than 1 Ma) changes in atmospheric CO_2 to the extent of doubling or quadrupling present atmospheric levels are unlikely to induce sufficient selection of the genome to bring about wholesale changes in leaf morphology at the species level. To generate the extent of changes in assemblage composition that are observed would require rapid simultaneous evolution of species. Nevertheless, the possibility that the proportions of taxa might change is a real one but here the magnitude of the effect is unlikely to alter significantly the conclusion that precipitation increased.

Fortunately, one does not have to rely solely on data from individual plant specimens. It has long been recognized that the onset of coal deposition in the Western Interior of the U.S.A. is approximately coincident with the K–T boundary (Brown 1962) which, together with an increase in the abundance and size of river channels, also suggests more humid conditions in the Palaeocene than the Maastrichtian. The more parsimonious explanation of the biotic and abiotic changes across the boundary is that the precipitation:evaporation ratio increased, but this does not exclude changes in CO_2 levels.

References

Bazzaz, F. A. 1980 Consequence of elevated CO_2 concentrations for plant photosynthesis, growth and competition. In *Abstracts of the 5th International Congress on Photosynthesis, September 7–13. Halkidiki, Greece.*

Brown, R. W. 1962 Paleocene flora of the Rocky Mountains and Great Plains. *U.S. geol. Surv. prof. Pap.* no. 375. (119 pages.)

Jarvis, P. G. & McNaughton, K. G. 1986 Stomatal control of transpiration: scaling up from leaf to region. *Adv. ecol. Res.* **15**, 1–49.

Moore, P. D. 1983 Plants and the palaeoatmosphere. *J. geol. Soc. Lond.* **140**, 13–25.

Phil. Trans. R. Soc. Lond. B **325**, 307–326 (1989)

Printed in Great Britain

Ammonoid extinction events

By M. R. House

Department of Geology, University of Southampton, Southampton SO9 5NH, U.K.

The ammonoid cephalopods range from the early Devonian to the late Cretaceous, a period of some 320 Ma. Because of their importance for biostratigraphic discrimination and their use in practical age dating for this period they have been intensively studied. Major extinctions at the close of the Devonian, end Permian, end Triassic and end Cretaceous have long been recognized and linked with regressional palaeogeographical events. The recognition of smaller-scale extinction events is relatively new and is especially well shown in the Palaeozoic, when there was a simpler distribution of land and sea pathways than in later periods when the influence of latitudinal distributions and local provinces was more severe. Extinction events in the Devonian show the nature of the process. Usually a gradual decline in diversity is followed by extinction; then there is a period of low diversity but often individual abundance. Then novelty appears and is seen in new characters of the early stages; elaboration and diversification follow. These fluctuations can often be correlated with changes in other groups and also with sedimentological and palaeogeographical changes. Usually a regression–transgression couplet is involved with evidence of ocean turnover indicated by anoxic or low-oxygen events. A new family, Sobolewiidae, is diagnosed.

A new analysis of diversity, appearances and extinctions is made at the family level for 2 Ma time units throughout the history of the Ammonoidea. This record is compared with modern attempts to portray sea-level fluctuations and onlap and offlap movements of marine seas. The correlation, even in detail, is impressive and gives support for the species/area theory. But it is argued that temperature, as well as sea-level factors, is important.

The evidence, on both large and small scales, shows an association of evolutionary change with palaeogeographical change. The new evidence does not suggest a role for periodicity above the Milankovitch Band level. Whether or not periodicity is involved, such factors seem more readily explained in endogenic earth causations and for the present these provide the most parsimonious explanations.

1. Introduction

The coiled ammonoids first appear in the late Lower Devonian, perhaps about 390 Ma BP on the radiometric scale. They became extinct at the close of the Cretaceous about 66 million years ago, near the end of the Maastrichtian. Although similar in basic morphology to other chambered cephalopods the group is characterized by a distinctive globular to egg-shaped protoconch and in all but the very oldest forms by a distinctive coiled nepionic stage, the ornament of which differs from later coils. Their origin is thought to be from straight orthoconic and ectocochleate bactritids, which had already adopted the distinctive protoconch. Their evolution is characterized by the production of an incredible variety of small forms, which can vary in degree of coiling, in the shape of whorls and the degree and nature of shell ornament. Characteristic also is the great variety in the pattern of folding of the septa that separate the shell chambers: when found fossilized as internal moulds these give the distinctive suture lines of the group.

For as long as they have been studied, the ammonoids have proved very useful to the geologist because the detailed changes of their evolution have enabled a timescale of zones and subzones to be established that is the most precise age-dating tool available for that period. As a result their history is probably as well or better documented through time than any other group.

This contribution is concerned with extinction periods in the Ammonoidea. During the compilation of the latest review of the group (House & Senior 1981) the contributors to the forthcoming revision of the *Treatise on invertebrate paleontology* volume on the Ammonoidea provided detailed ranges of families that enabled the accompanying evolutionary tree for the group to be constructed (figure 1). The data here are that of M. R. House (Devonian), J. Kullmann (Carboniferous), B. F. Glenister and W. M. Furnish (Permian), E. T. Tozer

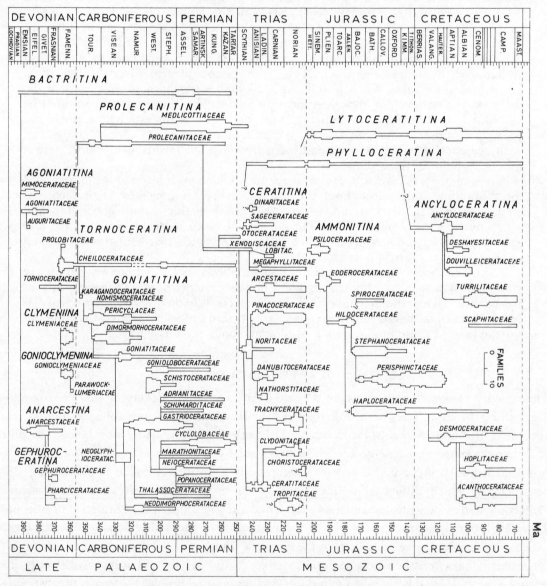

FIGURE 1. The evolutionary tree of the Ammonoidea from their appearance in the early Devonian to their extinction at the end of the Cretaceous. Based on House & Senior (1981).

(Triassic), D. T. Donovan, J. H. Callomon and M. K. Howarth (Jurassic) and C. W. Wright (Cretaceous), who have given details of the classification used. I was responsible for eliciting agreement over boundary problems. Thus this review is of a snap-shot of the available data at that time and there has been no attempt at a wholesale revision in the light of more recent data.

For this review the detailed manuscript ranges provided by the authorities have been

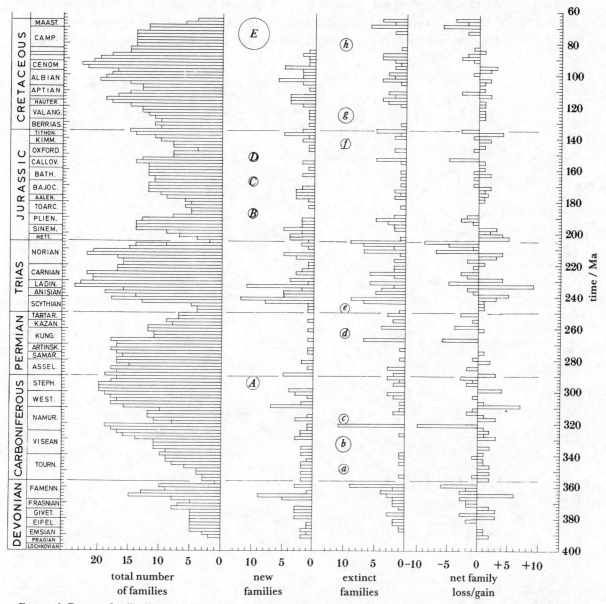

FIGURE 2. Data on family diversity through time of the Ammonoidea (excluding Bactritina) from the early Devonian to the end Cretaceous. Based on data from House & Senior (1981). Time units of two-million-year intervals largely based on Snelling (1985). Total number of families, which is an indication of morphological diversity at particular times, is taken as an indication of success or failure in exploiting a wide range of ecological niches. Totals of new families appearing within a two-million-year unit are an indication of degree of innovation and high values suggest ecological relaxation and exploitation; periods particularly noteworthy for a continued period lacking innovation are marked by circles (A–E). Totals of extinct families within a unit show families lost and high values are taken to indicate times of ecological stress; circles (a–h) mark continued periods when no families became extinct. More informative is the net family loss–gain plot, which indicates periods when innovation exceeds or is less than extinction.

analysed to give a review of the fortunes of the group at 2 Ma intervals for the time of known existence of the group (figures 2 and 4). The main concern is diversity, as indicated by the number of families at any 21 Ma unit, the number of new families appearing, and families lost and various aspects of the ratio of change and its relation to longevity of taxa involved (figure 4). Division of appropriate scales by two gives the rate for these factors in millions of years. This collation is used as the basis of a discussion on the extinctions but it should be stressed that the main concern is in loss of diversity, by which is inferred loss of niche occupancy through time. Thus whether or not families may be paraphyletic, and whether other

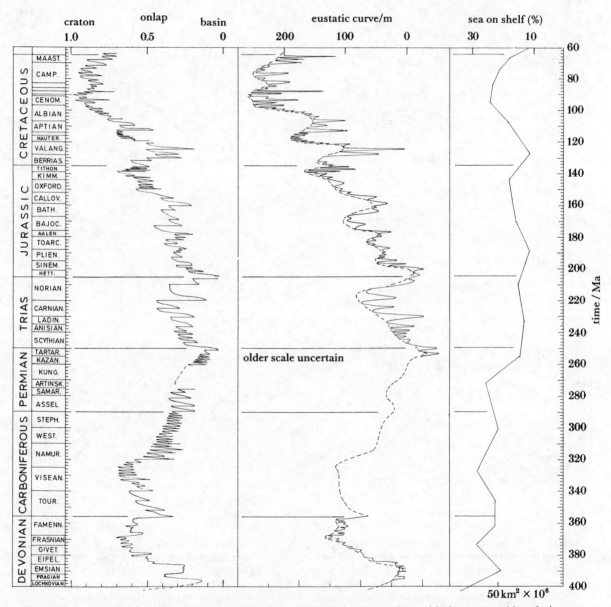

FIGURE 3. Data on changes in sea level through time plotted to enable comparison with the ammonoid evolutionary analysis. Onlap and eustatic data based as follows: Devonian, House (1983, Johnson *et al.* (1985, 1986), Krebs (1979); Carboniferous and Permian, Ramsbottom (1979), Saunders & Ramsbottom (1985), Ross & Ross (1985), Smith *et al.* (1974); Triassic, Holser & Magaritz (1987); Triassic to end Cretaceous, Haq *et al.* (1987). Total sea on continental shelf estimates based on Wise (1974).

interpretations of taxonomic procedure may be possible, are not of concern. The biological effects considered refer to the actual change in standing diversity. The aspects of what this meant in numerical abundance and 'success' can only be approached in a rigorous way by the known geographical distribution of faunas and, far more speculatively, as indicated by their fossilization record, interpreting common occurrence as indicating former abundance. Whereas with ranges reasonable objectiveness is possible, on other matters judgement and experience

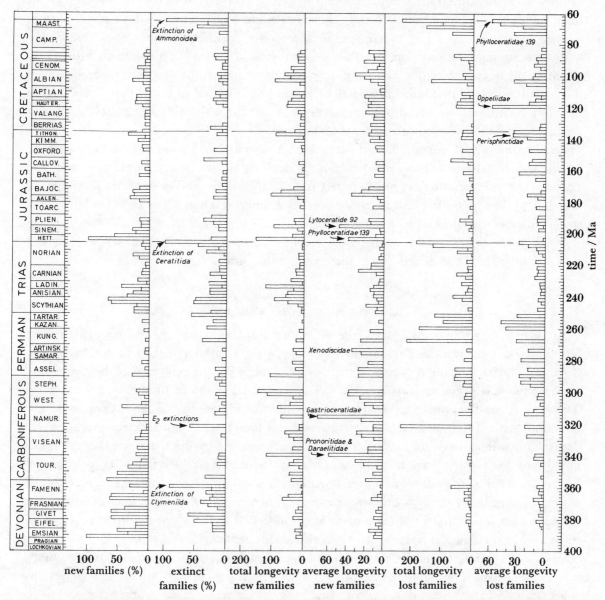

FIGURE 4. Further data on family diversity through time of the Ammonoidea (excluding Bactritina) from the early Devonian to the end Cretaceous (data as for figure 2). Plots of new/total and extinct/total give the ratios as a percentage for each of the two-million-year units used. The longevity, or time range, of families of Ammonoidea is plotted in relation to both the total longevity and the average longevity of families involved in appearance and extinction. Families are named in significant cases where a single family either gives the signal or is responsible for more of it. The loss of groups that have survived long periods of the vicissitudes of life is probably more significant than the loss of short-range taxa.

[71]

enter. At present, a similar review would not be possible at the generic level because for some time periods the data have not been published.

Those working on ammonoids have long seen a relation between palaeogeographical changes and periods of extinction and radiation (House 1963). This is analysed in more detail by using a compilation of data on marine transgressions and regressions, and sea-level changes thought to be worldwide and eustatic (figure 3). But several other hypotheses have been suggested as causes of extinction and diversification periods, so some of the important events are discussed in stratigraphic order so that evidence relating to particular events can be separated. The time scale used is largely that of Snelling (1985) and it will be apparent that for the Palaeozoic, if not so much for the Mesozoic, considerable inaccuracy results and variations in interpretation are possible. Nevertheless it seems worthwhile to attempt to assemble the data on a consistent time base so that rates and tempo can be compared, even if it is to be understood that refinements will be inevitable. It should be noted that figure 3 is not based on such rigorous correlation with the biostratigraphic scale as figures 2 and 4, and very detailed correlation of events should not be inferred.

Because my special interests lie with the older Ammonoidea, I shall discuss these in more detail. This will illustrate how, when even more detailed biostratigraphic data are taken into account, the interpretation of crudely totted and clumped data even at the 2 Ma level may be misleading. This will also illustrate that attempts at interpretation at an even cruder level of stages or series is subject to even more suspicion.

Much of the content of this contribution has been discussed with R. Thomas Becker and I also acknowledge the comments he made on the initial draft.

2. The ammonoid record

At several stages in their history the Ammonoidea almost became extinct. In the late Devonian, at the Frasnian–Famennian boundary, only a small group of tornoceratids and *Archoceras* survived. At the very close of the Devonian following the extinction of the clymeniids, at the Famennian–Tournaisian boundary, only restricted group of prionoceratids survived. The major mid-Carboniferous E_2 extinction, the second largest family extinction in the ammonoid record (figures 2 and 4), was not near to a total loss as eight superfamilies survived. The major extinctions by the close of the Permian witness to a gradual attrition through much of the period but with enough survivors to give cause for debate on the origin of Triassic groups. The end-Triassic extinctions come out in the present compilation (figure 2) as the largest family extinction event and perhaps only two or three genera survived it. The final demise of the Ammonoidea is the product of the longest attrition period in the history of the group and for some 18 Ma preceding it no new family-level taxa are reported.

The broad correlation of these events with regressional occasions (figure 3) has been much commented upon. The nature of the process, however, is less clear. What is clear is that any stock that did survive soon developed a wide range of diversifications. Conservatism appears to be a feature of survivors.

Devonian

In the Devonian diversification was largely a matter of inovation in patterns of folding of the septa. As sutural ontogeny has generally been held to be of great importance, and because sutures can readily be used to distinguish groups, high taxon grades have been assigned to such forms. Thus high-taxon extinction is highest in the Devonian with five suborders becoming extinct. This does not, of course, imply that these extinctions are in any way more 'significant' than others, but in that they led to the disappearance of certain special sutural possibilities they served to constrain later evolution.

The major extinction events during the Devonian number about eight and have been named in relation to localities or lithological units showing sedimentary perturbations with which they are associated (House 1985a). There is no implication that these are the only such events, but the experience of many years suggests that they may be the most significant. The relation between extinction events and environmental changes has been long recognized but only recently codified in this way. The first records of the coiled Ammonoidea are in the conodont *dehiscens* Zone, which the Devonian Subcommission is likely to use to define the base of the Emsian stage. Goniatites of the *dehiscens* Zone seem known in Alaska and the Yukon, probably Morocco, the Montagne Noire and Australia. This appearance is associated with a Zlichovian (approximately Lower Emsian) transgressive event that has recently been recognized even in Australia and Siberia (Talent & Yolkin 1987). By the following *gronbergi* Zone there was a considerable international distribution of goniatites. During this time loosely coiled and eventually tightly coiled and convolute forms are known, mostly members of the very simple-sutured Mimosphinctidae; but bizarre sutures are shown, by the Auguritidae, for example.

Daleje Event

The Zlichov Formation of Czechoslovakia represents a gradual deepening phase (Chlupáč & Kukal 1986) even reaching black calcareous shale facies in the upper part. This followed a regression at the base of the Zlichovian. The change to the deeper facies of the Daleje Formation is in detail gradual and initiated in the late *gronbergeri* Zone. Facies changes at this level are documented in Europe, North Africa, U.S.S.R., and eastern and western North America (House 1985a; Chlupáč & Kukal 1986). It appears to be an international event. In several areas, but especially in Europe, pyritic preservation indicates anoxic sea floors so an ocean overturning, spreading oxygen-low waters over shelves, seems a likely scenario. The loss of the main group of mimosphinctids and the auguritids is associated with this. Higher in the Dalejian two groups of great later significance appear, the Agoniatitidae and the Anarcestacea. Both soon show a closure of the primitive imperforate umbilicus and the latter shows a distinctive sutural pattern in which, in many, there is a lateral migration of an early subumbilical lobe into a more lateral position in the adult. However, it will be noticed that this event is not easy to resolve by using the crude 2 Ma unit plots of figure 2. This is because overall diversity is low and the quick rise of new groups masks the extinctions. Plotted in more detail, with generic range shown, the reality is rather more clearly seen (figure 5a). Similarly, the sea-level fluctuation corresponding to this is not well shown on figure 3, where high resolution is not possible, but it corresponds to deepening within Cycle 1b shown by Johnson *et al.* (1985). Other faunas show a decrease in provincialism associated with the transgressive event.

Basal Chotec Event

Chlupáč & Kukal (1986) have distinguished a Basal Chotec Event close to the new boundary between the Lower and Middle Devonian. Not marked explicitly on figure 5a, the event is indicated by the extinction of the Mimagoniatitidae (which in reality range higher in reduced importance) and of the Mimoceratidae. This event has been recognized as the *jugleri*

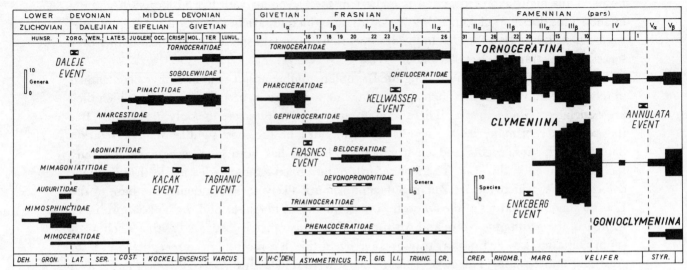

FIGURE 5. Family and generic data for certain groups of Devonian ammonoids (especially goniatites),
illustrating particular extinction events discussed in the text. Modified from House (1985a).

event by Walliser (1985), referring to the goniatite *Pinacites jugleri*, which appears following the
break. The genus *Pinacites* is not common but is highly distinctive and easily recognized. It is
widespread in North Africa and Europe and has recently been found in Alaska. That there is
an environmental and palaeogeographical effect here is suggested by the close relation with the
great changes in spore floras known close to this level (Riegel 1974). In Europe it corresponds
rather closely to the deepening following the Emsian near-shore clastic facies and in North
America to the progressive transgression of the Onondaga Limestone, one of the two most
significant onlap events in the North American Devonian.

Kačak Event

In the late Eifelian (as now defined), there is a distinct break when *Pinacites* disappears and
Cabrieroceras becomes internationally abundant. Again this is connected with a sedimentary
perturbation. The names *otomari* Event, after the common tentaculite, *Nowakia otomari*, or
rouvillei Event, after one species of *Cabrieroceras*, have been applied to it (Walliser 1984, 1985).
At this event goniatites with perforate umbilici appear to be lost. The distribution of
Cabrieroceras (House 1978) is considerable in Europe, North Africa and eastern North America
but it also occurs in western North America. In Morocco there is a pyritic level with the name-
bearer indicating local sea-floor anoxia, and the event is recognized in the Odershausen
Limestone in deeper-water facies. In New York State, U.S.A., it is seen as the Werneroceras
Bed associated with black shales. However, it is only recently that the environmental effects at
this level have been well documented in more shallow facies, and in particular K. Weddige and
W. Struve of the F.R.G. have recognized a regressional gap at this level in the Eifel succession
between the Giesdorf and Freilingen levels.

Taghanic Event

The extinction of the *Sobolewia* group (named on figure 5 as the Sobolweiidae *fam. nov.*,
diagnosed by the simple sutured, involute goniatites with convex lines of the genus),

Agoniatitidae, Maenioceratidae and 'Pinacitidae' (forms such as *Foordites* and *Wedekindella*), used to define the end of the Middle Devonian for Devonian workers (as almost universally in Oswald (1967), but the boundary is now taken higher. These earlier groups are shortly replaced by the distinctive multilobed Pharciceratidae and the globally widespread Gephuroceratidae. The new forms are associated with transgressions almost internationally but the North America sub-Taghanic regression is the main clue to the extinctions. The later extinction of most of the Pharciceratidae has been named the Frasnes Event.

Kellwasser Events

The upper of the two late Frasnian black limestones, known in the Schiefergebirge as the Lower and Upper Kellwasser Kalk, marks the extinction of the Gephuroceratidae and Beloceratidae. Only the anarcestid *Archoceras* and a small group of tornoceratids survive. There is an anomaly here, as the anoxic Upper Kellwasserkalk level is where the last gephuroceratids (especially *Crickites*) are seen abundantly. The anoxia itself seems unlikely to have been an immediate cause of extinction but there are several lines of evidence suggesting shallowing, or at least a spread of shallow seas, especially the increase in the conodont *Icriodus* (Sandberg *et al.* 1988). Bolide impact is another favoured theory (McLaren 1982, 1983) although disparate ages for iridium anomalies (Playford *et al.* 1984; McLaren 1985), or absence of any anomaly at critical points (McGhee *et al.* 1984, 1986), is making this increasingly unlikely. The sedimentological perturbations (Sandberg *et al.* 1988) and regressional lag deposits (Geldsetzer *et al.* 1987) interpreted as indicating bolide impact seem equally explained as epeirogenic or eustatic changes linked with a primary cause in tectonism associated with plate movement (Johnson 1988).

When studied in detail (Becker 1986; House 1985a), goniatite diversity shows a decline approaching the Frasnian–Famennian boundary and the base of the Lower *triangularis* Zone (figure 5b). The extinction is similar to those already documented here for which no extraterrestrial cause is invoked. In a general analysis of all faunas around this boundary (House 1975) it was argued that extinctions are spread around the boundary and sea-level fluctuations were thought to be a cause. Only corals (Pedder 1982), atrypids (Copper 1984, 1986), conodonts (Sandberg *et al.* 1988) and stromatoporoids (Stearn 1987) have been looked at in any detail and the records suggest a staged or gradual extinction rather than a sudden event. Copper (1986) prefers a cold-water cause; Stanley (1988) also prefers this, but by spreading of deeper anoxic water over a shelf by changes in the level of the pycnocline. This might be caused tectonically by epeirogenetic or eustatic changes as readily as by global climatic change. There is now no convincing evidence for Devonian glaciation anywhere, even in the Famennian of South America close to the supposed pole, so it is not possible to invoke glacial events.

Hangenberg Event

There are other extinction events within the Famennian (House 1985a; Korn 1986) but that at the close, just below the Devonian–Carboniferous boundary, is the most significant. Details have been published elsewhere (Price & House 1984; House 1985a; Becker 1988). In the uppermost Famennian the Tornoceratidae, Posttornoceratidae and Sporadoceratidae, as well as eleven families, become extinct as do the bizarre clymeniids, the only group of ammonoids with a dorsal siphuncle. During the diversity low only a limited group of Prionoceratidae are known, but early stages are quite diverse and they are usually abundant in the black and anoxic shales equivalent to the Hangenberg Shales in several parts of the world. Diversification leading to the major new groups, Prolecanitina and Goniatitina, follows gradually in the early Lower Carboniferous.

These selected extinction events in the Devonian are characterized by an association with sedimentary perturbations, especially in the form of anoxia. They share a common pattern of gradual decline, minimum of diversity and gradual diversification. These suggest a common cause. One of the reasons why I have favoured terrestrial explanations for such extinctions is that ocean upwelling associated with sea-level changes triggered by tectonic events and/or plate tectonic activity seems a reasonable cause. The events vary slightly among themselves, and this too would be expected from terrestrial events. It has been argued that the triggering of the Kellwasser Event specifically is extraterrestrial (McLaren 1983; Sandberg *et al.* 1988) and by bolide impact. However, the pattern of the Kellwasser Event, as has been shown, differs only in degree from the other events discussed, and unified and parsimonious interpretation is to be preferred. That is sought here in terrestrial causations.

When the detail known on Devonian extinction events is looked for in compilations such as figure 2 it is clear that the specialist biostratigrapher has at his or her command far more detailed information than is available to a compiler. The events in subsequent ammonoid history which will be discussed will be based largely on the compilation that has been attempted (figures 2 and 4) and not on such detailed knowledge.

Carboniferous

Carboniferous ammonoid evolution indicates an initially progressive radiation, arising from survivors of the Hangenberg Event, into a great proliferation in Goniatitina and in the less diverse but highly distinctive Prolecanitina, which show elaboration on umbilical lobation. The steady and consistent rise in total diversity during the Tournaisian and Visean is remarkably systematic (figure 2). Also noteworthy are the periods (352–346, and especially 338–328 on the units of figure 2) when no families became extinct at all. The Visean period is the largest timespan known in the Palaeozoic when this occurs for so long and this is equalled only once in the later history in the early Cretaceous. A relation with the widespread early Carboniferous transgression, giving broad shallow seas combined with indications of equable climates, may be the cause.

The major extinction event is in the early Namurian at the end of E_2. The data on the accompanying diagram are from Kullmann (1981) but he has published more specific accounts of changes in evolutionary rates during the period (Kullmann 1983, 1985) based on an analysis of genera at approximately genus-zone divisions rather than zonal level or the attempt at standardized time units adopted here. Generic diversity is at its maximum at about the Visean–Namurian boundary which is second in importance on the family plot (figure 2). The last occurrence of genera is at its peak in E_2 and this does correspond with the family data (figure 2); it is the Palaeozoic extinction peak for total families but not the ratio of extinct to total present. However, E_2 is clearly the major extinction event for the period.

Palaeogeographical correlation with evolutionary diversity increase or reduction is less clearly established. The long-known North America regression between the Kaskaskia and Absaroka cratonic sequences (Sloss 1963) closely corresponds to this and the result emphasizes the Mississippian–Pennsylvanian boundary. In Europe the H genus zone corresponds to a significant influx of Millstone Grit clastics as the prelude to Coal Measure deltaic facies. So in general a palaeogeographical change is likely to be a significant cause of the extinctions.

But in detail it is clear that there are many small-scale transgression/regression couplets through the whole Carboniferous (see Ramsbottom 1979; Heckel 1986; Ross & Ross 1985). Comparison of these between Europe and North America has been attempted by Saunders *et al.* (1979) and the British succession, which is the most detailed, indicates maxima around E_{2c} to H_{1ab} and H_{2a} to H_{2c} suggesting confirmation of the correlation. But the data do not give really precise plots for figure 3 and, of course, no palaeogeographic maps of this detail have been constructed to allow precise estimates of onlap.

The relation between cyclothemic pulses and the entry of goniatite shales with distinctive faunas is the basis for much European Carboniferous zonation, but the data refer to internationally restricted taxa and would not show on a plot such as figure 2. A much more detailed study of the British data by Holdsworth & Collinson (1988) is more helpful in indicating the intimate relation between goniatite faunas and facies changes but international correlation is not advanced enough for this to be documented on a larger scale. Part of the reason this is possible for the Devonian data is the existence of a parallel conodont biostratigraphy.

For the later Carboniferous there is an interesting conflict of evidence. Kullmann's (1985) generic analysis shows a progressive and asymptotic decline in total number of genera from R_1 to the end of the period. The family data (figure 2) on the other hand, show a steady increase to the mid-Stephanian, then a final 8 Ma period (298–290) when no new families appear. This is the only period of such a standstill in the Palaeozoic record and it is equalled only by that preceding the extinction of the ammonoids at the end of the Cretaceous. With the already established glaciation in the Southern Hemisphere, and Coal Measure or terrestrial environments in much of Laurussia, probably rather special conditions characterized the areas of goniatite colonization. Although early collisions between Gondwanaland and Laurussia were initiated in the Devonian with the Acadian orogeny, similar effects were long continued, culminating in the late Visean with widespread European olistostromes and thrusting and later in the main paroxysms of the Variscan orogeny. These events will have been the major cause of the sea level fluctuations which, like a bioseimograph, are reflected in ammonoid evolution, but the detailed correlation is not established. Tectonic and sea-floor spreading effects may also cause the prominent sedimentary rhythmicity at a smaller scale but by the early Namurian the increasing effects of the growth of the Gondwanaland ice cap give also the probability of additional climatically forced sea-level changes.

Permian

The transition between the Carboniferous and Permian is not marked by a major extinction event. Rather, the introduction of the new families Shikhanitidae, Perrinitidae, Metalegoceratidae, Paragastrioceratidae and Popanoceratidae gives a distinctive defining element, although work on the international definition of the actual system boundary is still proceeding. The family data show a gradual decline to the end of the Permian (figure 2), with a significant fall away only after the mid-Kungurian where there is the major extinction event for the system, seven families being lost, corresponding to the late Wolfcamp. Although in general this corresponds to a regression (figure 2) the data are far from clear and international definition and correlation leave much to be desired.

The Permian–Triassic boundary represents a low point in ammonoid family diversity, only surpassed by those at the end of the Devonian, Triassic and Cretaceous. At this point the helpful onlap–offlap and eustatic curves of Haq *et al.* (1987) give evidence of possible causation because the boundary represents the lowest point for both offlap and sea-level lowering in the whole period from the latest Permian to the end Cretaceous. Again a relation between decrease in area of shallow seas and decrease in diversity seems an inescapable conclusion. Holzer & Magaritz (1987) have summarized other changes associated with the boundary regression/ transgression couplet, particularly perturbations in $^{87/86}$Sr, δ^{34}S and δ^{13}C. This has been recognized for a long time as the most significant extinction event for marine animal taxa (see Newell 1952, 1963, 1967; Raup & Sepkoski 1984).

It is of particular interest that the end Permian also saw the extinction of the Tornoceratina, the range of which has now been extended into the latest Permian with *Qinglongites* (Zheng 1981). This group alone had survived all the earlier extinction events from the Middle Devonian. For the Devonian there is evidence that the group may have favoured colder or deeper waters (House 1985 a).

Triassic

The Permo-Triassic boundary extinctions for the ammonoids are sufficiently severe for there to have been differences of opinion on the relationship of succeeding groups. The evolution has been reviewed by Tozer ((1971 a) and in House & Senior (1981)) and he has also given a splendidly readable review of the major revisions resulting upon new views of the sequences in the Alps (Tozer 1984), which starts with a quotation at the beginning: 'Were it not for fossils geology would be mere vulgar engineering'.

From the detailed zonal data of Tozer, the evolution of superfamilies of Triassic ammonoids are illustrated here (figure 5) by plots giving total number of genera in each zone. This illustrates a phased pattern of innovation. The earliest Triassic appears to have been a period of standstill (shown as '*e*' on figure 2). The Noritaceae diversify in the Nammalian (Na. on figure 5), followed by six superfamilies appearing in the Spathian (Sp.), a further four in the Ladinian and a final one in the late Norian (or Rhaetian). This shows also how the dominant period or acme for a group is usually markedly later than its first appearance. The decline is also initiated long before the extinction of a group and the extinctions indicated by generic data may differ from those at family or superfamily level. There is also a relay-like replacement of faunal 'packets'; these are probably replacing each other in a similar ecological niche, but it is too early to quantify this. Noticeable also are the high rates in the production of new and extinction of old families. This flourishing evolution emphasizes why the period has been called the 'Age of ammonites'.

The new sea-level curves (figure 3) start to suggest a striking correlation of radiation periods for the Triassic ammonoids with transgressive pulses and of extinctions with regressive periods. Thus it would appear that there are several extinction periods (Benton 1986).

The extinction at the close of the Triassic is, in terms of number of families becoming extinct, the greatest in the history of the Ammonoidea (figure 2). So profound was the effect that it has resulted in great uncertainties in finding the ancestors for Jurassic groups (see Tozer (1971) and Wiedmann (1973)).

Jurassic

Post-Triassic ammonoid evolution is set in a palaeogeography showing the effects of the fracture and break-up of Laurussia and in stages of Gondwanaland. The easterly expanding

Tethys Ocean was mostly near-tropical and led to the development of specialist warmer Tethyan faunas. In higher latitudes the proto-Atlantic and associated fractures, and geographically contained polar areas, lent themselves to the development of colder Boreal faunas and of local provincial evolution in restricted basins.

The two longest-surviving Jurassic and Cretaceous groups, the Phylloceratina and Lytoceratina (figure 1), appear to have characterized deeper levels of more equatorial waters. Formerly it was thought that these groups gave off stocks of coarsely ribbed Ammonitina in evolutionary waves that reiteratively colonized boreal waters, but this simplicity is not now accepted. The current view is that the Ammonitina are independently derived from the Triassic, but this includes groups colonizing Tethyan as well as Boreal shallow waters. Evolution resulted from the success or failure of these related to migrations driven by sea-level changes (eustatic or epeirogenic) and by climatic changes. Because the palaeogeography was complex, so was the resulting evolution. But beginnings are being made in relating evolution to environmental changes.

For the earliest Jurassic evolution Donovan (1988) has suggested that the Arietitidae originated in low latitudes to colonize other areas. He notes significant extinctions within and at the end of the Sinemurian, the former possibly being linked to a regression and the latter corresponding to a switchover in dominance of psilocerataceans by eoderocerataceans (figure 6).

The end Pliensbachian is a marked extinction event for marine groups (see Raup & Sepkoski 1984) and this has been documented in detail for the bivalves and other groups by Hallam (1986) who considers that the main extinction falls within the early Toarcian (Hallam 1987a). This event shows in the international ammonoid record by the loss of the Eoderoceratacea (figure 6), especially the amaltheids. This Hallam links with regression.

The Toarcian *falciferum* Zone represents a widespread deepening of facies in Europe associated with anoxia, giving the famous exotic preservation in the Lagerstätten of Holzmaden in Germany and the Jet Rock facies of Yorkshire. This is thought to be related to a global rise in sea level and climatic cooling cannot be ruled out as an additional factor. It corresponds to the second-lowest count of total familes on figure 2 and the first of the significant periods when no new families were produced (shown as B on figure 2). Tethyan phylloceratids extended far to the north at this time, and with them extended an evolutionary radiation of Tethyan derived dactylioceratids and hildoceratids giving rise to dominance in the later Toarcian of the Hildocerataceae (figures 1 and 5).

The Aalenian transgression, and transgressive events within the Bajocian and Bathonian (figure 3) are associated with the rise of Stephanocerataceae, Perisphinctaceae and Haplocerataceae (figure 1), which then dominate the later Jurassic. The first proliferated especially in Boreal waters. However, this is not a period of great family diversity. Ager (1981) has suggested this was a Jurassic warm period. In this interval Bayer & McGhee (1984) have documented iterative morphological sequences linked with sedimentological rhythms.

The celebrated Callovian transgression was a staged event but at its peak it probably deserves much greater emphasis than given to it by Haq *et al.* (1987) and shown on figure 3. Again in Europe there is an anoxic event and after this, at the close of the stage, the extinction of the warmer-water reineckiids and cooler-water kosmoceratids give the marker taken to define the Middle/Upper Jurassic boundary.

An early Oxfordian global deepening shows by the extension of Boreal stephanocerataceans

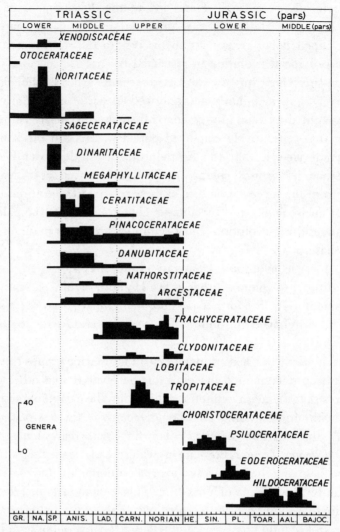

FIGURE 6. Diversity in ammonoid superfamilies for the Triassic and early Jurassic based on data of Tozer (1981) and Donovan *et al.* (1981). The time units are the zonal divisions of stages. This illustrates how evolution of certain morphologically defined clades often shows a slow growth of diversity, a peak and then a decline. It is the decline rather than the extinction that may be the more indicative of selection stress for the group. Noticeable is the 'packet' type evolution in which, relay-style, one group may replace another, possibly in the same ecological niche. From House (1985*b*).

southward in Europe and especially southward through the western interior of North America, with sufficient sea-floor anoxia to give the famous pyritic ammonites of both areas. The later Oxfordian has shallower environments and probably climatic amelioration.

Progressive deepening through the Kimeridgian appears to be having a control because just before the maximum in anoxia is the extinction of the cardioceratids (and with them the Stephanoceratacea), a level that is taken as the boundary between the Kimmeridgian and Tithonian. The large rise in families in the late Tithonian corresponds to a blooming of the perisphinctacians, which include many giants. This culmination would link with another period of warming recognized by Ager (1981). There is an extremely good correlation between late Jurassic regression and the decline in diversity near the Jurassic/Cretaceous boundary (figures 2–4) (Hallam 1986).

Cretaceous

The broad analysis of family diversity (figure 2) shows a maximum in the Upper Cretaceous. This shows a close correspondence with the encroachment of marine waters over cratonic shelves (figure 3) and by the Cenomanian/Turonian interval both reach their maximum for the whole period of existence of the Ammonoidea (figures 2 and 3). Subsequent regressive phases led to the extinction of the group at the close of the Maastrichtian. The novel feature of the early Cretaceous is the rise of the Ancyloceratina, which include the heteromorphic ammonites so distinctive of the period. Otherwise Ammonitina evolution results from Jurassic derivative stocks of the Haploceratraceae and the declining Perisphinctaceae (figure 1). A revised family tree for Cretaceous ammonoids has been published by Kennedy & Wright (1985).

The Berriasian faunas comprise holdover stocks from the Jurassic without the type of inovative radiation that might have been expected after the Tithonian extinctions. The sea-level data (figure 3) suggest a cause in continued pulses of regression. Nevertheless, the Valanginian is one of the significant periods when few families become extinct (marked as *g* on figure 2).

It is with the late Lower Cretaceous transgressions that renewed evolution is seen and this reaches a maximum in the Albian interrupted by late Aptian regressions. The Haq *et al.* (1987) curve here owes much to European data where the relation between sedimentary perturbations and ammonoid evolution has been well documented. Early Albian deepening may be the cause of the demise of the Douvilleiceratraceae, Deshayesitaceae and Ancycloceratraceae.

The Cenomanian transgression is generally considered to represent the greatest onlap of marine waters over continental shelves in the Phanerozoic (Hallam 1977) (figure 3) but the term is used in a general way. Hancock & Kauffman (1979) found a Turonian sea-level high-point for the western interior of the U.S.A. and a Campanian–Maastrichtian maximum for northern Europe. Much of the late Cenomanian peak for ammonoid families (figure 2) comprises families that had their origin in the Albian.

After the Coniacian the ammonoid history is one of progressive family decline (figure 2) and generic decline. No new families were produced during this period and the interval, perhaps of 19 Ma, represents the longest such period in the whole history of the Ammonoidea (marked *E* on figure 2). The evidence of increasing selection pressures that this indicates, together with the associated palaeogeographic changes, are the reasons why palaeontologists do not concede that meteorite impact can be the only cause of the end-Cretaceous extinction (views summarized by Hallam 1987*b*).

The actual documentation of the final demise of the Ammonoidea shows a progressive geographical and diversity restriction (Hancock & Kennedy 1981; Ward 1983, 1988; Wiedmann 1988*a, b*). The latest evidence suggests a cutoff of several genera in the last 10 m of the best-documented sections along the Bay of Biscay coast. It is, however, impressive that the very last ammonite known at Zumaya, *Neophylloceras ramosum*, is a member of the Phylloceratitina, the one stock that survived all the earlier extinction events of the Mesozoic. Was this because of evolutionary robustness of the stock or because it lived in an environment that was least subject to rigorous change? If the latter it may hold this in common with the Tornoceratina, the longest-surviving group of the Palaeozoic.

Misgivings

No palaeontologist can be happy about the premises of using taxonomic categories for numerical analysis, as there is no consistency or standard in the definition of such grades nor any likelihood of that being achieved. As the Devonian and Triassic analysis shows, when generic-level data are available a more useful precision results but for the Ammonoidea a comprehensive survey of the whole group at such a level is not yet possible. There are also problems related to the radiometric time units used, for these become vague in the extreme by the Palaeozoic and the scheme used here is but one of many possible. For evolutionary analysis undoubtedly the record of actual successive beds is better than that of subzones, which is better than that of zones, which is considerably better than that of stages. The attempt here at 2 Ma time units is again probably the best that can be done at present for the group as a whole. Parts of the record are known, by contrast, in very great detail but a review at that level could not be attempted by a single individual or for the whole group.

General conclusions

The correlation demonstrated above between the detailed diversity record of the Ammonoidea, onlap–offlap and eustatic curves shows a vivid confirmation of the long-recognized species–area relation in evolution. Indeed, it may well be that the bioseismograph of the ammonoid record, and that of other groups, will provide a key to environmental controls of evolution in this way. This differs from views expressed by Stanley (1984) and Wilde & Berry (1984).

It has been argued throughout this paper that, when studied in detail, most evolutionary perturbations seem to be related to environmental and palaeogeographical changes and that terrestrial causations seem the most parsimonious to explain the record. Furthermore, in the Palaeozoic, a relation between certain sea-level changes and often anoxic events is increasingly being recognized to be related to specific tectonic events. Yolkin & Talent (1988) have demonstrated this for parts of the early Devonian. The effects of the Acadian and Antler orogenies seem part of this and there is considerable evidence of tensional tectonics associated with the Frasnian–Famennian Kellwasser Event. The next stage will be a more precise analysis of sea-level changes as a means of estimating these effects. It is singularly unfortunate that international agreement on stratigraphic units and their correlation is at such an early stage that the production of palaeogeographical maps is at best piecemeal and cannot contribute yet to the more accurate calculation of land/sea area which is required.

There is also a climatic factor more generally involved. It is at present impossible to quantify this effect and it is clearly an important factor, but not the only factor as some would argue. Climatic control of sea level and of sediment types adds a complication to a simplistic consideration of area/species effects. This is also true of anoxic levels, which are here interpreted as more likely caused by ocean upwelling that is tectonically driven than by climatic 'greenhouse' effects.

There is an impression from the data in figure 2 that extinctions are rather more spasmodic and irregular than innovation, which has been steadier. This has been tested by assembling histograms of the numbers of families known to appear ('new') and become extinct in the 2 Ma class intervals used (figure 7). The 'new' plot is the more regularly asymptotic. A similar

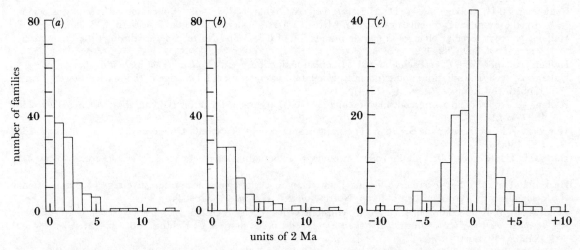

FIGURE 7. Histograms illustrating the frequency of (a) 'new', (b) 'extinct' and (c) 'gain/loss classes of total families, based on data shown in figure 2. The more regularly Gaussian nature of the 'new' and 'gain' plots is taken to indicate that the character of extinction and loss events is more spasmodic and irregular.

plot for loss–gain also shows the 'gain' plot to be more asymptotic. This suggests that appearances are more regularly programmed and losses more disjunct and irregular.

As for periodicity, there is little doubt that at the smallest scale Milankovitch Band orbital forcing and consequent climatic control is one of the factors important in detailed evolutionary change. This has also been argued for large-scale patterns. The great weakness here is that the actual radiometric timescales available, especially for the Palaeozoic, are far too inaccurate to show whether these really are periodic. The same is probably true of periodicity at the 26 Ma level and no evidence of that periodicity can be traced in the data assembled here (figures 2–4), notwithstanding the importance of certain extinction events that have long been known. Perhaps the advent of a timescale based on Milankovitch cycles will be the only way ahead here.

Finally, I must express the view that it is only by the detailed accumulation of factual data that improvements can be made. It behoves every scientist who wishes to be called a palaeontologist to add as much as possible to this body of knowledge. For this alone will lay a firm foundation for future advances.

REFERENCES

Ager, D. V. 1981 Major marine cycles in the Mesozoic. *J. geol. Soc. Lond.* **138**, 159–166.
Bayer, U. & McGhee, G. R. 1984 Iterative evolution of Middle Jurassic ammonite faunas. *Lethaia* **17**, 1–16.
Becker, R. T. 1986 Ammonoid evolution before, during and after the 'Kellwasser-event' – review and preliminary results. In *Global bio-events* (ed. O. H. Walliser), pp. 181–188. Berlin and Heidelberg: Springer-Verlag.
Becker, R. T. 1988 Ammonoids from the Devonian–Carboniferous boundary in the Hasselbach Valley (Northern Rhenish Slate Mountains). *Cour. Forsch.-Inst. Senckenberg* **100**, 193–213.
Benton, M. 1986 More than one event in the late Triassic mass extinction. *Nature, Lond.* **321**, 857–861.
Chlupáč, I. & Kukal, Z. 1986 Reflections of possible global Devonian events in the Barrandian area, C.S.S.R. In *Global bio-events* (ed. O. H. Walliser), pp. 169–179. Berlin and Heidelberg: Springer-Verlag.
Copper, P. 1984 Cold-water oceans and the Frasnian–Famennian extinction crisis. *Geol. Soc. Am.* Abstract programs, **16**, 10.
Copper, P. 1986 Frasnian/Famennian mass extinction and cold-water oceans. *Geology* **14**, 834–839.
Donovan, D. T. 1988 Evolution of the Arietitidae and their descendants. *Cah. Inst. Cath. Lyon, ser. Sci.* **1**, 123–138.
Donovan, D. T., Callomon, J. H. & Howarth, M. K. 1981 Classification of Jurassic Ammonitina. In *The Ammonoidea* (ed. M. R. House & J. R. Senior), pp. 101–155. London and New York: Academic Press.

324   M. R. HOUSE

Geldsetzer, H. H., Goodfellow, W. D., McLaren, D. J. & Orchard, M. J. 1987 Sulfer-isotope anomaly associated with the Frasnian–Famennian extinction, Medecine Lake, Alberta, Canada. *Geology* **15**, 393–396.

Hallam, A. 1977 Secular changes in marine inundation of USSR and North America through the Phanerozoic. *Nature, Lond.* **269**, 769–772.

Hallam, A. 1986 The Pliensbachiabn and Tithonian extinction events. *Nature, Lond.* **319**, 765–768.

Hallam, A. 1987*a* Radiations and extinctions in relation to environmental change in the marine Lower Jurassic of northwest Europe. *Paleobiology* **13**, 152–168.

Hallam, A. 1987*b* End-Cretaceous mass extinction event: argument for terrestrial causation. *Science, Wash.* **238**, 1237–1242.

Hancock, J. M. & Kauffman, G. 1979 The great transgressions of the late Cretaceous. *J. geol. Soc. Lond.* **136**, 175–186.

Haq, B. U., Hardenbol, J. & Vail, P. 1987 Chronology of fluctuating sea level since the Triassic. *Science, Wash.* **235**, 1156–1167.

Heckel, P. H. 1986 Sea-level curve for the Pennsylvanian eustatic marine transgressive-regressive depositional cycle along midcontinent belt outcrop, North America. *Geology* **14**, 330–334.

Holdsworth, B. K. & Collinson, J. D. 1988 Millstone Grit cyclicity revisited. In *Sedimentation in a synorogenic basin complex: the Upper Carboniferous of Northwest Europe* (eds. B. M. Besly & G. Kelling), pp. 132–151. Glasgow and London: Blackie.

Holzer, W. T. & Magaritz, M. 1987 Events near the Permian-Triassic boundary. *Mod. Geol.* **2**, 155–180.

House, M. R. 1963 Bursts in evolution. *Adv. Sci.* **19**, 499–507.

House, M. R. 1975 Faunas and time in the marine Devonian. *Proc. Yorks. geol. Soc.* **40**, 459–490.

House, M. R. 1978 Devonian ammonoids from the Appalachians and their bearing on international zonation and correlation. *Spec. pap. Palaeontol.* **21**, 1–70.

House, M. R. 1983 Devonian eustatic events. *Proc. Ussher Soc.* **5**, 396–405.

House, M. R. 1985*a* Correlation of mid-Palaeozoic ammonoid evolutionary events with global sedimentary perturbations. *Nature, Lond.* **313**, 17–22.

House, M. R. 1985*b* The ammonoid time-scale and ammonoid evolution. In *The chronology of the geological record* (ed. N. J. Snelling), pp. 273–283. Oxford: Blackwells.

House, M. R. 1988 Extinction and survival in the cephalopoda. In *Extinction and survival in the fossil record* (ed. G. P. Larwood), pp. 139–154. Oxford: Clarendon Press.

House, M. R. & Senior, J. R. (eds) 1981 *The Ammonoidea* (593 pages.) London and New York: Academic Press.

Johnson, J. G. 1988 Volcanism, eustacy, and extinction. *Geology* **16**, 573–587.

Johnson, J. G., Klapper, G. & Sandberg, C. A. 1985 Devonian eustatic fluctuations in Euramerica. *Bull. geol. Soc. Am.* **96**, 567–587.

Johnson, J. G., Klapper, G. & Sandberg, C. A. 1986 Late Devonian eustatic cycles around the margin of Old Red Sandstone Continent. *Ann. Soc. géol. Belgique* **109**, 141–147.

Kennedy, W. J. & Wright, C. W. 1985 Evolutionary patterns in late Cretaceous ammonites. *Spec. Pap. Palaeontol.* **33**, 131–143.

Korn, D. 1986 Ammonoid evolution in late Famennian and early Tournaisian. *Ann. Soc. géol. Belgique* **109**, 49–54.

Krebs, W. 1979 Devonian basinal facies. *Spec. Pap. Palaeontol.* **23**, 125–139.

Kullmann, J. 1983 Maxima im Tempo der Evolution karbonischer Ammonoideen. *Palaontol. Zeitschr* **57**, 231–240.

Kullmann, J. 1985 Drastic changes in Carboniferous rates of evolution. In *Sedimentary and evolutionary cycles* (ed. U. Bayer & A. Seilacher), pp. 35–47. Berlin and Heidelberg: Springer-Verlag.

McGee, G. R. Jr, Gilmore, J. S., Orth, C. J. & Olsen, E. 1984 No geochemical evidence for an asteroid impact at late Devonian mass extinction event. *Nature, Lond.* **308**, 629–631.

McGhee, G. R. Jr, Orth, C. J., Quintana, L. R., Gilmore, J. S. & Olsen, E. J. 1986 Late Devonian 'Kellwasser Event' mass-extinction horizon: no geochemical evidence for a large-body impact. *Geology* **14**, 776–779.

McLaren, D. J. 1982 Frasnian–Famennian extinctions. *Geol. Soc. Am. Spec. Pap.* **190**, 477–484.

McLaren, D. J. 1983 Bolides and biostratigraphy. *Bull. geol. Soc. Am.* **94**, 313–324.

McLaren, D. J. 1985 Mass extinction and iridium anomaly in the Upper Devonian of Western Australia: a commentary. *Geology* **13**, 170–172.

Newell, N. D. 1952 Periodicity in invertebrate evolution. *J. Paleontol.* **26**, 371–381.

Newell, N. D. 1963 Crises in the history of life. *Scient. Am.* **208**, 76–92.

Newell, N. D. 1967 Revolutions in the history of life. *Spec. Pap. geol. Soc. Am.* **89**, 63–91.

Oswald, D. H. (ed.) 1968 *International Symposium on the Devonian System, Calgary, 1967*. Alberta: Association of Petroleum Geology.

Pedder, A. E. H. 1982 The rugose coral record across the Frasnian/Famennian boundary. *Geol. Soc. Am. spec. Pap.* **190**, 485–489.

Playford, P. E., McLaren, D. J., Orth, C. J., Gilmore, J. S. & Goodfellow, W. D. 1984 Iridium anomaly in the Upper Devonian of the Canning Basin, Western Australia. *Science, Wash.* **226**, 437–439.

Price, J. D. & House, M. R. 1984 Ammonoids near the Devonian–Carboniferous boundary. *Cour. Forsch.-Inst. Senckenberg* **67**, 15–22.

Ramsbottom, W. H. C. 1979 Rates of transgression and regression in the Carboniferous of NW Europe. *J. geol. Soc. Lond.* **136**, 136–153.

Raup, D. & Sepkoski, J. Jr 1984 Periodicity of extinctions in the geological past. *Proc. natn. Acad. Sci. U.S.A.* **81**, 801–805.

Riegel, W. 1974 Phytoplankton from the Upper Emsian and Eifelian of the Rhineland, Germany – a preliminary report. *Rev. Palaeobot. Palynol.* **18**, 29–39.

Ross, C. A. & Ross, J. R. P. 1985 Late Paleozoic depositional sequences are synchronous and worldwide. *Geology* **13**, 194–197.

Sandberg, C. A., Ziegler, W., Dreesen, R. & Butler, J. A. 1988 Late Frasnian mass extinction: conodont event stratigraphy, global changes, and possible causes. *Cour. Forsch.-Inst. Senckenberg*, **102**, 263–307.

Saunders, W. B., Ramsbottom, W. H. C. & Manger, W. L. 1979 Mesothemic cyclicity in the mid-Carboniferous of the Ozark shelf region? *Geology* **7**, 293–296.

Sepkoski, J. Jr 1986 Global bioevents and the question of periodicity. In *Global bio-events* (ed. O. H. Walliser), pp. 47–61. Berlin and Heidelberg: Springer-Verlag.

Sloss, L. L. 1963 Sequences in the cratonic interior of North America. *Bull. geol. Soc. Am.* **4**, 93–114.

Smith, D. B., Brunstrom, R. G. W., Mannin, P. I., Simpson, S. & Shotton, F. W. 1974 *A correlation of Permian rocks in the British Isles*, special report of the Geological Society of London no. 5. (46 pages.)

Snelling, N. J. (ed.) 1985 The chronology of the geological record. *Mem. Geol. Soc. London* **100**, 1–343.

Stanley, S. M. 1984 Marine mass extinction: a dominant role for temperature. In *Extinctions* (ed. M. H. Nitecki), pp. 69–117. University of Chicago Press.

Stanley, S. M. 1988 Palaeozoic mass extinctions: shared patterns suggest global cooling as a common cause. *Am. J. Sci.*, **288**, 334–352.

Stearn, C. W. 1987 Effect of the Frasnian–Famennian extinction event on the stromatoporoids. *Geology* **15**, 677–679.

Talent, J. A. & Yolkin, E. A. 1987 Transgression–regression patterns for the Devonian of Australia and south west Siberia. *Cour. Forsch.-Inst. Senckenberg* **92**, 235–249.

Tozer, E. T. 1971 Triassic time and ammonoids: problems and proposals. *Can. J. Earth Sci.* **8**, 989–1031.

Tozer, E. T. 1984 The Trios and its ammonoids: the evolution of a timescale. *Geol. Surv. Canada Miscell. Rep.* **35**, 1–171.

Walliser, O. H. 1984 Geologic processes and global events. *Terra Cognita* **4**, 17–20.

Walliser, O. H. 1985 Natural boundaries and Commission boundaries in the Devonian. *Cour. Forsch.-Inst. Senckenberg* **75**, 401–408.

Ward, P. D. 1983 The extinction of the ammonites. *Scient. Am.*, **249**, 136–147.

Ward, P. D. 1988 Maastrichtian ammonite and inoceramid ranges from Bay of Biscay Cretaceous–Tertiary boundary sections. *Revista España de Paleontologia*, No. Extraord. (October 1988), 119–126.

Wiedmann, J. 1973 Evolution or revolution of ammonoids at Mesozoic system boundaries. *Biol. Rev.* **48**, 159–194.

Wiedmann, J. 1988*a* Ammonoid extinction and the 'Cretaceous–Tertiary Boundary event'. In *Cephalopods, past and present* (ed. J. Wiedmann & J. Kullmann), pp. 117–140. Stuttgart: Schweizerbartsche.

Wiedmann, J. 1988*b* The Basque coastal sections of the K/T boundary – a key to understanding 'mass extinction' in the fossil record. *Revista España de Paleontologia*, No. Extraord., (October 1988), pp. 127–140.

Wilde, P. & Berry, W. B. N. 1984 Destabilisation of the oceanic density structure and its significance to marine 'extinction' events. *Palaeogeogr. Palaeoclimatol. Palaeoecol.* **48**, 143–162.

Wise, D. U. 1974 Continental margins, freeboard and the volumes of continents and oceans through time. In *The geology of continental margins* (ed. C. A. Burk & C. L. Drake), pp. 45–58. New York: Springer-Verlag.

Zheng, Z. 1981 Uppermost Permian (Changhsingian) ammonoids from Western Guizhou. *Acta palaeont. Sinica* **20**, 107–114.

Discussion

W. A. KERR (*The Thatched Cottage, Whitchurch Hill, U.K.*). The decapod *Spirula* has an internal skeleton, as do the belemnites. Might not the ammonoids have the same? *Spirula* was the subject of research given to Professor J. Graham Kerr by the Danish Government.

M. R. HOUSE. The ammonoids are ectocochleate. The shells sometimes show colour banding and periostracum, with no evidence of tissues outside the shell as shown on the guard of some belemnites, which are endocochleate. Nevertheless there are suggestions that there may have been tissue outside the aperture in some cases. There are many ammonoids with a very constricted aperture in the adult, often caused by periodic apertural constriction during growth. In some of the heteromorphs the shell aperture closely abuts earlier whorls, giving very little space for tentacles and the food-gathering facility. Although both radulae and jaw structures are now known in ammonoids, for forms with constricted apertures it is difficult to envisage a 'normal' feeding pattern for them and it may be they developed some type of filter feeding or techniques enabling them to live on small organisms; in this case extra-apertural tissues may well have been involved.

Phil. Trans. R. Soc. Lond. B **325**, 327–355 (1989)

Printed in Great Britain

There are extinctions and extinctions: examples from the Lower Palaeozoic

By R. A. Fortey

*Department of Palaeontology, British Museum (Natural History), Cromwell Road,
London SW7 5BD, U.K.*

The extinction events at the Cambrian–Ordovician and Ordovician–Silurian boundaries are compared and contrasted. A simple theoretical model shows that times of increased cladogenesis produce elevated rates of taxonomic pseudo-extinctions, according to the recognition of paraphyletic groups. Taxonomists have traditionally placed stratigraphically early and morphologically primitive members of clades into paraphyletic groups. The Cambrian–Ordovician boundary coincided with such a period of cladogenesis. Extinctions occurred among shelf taxa: deeper-water taxa were mostly unaffected. The various explanations that have been proposed to explain Cambrian–Ordovician extinctions are evaluated. The Cambrian–Ordovician boundary event was probably similar to 'biomere-type' events that preceded it in the Cambrian and followed in the Ordovician. However, the rapid, but apparently staggered appearance of major new taxa at this time elevated taxonomic pseudoextinctions. In contrast, the Ordovician–Silurian extinction event terminated many major clades. An important 'oceanic' event (or events) profoundly affected outer- to off-shelf taxa (including plankton), some having had long and stable histories. The late Ordovician glaciation produced changes in shelf taxa, but changes in brachiopod faunal composition were spread over a long time compared with that for oceanic events. The likely role of anoxia in explaining deeper water end-Ordovician extinctions at the time of deglaciation is discussed.

1. Introduction

When Newell (1967) identified major extinction events, those at the Cambrian–Ordovician and Ordovician–Silurian boundaries were regarded as of major consequence. However, these intervals have not attracted the same amount of attention as events later in the geological column, particularly those associated with supposed catastrophic extinction (Raup & Sepkoski 1982, 1986; Jablonski 1986). If Sepkoski's (1981, 1984) account of the fossil record is correct, the Cambrian–Ordovician boundary marks the onset of expansion of the Palaeozoic evolutionary fauna, whereas the Ordovician–Silurian boundary marks both the decline of the Cambrian evolutionary fauna and a relative rise in importance of the Modern evolutionary fauna. It might be claimed that events in the early Palaeozoic set the course of subsequent evolution in the marine realm, notwithstanding the major extinction events that were to follow. It has also been claimed that the Cambrian soft-bodied biota, such as those represented in the Burgess Shale, reveal a phase of evolutionary 'experiment' (Conway Morris 1979), even with distinct phyla; most did not survive the Cambrian Period. If all, or any, of these conjectures are true, it is clear that the extinction events at the Ordovician–Silurian and Cambrian–Ordovician boundaries are of some interest. There have been many recent papers on the Ordovician–Silurian extinction events; usually from the perspective of a single group (e.g. brachiopods (Sheehan 1975, 1986), graptolites (Koren 1988; Melchin & Mitchell 1988)). The pattern of arthropod extinctions through both boundaries was described by Briggs *et al.* (1988). The factors that might have been involved at the Cambrian–Ordovician boundary have mostly been discussed in relation to trilobite evidence (Fortey 1983; Westrop & Ludvigsen 1987). This paper reviews some of the problems special to these extinction horizons, and examines the mechanisms that might have brought about the extinctions themselves.

2. Problems peculiar to lower palaeozoic extinction horizons

There are some problems with the Cambrian–Ordovician and Ordovician–Silurian extinction events that should be stated at the outset. Not least is the problem that in many cases stratigraphic knowledge is still at the descriptive stage. For example, the detailed Ordovician–Silurian stratigraphy of the Oslo region, Norway (Brenchley & Cocks 1985), or the Eastern Great Basin, U.S.A., has only been fully described recently; there are still parts of the world where stratigraphic knowledge is lacking or incomplete. Some of the most important sections for studying the Cambrian–Ordovician boundary have only been closely studied in the past decade, and some have only just been described (Chen *et al.* 1985). None the less much new knowledge has become available over the past few years (Cocks 1988; Norford 1988). The stratigraphic emphasis has been the result of the need to define the bases of the Ordovician and Silurian Systems; information directly pertaining to extinctions has mostly been derived as a by-product.

Correlation problems

Although the international correlation of the Cambrian–Ordovician and Ordovician–Silurian boundaries is much more precise than before, many problems remain. International agreement has been reached on the base of the Silurian (at the base of the *Akidograptus acuminatus* biozone), but the base of the Ordovician has yet to be defined formally, and any discussion of the Cambro–Ordovician event has to take an arbitrary decision on the matter. The problems can be summarized in two categories.

(i) Correlation between different biofacies and lithofacies. This has always been a problem in the Ordovician with regard to the graptolitic and shelly facies. Much of the discussion on fixing the base of the Silurian centred on such problems. Graptolites give a very refined stratigraphy in some facies through the critical late Ordovician to early Silurian interval. Sequences in conodont faunas that might be used to effect correlation into shelf carbonate sequences are less refined. The regressive sequences at the top of the Ordovician are usually lacking in graptolites, and correlation of such rocks with the graptolitic facies is difficult. Furthermore, the Hirnantian-style shelf faunas are typical facies faunas, and as such are subject to problems of diachronism (Rong 1984). All these problems combine to give a simplified view of the chronology of the late Ordovician glaciation, and it may be the case that a polyphase glacial episode will be recognised when some of these stratigraphic problems are resolved. There are comparable but less acute biofacies problems in the late Cambrian to early Ordovician (Fortey *et al.* 1982; Shergold 1988) with contrasting platform and off-shelf faunas.

(ii) International correlation. Discussion of a supposed event requires a reliable basis for correlation between what were separate continents in the early Palaeozoic. Particularly in the Cambro–Ordovician interval there was a high degree of provincialism in inshore faunas as a result of relatively dispersed continental masses at that time (Ziegler *et al.* 1979), coupled with a comparatively strong climatic gradient. This means that certain events (like the early history of cephalopods – most fully known from China) are manifest only in confined areas, and that it is not easy to establish synchroneity of events without circular argument. The recent discovery of a magnetic reversal at this horizon could be a useful adjunct to biostratigraphy, if it is confirmed in other sections. None the less, conodonts (Miller 1988), graptolites and trilobites are a reasonable control internationally, even if there are still arguments about the exact level appropriate for the definition of the base of the Ordovician.

3. Systematics, paraphyletic groups, and extinction events

Systematics and the study of extinctions are inextricably connected. If extinction is the termination of lineages, the meaning of the taxa concerned defines the nature of the event. The problems involved are not trivial ones. For example, Briggs *et al.* (1988) identified several cases where taxonomic practice itself could give rise to spurious extinctions (taxonomic pseudo-extinctions). This applied where a particular horizon served to separate the specializations of different taxonomists who became disposed to apply different names on either side of a boundary. A supposed extinction may merely record the disappearance of a paraphyletic taxon, which is a consequence of the inadequacy of phylogenetic resolution, and not the termination of a clade. Similar criticisms by Patterson & Smith (1988) were applied to family level data used by Raup & Sepkoski (1986). These systematic problems may have a major influence on the kind of extinction pattern, or even call its reality into question. Similar problems have some influence on the differences between the Cambrian–Ordovician and Ordovician–Silurian boundary events as discussed below.

Paraphyletic groups and extinction horizons

Patterson & Smith (1988) noted that a significant part of the family-level data used by Raup & Sepkoski (1986) was flawed. Some of these limitations, such as the use of monospecific families, which are simply those for which relationships have not been determined, are obvious. But the observation that many of the family taxa were paraphyletic groups is much more interesting. Is a cyclicity of extinction actually a cyclicity of paraphyletic groups?

Patterson & Smith's approach to the systematics of these families was cladistic, and correctly so, for such methods of analysis 'permit the sensible marriage of history with models involving process' (Levinton 1988, p. 492). It is the meaning of paraphyletic groups that is important in the context of extinction events (Smith 1988). From a cladistic taxonomic standpoint, paraphyletic groups are meaningless in the sense that they are based on shared primitive characters (symplesiomorphies) rather than synapomorphies, which define 'good' groups (Wiley 1981). However, this does not mean that their appearance is historically meaningless. Why, after all, should paraphyletic taxa be concentrated at certain horizons in the geological record? The employment of paraphyletic groups by palaeontologists has usually been based on shared general resemblance, regardless of whether the characters were primitive or derived. Paraphyletic groups of this kind share one or more characters that link them with derived groups, but their overwhelming plexus of primitive characters often makes it impossible to decide to what derived taxon any member of the paraphyletic group is related. They are 'bags' to which primitive members of a clade are consigned. Although this makes them difficult to deal with from a cladistic point of view (meaningless, even) the presence of such groups is not without meaning historically.

I have attempted to model a possible explanation for the presence of paraphyletic groups in association with extinction events in figure 1.

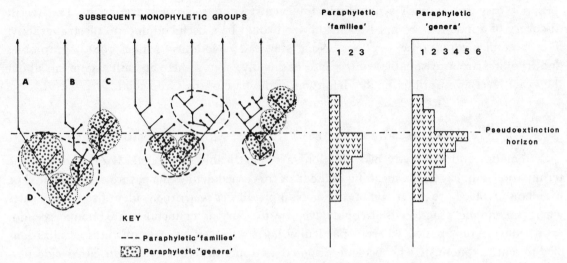

FIGURE 1. Early phases of cladogenesis, if concentrated in a short time interval, produce a temporal concentration of paraphyletic groups as recognized in traditional taxonomy and a taxonomic pseudoextinction of some magnitude. Points represent hypothetical species, arrows signify groups that would have had a long subsequent history and be recognized as clades by most taxonomists.

It has been widely claimed that periods of cladogenesis follow extinction events, and hence that the early history of clades will be recorded in the rock sequences immediately overlying horizons of extinction. The early species may retain characters from their immediate ancestors, and phyletic divergence may not have proceeded far enough to be certain to what (if any) derived taxon any given species should be assigned. These are precisely the circumstances when early representatives are grouped together as paraphyletic taxa. Even with the added precision provided by cladistics it may not be possible to dismantle the constituents of such taxa into derived clades, simply because character complexes defining subsequent clades may not have been fully assembled. A cluster of early species may differ only in the most trivial characters; ornament or size. A taxonomist may have placed such species within a single genus regardless of whether that genus may subsequently prove paraphyletic. Even if the phylogenetic relationships of derived taxa are known it may be that any one of the early species may substitute for any other as the sister group of the derived clades; they may not show characters that enable their relationships to be resolved more finely. The more complete the fossil record, the more likely it becomes that such early stem species will be discovered, and the more likely it is that there will be numbers of such species that may be grouped together in a larger paraphyletic taxon. This implies that later taxa will be identified into monophyletic clades (A, B, C, in figure 1), and that the paraphyletic taxon (D) will include those ancestral taxa where relationships to the derived groups are not clear. The size of D is limited by the extent that derived taxa can be 'pruned' from it. The disappearance of taxon D (that might be claimed as a family) is a taxonomic pseudoextinction. A repetition of the same argument leads to the conclusion that within the paraphyletic family, the earliest and most primitive members may themselves be congregated into a paraphyletic genus (figure 1) including those species plesiomorphic in all fossilisable respects to the rest of the clade A + B + C + D.

It may be concluded that a period of cladogenesis (e.g. that following an extinction event) may yield a high rate of taxonomic pseudoextinction. Families and genera, at least as presented in tabular summaries of ranges, will 'disappear'. Could some such factor have been responsible for the cyclical appearance of paraphyletic groups in Patterson & Smith's (1988) analysis of Raup & Sepkoski's (1986) data?

Stratigraphic refinement comes into this argument. Most extinction data are derived from rocks representing a given period, usually those of a stage. This is often the most feasible unit for worldwide correlation. An average stage in the Jurassic has a duration of about 7 Ma, but even if it were less than half this figure this would presumably represent enough time to gather together the effects of extinction (genuine termination of clades) and the earlier part of subsequent re-diversification (taxonomic pseudoextinction of paraphyletic clades). In a particular case this would no doubt depend on how and where stage boundaries were drawn. It seems possible that some extinction events were composite, with different groups having terminated at different times, perhaps even staggered over several million years. This might prove to be the case with the Ordovician–Silurian boundary event. If this were so, true extinction and taxonomic pseudoextinction at early cladogenesis could overlap in time. Both would produce an extinction peak.

This is not mere quibbling over taxonomy. The basal Ordovician has long been recognized as a period whence major taxa originate. Many of them are still extant. The same phenomenon would be reflected in the steep climb in Sepkoski's (1984) curve of overall familial diversity at this level. Hence one might expect this time interval to include many paraphyletic taxa (the ancestors of later clades), and a correspondingly high rate of taxonomic pseudoextinctions.

There are few full phylogenetic analyses of taxa at their times of origin, close to the base of the Ordovician. However, some case histories indicate that there is a relative preponderance of paraphyletic taxa early in the history of clades at this time. Figure 2 shows the proportion of paraphyletic taxa within the Graptoloidea during the early Ordovician based on a phylogenetic classification by Fortey & Cooper (1986) and incorporating some further indications of paraphyletic taxa introduced by Erdtmann (1988). The early graptoloids are all placed in an avowedly paraphyletic family, the Anisograptidae. Members of this family have not yet been unequivocally ascribed to any derived clade. Anisograptids retain a primitive characteristic, bithecae, that is lost in later clades. Within the Anisograptidae, the earliest and most primitive genera are themselves considered to be paraphyletic taxa, especially on the basis of Erdtmann's account (1988). Later anisograptids probably include some genera that are true

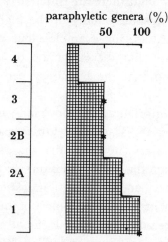

FIGURE 2. Decrease in proportion of paraphyletic genera in the early history of the Graptoloidea through time. The Tremadoc zones of Erdtmann (1988) are shown on the left.

clades. Overall there is a preponderance of paraphyly early in the history of the group. Disappearances of such taxa are pseudoextinctions and these will also be concentrated earlier in the group's history, with another event at the base of the Arenig. Among trilobites too there are more paraphyletic groups in the region of the Cambrian–Ordovician boundary. Subfamily or family level has to be considered because generic relationships are not worked out. The paraphyletic nature of some of these trilobites is not obvious from the current literature. These are mentioned here. For example hystricurids, so characteristic of the early Ordovician in palaeoequatorial regions, were regarded by Fortey & Owens (1975), in a treatment antedating a cladistic viewpoint, as the stock that gave rise to Proetida by more than one phylogenetic line. In other words, they are a paraphyletic group, although they have not formally been described as such. Symphysurinids are typical of early Ordovician platform limestones. In a phylogenetic analysis of Asaphina, Fortey & Chatterton (1988) noted that symphysurinids were not asaphids (as they had been regarded) but exhibited a combination of primitive and advanced characters that made it likely that they were ancestral to both cyclopygids and nileids. Hence symphysurinids probably constitute another paraphyletic taxon. It has been claimed (Ludvigsen & Westrop 1984) that Missisquoiids are ancestral to the scutelluid clade. According to Fortey & Chatterton (1988), jegorovaiids are the trinucleacean equivalent. These taxa all become extinct within a biozone or two of the Cambrian–Ordovician boundary. Many trilobite families passed through the Cambrian–Ordovician boundary unscathed, and hence paraphyletic taxa constitute only a proportion (perhaps 20%) of the whole sample.

This provides some indication that a preponderance of paraphyletic taxa in some parts of the geological column may elevate 'extinction rates' as recorded in faunal lists. Termination of clades must be distinguished from the more arbitrary 'disappearance' of paraphyletic taxa. Species-level extinction may still be a real phenomenon within a paraphyletic taxon of higher rank, but the disappearance of paraphyletic taxa above species rank could be intimately connected with an immediately preceding extinction event, or with a spurt in cladogenesis.

4. THE CAMBRIAN–ORDOVICIAN EVENT – CLADOGENSIS OR EXTINCTION?

Newell (1967) and Sepkoski (1986) have recognized an extinction event of importance at the Cambrian–Ordovician boundary. That there are differences between Ordovician faunas and those of the Cambrian is not in dispute; many of the higher taxa still living (among the bryozoans, cephalopods and bivalves, for example) first appear close to the Ordovician boundary. It is natural to assume that the extinction of a Cambrian fauna predated the origination of 'modern' taxa, and that the boundary between the systems was when this happened.

The boundary between the Cambrian and Ordovician systems is under scrutiny from the International Working Group on the Cambrian–Ordovician boundary; this is seeking to determine at what level the boundary should be drawn, and in what section the boundary horizon should be defined. Although the problem has not yet been resolved, a positive outcome of this interest has been the discovery and investigation of several sections spanning the critical interval (Norford 1988). These include sections in most of the Ordovician palaeogeographical regions. The boundary is now reasonably defined globally. Thanks to detailed studies on conodonts, trilobites and graptolites the international correlation is now on a better footing, although there are still disagreements. For this paper the boundary is taken at the base of the

North American *Parakoldinioidia* (= *Missisquoia*) biozone and its equivalents that marks the appearance of trilobite faunas with an Ordovician cast.

If the Ordovician was a time when many groups with a long subsequent history appeared it is likely that the species that are recovered from the earliest rocks would have included the ancestors of at least some later clades. This makes it likely *a priori* that they will have been placed in paraphyletic groups, especially by palaeontologists basing their classifications on general resemblance. Even with the refinements of cladistic analysis such groups may not be resolved into derived clades because of a paucity of synapomorphies capable of solving their phylogenetic relationships. The question then arises whether the Cambrian–Ordovician event could have been dominated by taxonomic pseudoextinctions as described above. Is it an important event because of the cladogenesis going on at the time, rather than as a result of the extirpation of major clades?

Briggs *et al.* (1988) summarized trilobite data across the boundary. At the family level several of the late Cambrian groups were unsatisfactory, being polyphyletic or paraphyletic taxa. After subtracting these, there was greater family continuity across the boundary than had been supposed. One major group (Dikelocephalacea) did become extinct at or close to the boundary. Extinction was more prevalent at the generic level. This was preceded by a decline. The taxa affected were inhabitants of platform areas; deeper water taxa appeared to be largely unscathed. Some of these deep water taxa had long ranges extending from late Cambrian until well into the Ordovician (figure 6). The Cambrian–Ordovician boundary is thought by many workers to coincide with a eustatic event (Lange Ranch event of Miller (1985)). All workers agree that there was a Tremadoc overstep, although Ludvigsen *et al.* (1988) regard the evidence for a Cambrian regression preceding it as inadequate.

The late Cambrian genera that became extinct were much less likely to represent paraphyletic taxa; data on trilobite distribution support an extinction event – but how major an event? The Lower Palaeozoic has other, minor, extinction events, such as the Cambrian 'biomere' events (Palmer 1984). Similar events probably delimit the Ordovician series (Fortey 1984). Shelf trilobites were apparently more vulnerable to extinction at biomere boundaries, and the picture generally drawn of a biomere (Stitt 1975) is one of a radiation on-shelf followed by a stabilization of diversity before a comparatively slow decline, and eventual sudden extinction. One way to examine whether the Cambrian–Ordovician boundary is different from one of these minor events is to compare their extinction patterns. The Cambrian biomere events and their Ordovician equivalents have not been claimed as major extinction events. In figure 3 the trilobite evidence is summarized across the Tremadoc–Arenig boundary. This is the next horizon above the Cambrian–Ordovician boundary at which there appears to be a turnover in genera, and more 'Ordovician families' (trilobites: trinucleids, calymenids *sensu stricto*, for example) appear just above it, so it affords an appropriate comparison. There are difficult correlation problems at this level, however, and my Arenig base is taken at an horizon equivalent to the *Tetragraptus approximatus* graptolite biozone with an attempt at its equivalents elsewhere. This includes a correlation with the base of Zone G2 of the zonal scheme for platform North America of Ross (1951), and the base of the Moridunian (Fortey & Owens 1987) in the Anglo-Welsh area. The correlation problems are discussed in more detail by Bergström (1986), Cooper & Fortey (1982) and Fortey & Owens (1987).

There seems to be considerable similarity between the Cambrian–Ordovician and Tremadoc–Arenig patterns. Last appearances of genera rise at the end of the Tremadoc,

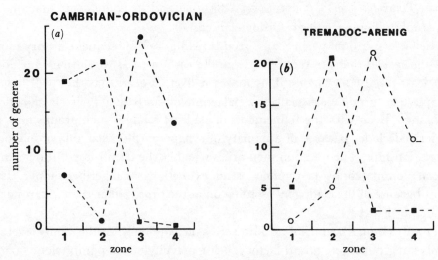

FIGURE 3. The similarity of pattern between last (■) and first (○) appearances of trilobite genera across the Cambrian–Ordovician boundary, and the Tremadoc–Arenig Series boundary within the Ordovician, the former from Briggs *et al.* (1988), the latter a new compilation. In both, an elevation in last appearances just before the boundary is complemented by a great increase in first appearances above the boundary. This has been claimed to correspond with a regressive–transgressive event in both cases.

whereas there is a peak in first appearances of genera at the base of the Arenig. The latest Tremadoc (Lancefield 2 in graptolite terms), like the end of the Cambrian, has been claimed as a regressive interval on the platform (Fortey 1984). The last appearance of the trilobite superfamily Ceratopygacea is in the late Tremadoc, this group has a history extending back to the middle Cambrian. Its disappearance can be compared with the extinction of the Dikelocephalacea close to the base of the Ordovician.

Overall, there is a greater number of families that become extinct at the Cambrian–Ordovician boundary than at the Tremadoc–Arenig boundary. The similarity in the pattern of disappearances and appearances of genera leads one to suspect that this represents a difference in degree rather than in kind. At the Tremadoc–Arenig boundary the graptolite family Anisograptidae probably became 'extinct', but this is a paraphyletic taxon (Fortey & Cooper 1986) and its disappearance is a taxonomic pseudoextinction. It seems that the Cambrian–Ordovician event might be no more than one more, perhaps rather exacerbated, biomere-type boundary. If this were the case, elevated extinction rates at this level may reflect an increased rate of cladogenesis commensurate with the general increase in diversity through the early Ordovician interval (Sepkoski 1984). The apparent extinctions may include taxonomic pseudoextinctions produced by the 'disappearance' of paraphyletic groups in the early history of clades.

The tests of such a theory reside in phylogenetics. Cladistic methods have hardly been applied to phylogenetic studies of the relevant groups through this interval. More traditional taxonomic methods tend to show paraphyletic groups as short-lived 'trunks' from whence various (presumably monophyletic) 'branches' arise. See, for example, Flower's (1964) view of cephalopod phylogeny where most such 'branches' arise from the early Ordovician Ellesmeroceratida. There is already evidence that the times of origin near the Cambrian–Ordovician boundary of such 'rootstocks', like paraphyletic taxa, are not simultaneous. Nor do they necessarily correspond with the Cambrian–Ordovician boundary used here (base *Parakoldinioidia* biozone) that is based on platform North American trilobite faunas (figure 4).

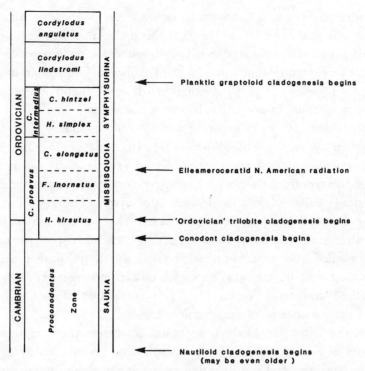

FIGURE 4. Timing of inception of major cladogenesis in different groups through the Cambrian–Ordovician boundary interval, showing that it is not simultaneous in different groups of organisms. Taxonomic pseudoextinction is to be expected; compare with figure 1.

For example, the inception of the radiation of the Graptoloidea began approximately one zone above this (probably within *Symphysurina* biozone equivalent). An initial radiation of the Cephalopoda has recently been discovered in the latest Cambrian of China (Chen *et al.* 1979) with little record elsewhere. The taxa described include ancestors of several derived groups; it is likely that phylogenetic analysis will show such groups to be paraphyletic. Typically, they are accorded high taxonomic rank because their stratigraphic position and characteristics make it awkward to accommodate them within the known phylogeny. The fossil record of other molluscan groups is too discontinuous through the critical interval for comment. The limitations on echinoderm data have recently been discussed by Smith (1988). Where Ordovician-style conodonts begin has not been agreed among conodont workers, but *Cordylodus proavus* and its associates have been regarded as the 'natural' base to the Ordovician (Miller 1988), including taxa ancestral to later Ordovician forms. The base of the *C. proavus* biozone lies below that of the *Parakoldinioidia* biozone. Unless there are major surprises to come in the international correlation, it seems that the onset of cladogenesis was different in different groups, and probably spanned several million years. This does not imply a catastrophic extinction event followed by renewed speciation.

Taking these facts together, we conclude that if cladogenesis was staggered in different groups through a comparatively broad time interval at or below the base of the Ordovician it is likely that a variety of groups ancestral to those typical of the Ordovician and later will have evolved in the latest Cambrian to earliest Ordovician interval. Traditional taxonomic practice will have erected stratigraphically-based and likely paraphyletic groups for such early taxa. The boundary event extinguished shelf taxa, at least at generic level, and is present as an unconformity in many sections. For the same reason this may be a natural 'cut off' for those groups that underwent early cladogenesis below the boundary: that is, taxonomic pseudoextinctions will tend to be concentrated at this horizon. This, combined with the effect of grouping stratigraphically separated events in broad-scale analyses, may account for the elevation in higher level taxon extinction near the Cambrian–Ordovician boundary. Much of it may be taxonomic pseudoextinction reflecting the 'diversity pump' effect at this time, which was probably the most rapid period of cladogenesis in the fossil record.

The extinction event at the Cambrian–Ordovician boundary affected shelf faunas particularly although by no means all of these were exterminated. Trilobite taxa that lived in outer shelf to slope environments were little affected (Fortey 1983, Ludvigsen & Westrop 1987). This is not different from Cambrian biomere events, nor, apparently, from those events at Ordovician series boundaries, such as the Tremadoc–Arenig boundary discussed above. For example, the long-lived trilobite taxa Leiostegiacea and Dikelokephalinidae, both of them shelf taxa and both with histories extending back to the mid Cambrian, were apparently terminated at or near the Llandeilo–Caradoc boundary having weathered earlier perturbations. One could argue that the Llandeilo–Caradoc event, which accompanied one of the greatest marine transgressions in the Phanerozoic (Vail *et al.* 1977), was the equal of the Cambrian–Ordovician event in terms of the elimination of major trilobite clades.

To summarize, the Cambrian–Ordovician boundary marks the inception of many higher taxa. Early phases of cladogenesis multiply paraphyletic groups; many are accorded high taxonomic status because of the difficulties in assigning them to advanced clades. The disappearance of these taxa represents taxonomic pseudoextinction. Such pseudoextinctions were of importance around the Cambrian–Ordovician boundary. The detailed time of inception of new clades was different between groups and a major catastrophic extinction seems improbable. The extinction of genera in platform sites was apparently similar to other events earlier in the Cambrian or later in the Ordovician.

Possible causes of Cambrian–Ordovician extinctions

If the notion of any catastrophic extinction is taken out of consideration, the stratigraphic record can be examined for evidence of possible causes of platform extinctions. If comparison with other Cambrian and Ordovician events is valid it is reasonable to seek common cause.

That biomere boundaries in the Cambrian and the Cambro-Ordovician boundary share common features has been observed by several workers (Stitt 1975, 1977; Ludvigsen 1982). Some biomere boundaries are not accompanied by dramatic facies changes (Palmer 1984). In the majority of Cambrian–Ordovician platform sequences where data is recorded there is a break, or paraconformity, at the boundary (as taken at the base of the *Parakoldinioidia* trilobite biozone) that has been regarded as representing a eustatic regression. The basal Ordovician is transgressive. Although the reality of the late Cambrian regressive event has been challenged locally (Ludvigsen *et al.* 1988) it is a striking fact that apparently identical features occur at exactly the same time in carbonate platform sequences on what were separate plates in the Ordovician (Miller 1984). These features occur widely in the U.S.A., Australia and in the North China platform. I find it difficult to explain this without assuming the existence of eustatic control. However, it is also true that such platform carbonates are generally characterized by small diastems that may be of no more than local significance, and there is the danger of circularity in diagnosing such phenomena as worldwide. Whatever the possible disagreements over the latest Cambrian regression, all authors seem to agree that the early

Ordovician marked a time of transgression. This can produce complicated results locally with biofacies migrations (Ludvigsen & Westrop 1983). One general effect is the shelfward movement of more exterior biofacies, limiting the extent of inshore biofacies (Fortey 1983; Westrop & Ludvigsen 1987). The sudden appearance of offshelf taxa, such as olenid trilobites, within shelf sequences is another consequence. This is analogous to the claimed repopulation of the shelf from offshelf forms that has long been described in biomeres (Palmer 1965). I claimed similar regressive-transgressive cycles in the Ordovician (Fortey 1984) at the Tremadoc–Arenig and Arenig–Llanvirn boundaries, and approximately commensurate with the Llandeilo.

There are several models available to explain how these events could have caused extinction of some platform faunas. They differ significantly, and are summarized in simple form in figure 5.

(i) *With regression, and peripheral refugia*

This model treats a combination of regression and transgression as causing the extinction of platform taxa. The term 'refugia' is used to mean small sites where species may persist during a time of crisis over their formerly widespread habitat area. The regressive phase accounts for the high proportion of last appearances of genera in the latest Cambrian. Some trilobite taxa survive as Lazarus forms (i.e. they reappear later in the Ordovician in shelf facies comparable with those in the late Cambrian, e.g. bynumiids, catillicephalids and lecanopygids). The survival of these is explained by shelf-edge refugia during the regressive interval (Fortey 1983). Subsequent transgression introduced offshelf taxa onto the shelf, and also permitted recolonization of the shelf from the marginal refugia. The implication is that there was greater phylogenetic continuity between Cambrian and Ordovician shelf genera than had been supposed. A rather similar picture can be painted for the Ordovician regressive and transgressive cycles, where there is generally greater taxonomic continuity across boundaries between supposed eustatic events than has been claimed in the Cambrian between biomeres.

(ii) *Transgression and biofacies changes*

Ludvigsen (1982), Ludvigsen & Westrop (1983) and Westrop & Ludvigsen (1987) associate extinctions with a transgressive event alone, producing biofacies shifts and biogeographical changes that are considered sufficient to account for the reduction in habitat variety available for on-shelf trilobites. Extinction was associated with loss of inshore habitats appropriate for particular taxa (especially 'thermophylic and stenogeographic' ones (Ludvigsen 1982, figure 18)) and the shoreward spread of 'eurygeographic and cryophylic' trilobites. The explanation of extinction is thus a combination of the species–area effect, biofacies redistribution and onshore migration of cold-water taxa.

(iii) *Cold-water excursions without necessary transgression*

This appears to be the favoured explanation for Cambrian biomere patterns (Stitt 1977; Palmer 1979, 1984). I do not wish to discuss the problem of where the top of the biomere lies, as opposed to the base of the succeeding biomere (Palmer 1979). Extinctions are associated with change in oceanic structure, especially the upward displacement of the thermocline in the water column such that the more oceanic, long-lived taxa can make their excursion onto the platform, the abrupt change in conditions extirpating its previous inhabitants. Transgressive

[97]

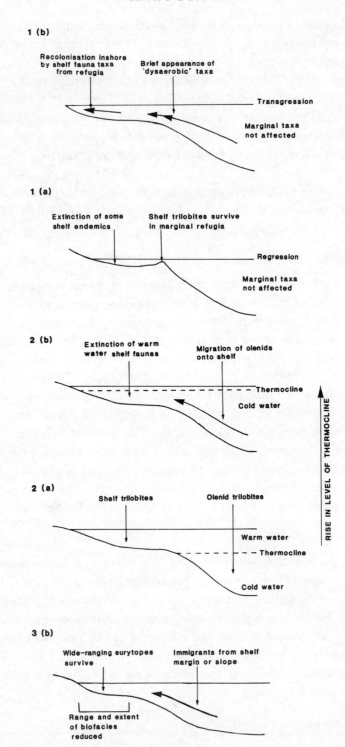

FIGURE 5. Three theories about causes of on-shelf extinction at the Cambrian–Ordovician boundary (a) and (b) represent immediately before and after the boundary interval. (1) Combined regression–transgression, with marginal refugia at regressive maximum to account for continuity of some taxa. (2) Raising of level of thermocline causing platform crisis and appearance thereon of olenid trilobites. (3) Shelfward displacement of exterior biofacies with loss of extent and variety of inshore biofacies; pre-boundary interval much like 2(a).

or regressive events are not necessary for this explanation, and an example of regression within the pterocephaliid biomere, but without major biological effect is cited. Palmer (1984, p. 609) accepts the possibility that the biomere boundaries may record 'widespread anoxic events'.

These models are based on what happened in shelf carbonate sequences, particularly those of North America. A similar pattern may be observed in the Sino-Korean platform (Chen *et al.* 1985). However, contemporary sequences in Scandinavia and Wales record nothing dramatic, and there is undisturbed phyletic continuity among olenid trilobites through the same interval (Henningsmoen 1957). Explanations of the Cambrian–Ordovician events have to take such continuity into account, as well as the observed change.

All the models share one common feature: the shelfward movement of exterior biofacies. This can be accomplished with or without transgression, according to the favoured hypothesis. Is there evidence for more widespread anoxia in the earliest Ordovician accompanying this biofacies shift? It would be true to say that in Norway, the Anglo–Welsh area, and North Africa the early Ordovician *Dictyonema* shales are widespread, and often only yield the planktic graptolite that gives the shales their name. Benthos is often completely lacking (Thickpenny & Leggett 1987). Where present, as in North Wales (Rushton 1982) or Random Island, eastern Newfoundland, the commonest trilobites are often olenids. Olenid trilobites also appear in platform sequences, or become abundant in peripheral platform sequences at the same time. For example, genera such as *Bienvillia*, *Jujuyaspis* and *Parabolinella* make their appearance, occasionally in abundance, in earliest Ordovician sequences (Winston & Nicholls 1967; Ludvigsen 1982). Nobody has suggested that these taxa constitute a 'rootstock' for subsequent platform evolution, in the manner of the biomere model (Palmer 1965). They are interlopers from more peripheral sites.

The olenids are particularly interesting, therefore, in interpreting boundary events. The family has a long history through the later Cambrian to the late Ordovician. Most authors agree that they were a specialized group, but there are different opinions as to what they were specialized for. Their common occurrence with dark shales, lacking other benthos, or with laminated, pyritic limestones or 'stinkstones', has long led to their being associated with dysaerobic environments (Henningsmoen 1957). This has led some authors to suppose that they were pelagic animals, living above a deoxygenated sea floor, but their strict facies association, and many aspects of their functional morphology, make this an unlikely assumption for the major part of the group (Fortey 1985). It is also claimed that one of the main constraints on their distribution was that they lived beneath the thermocline (Taylor & Forrester 1979). This accounts for an unusual interprovincial spread of olenid taxa under these uniform oceanic conditions. Such an explanation would naturally favour upward displacement of the thermocline at the Cambrian–Ordovician boundary. I favour adaptation to the dysaerobic environment as the most plausible explanation for olenid distribution and morphology. Because thermocline and oxygen minimum layer may correspond, it is not perhaps surprising to find the two factors operating together at former continent margins, and the olenid biofacies always represent some of the deepest occupied by trilobites. The Olenid biofacies may, however, become established in marginal basins, at some remove from the continent edge, as in the early Arenig of South Wales (Fortey & Owens 1987). There, the development of a local basin poor in oxygen seems to provide the most plausible explanation of the presence of the biofacies: deep water biofacies in adjacent areas have a different assemblage of trilobites. Truly pelagic trilobites accompany the olenid biofacies in oceanic settings, but may be lacking in sites with restricted access to the open sea, as in South Wales.

If this view of the habits of olenid life is correct, their incursion into platform sites at the Cambrian–Ordovician boundary was probably the result of a shelfward movement of water poor in oxygen, to which the olenids were adapted. This neatly explains the truncation of the ranges of trilobites with normal respiratory requirements that could linger on only in confined inner shelf areas. The majority of deep water taxa already had the necessary physiology to survive the boundary perturbation. Note that there are some sections (such as at Dayangcha, China) where the early Ordovician transgression fails to introduce olenids into the section. Although there is evidence for facies shifts in China compatible with a regressive–transgressive event, the evidence for an oxygen crisis is not clear from the faunas alone. In general, perhaps the shoreward movement of exterior facies with short-lived oxygen reduction, especially at the margins of shelves, provides a more satisfactory explanation of the facts than the elevation of the thermocline. Would it really be possible for cool water to persist on a flooded, shallow platform with tropical insolation for the duration of a trilobite zone? Geochemistry may eventually provide the evidence critical for arbitration between these alternatives. Then there is the question of whether refugia were necessary to explain the persistence of taxa across the Cambrian–Ordovician boundary. The answer will partly depend on confirming or denying the existence of a sub-Ordovician regression. As I discuss below it is clear that there had to be such refugia over the Ordovician–Silurian boundary, even though they are unknown in the fossil record. If there were a major regression across the Cambrian–Ordovician boundary, the record of these sites is likely to be equally poor, although they may be recorded in boulders (derived from sites near the continent edge) in such deposits as the Cow Head Group, western Newfoundland (James & Stevens 1987). I find it difficult to account for the sudden appearance of hystricurid and symphysurinid trilobites in the early Ordovician. There is no phylogenetic hypothesis to suggest that they originated from the outer shelf invaders as the biomere model supposes. It seems far more likely that their sister taxa were Cambrian, and inhabitants of the same kind of in-shelf biofacies. In general, trilobites appear to have remained in their preferred habitats for long periods of time (Robison 1972; Fortey 1980). It is not possible to produce shelf genera with characteristic autapomorphies without plausible antecedents. Simply invoking a vague 'radiation' is not good enough. In some cases the phylogenetic connections are known, at least in a general way, between Cambrian and Ordovician forms. For example, leiostegiids are typical trilobites of Ordovician zone C–D interval in Nevada and Utah. Probably related leiostegiaceans occur in the Upper Cambrian. This leads me to ask where they were in the earliest Ordovician. The Lazarus effect is even more pronounced with the lecanopygid *Benthamaspis* (Briggs *et al.* 1980). Refugia are one possibility. This could be proved by further search in sites marginal to former continents.

5. Ordovician–silurian extinction events

The Ordovician–Silurian boundary is now internationally defined as lying between the Ashgill Series and the Llandovery Series. The Ashgill, and in particular its final stage, the Hirnantian, is known to coincide with a major glacial episode centred on the African part of Gondwana (see Brenchley (1984) for literature). For trilobites, graptolites, echinoderms, molluscs and other groups the late Ordovician marks the last appearance of several major clades. The groups that disappear are good clades and hence taxonomic pseudoextinction was not a major problem in elevating higher taxon extinction rates. The Llandovery was a time of uniform faunas with low diversity. Therefore, at least as far as the fossil record allows us to see it, there was not a period of rapid cladogenesis early in the Silurian. To this extent the reality of the Ordovician–Silurian event is uncontroversial.

Necessity for early Silurian faunal refugia

An important point with regard to the Ordovician–Silurian event is that the known poverty of early Llandovery faunas does not fully record the history of several animal groups. This is because there are many Lazarus taxa that disappear in the later Ordovician only to reappear in the Silurian after a gap of several million years. I can find examples from most groups. The brachiopod subfamily Leptellininae is abundant in the mid-Ordovician, but disappears approximately in the middle of the Ashgill not to reappear, again widespread, until the late Llandovery. Cocks (1988) notes the temporary absence of the superfamilies Craniacea and Eichwaldiacea from the early Llandovery (Rhuddanian). Paul (1982) records only one cystoid family with a Llandovery record, out of eight families that survive the Ordovician–Silurian event. Five more reappear in the Wenlock, whereas there is an even longer gap before the remaining two reappear. There are certain well-defined trilobite genera known from the late Ordovician that disappear through the Llandovery, or only appear in the latest Llandovery. Examples include: the only surviving member of Asaphina, *Raphiophorus*; proetides such as *Scharyia*, *Panarchaeogonus* and *Xenocybe*; the cheirurids *Sphaerocoryphe* and *Staurocephalus* (P. D. Lane, personal communication). The family Brachymetopidae is known from the Caradoc and Ashgill, and again from the Wenlock, but has no Llandovery record. I do not believe that there is any risk of the resemblance between Ordovician forms and their Silurian congeners being homoplasy. Of trilobite families that survived the Ordovician, harpetids and illaenids are absent from the early Llandovery. If one accepts the view of Destombes and Henry (1987) of calmoniid phylogenetics, we have an even longer gap between their Ashgillian Gondwanan ancestors and Devonian descendants. Only in the graptoloids has the record been claimed to be complete enough to read continuously through this interval.

The implication is clear: there have to be refugia for many taxa through the early Silurian. We have no fossil record of these refugia. The early Llandovery record is highly imperfect. This means that one cannot be certain of the sharp termination of clades at the Ordovician–Silurian boundary. After all, if we know that certain taxa must have endured, despite the lack of evidence for this at the moment, how can we be sure that the same refugia did not include the last representatives of taxa known at the present, no later than Ashgill? In the account that follows the assumption is made that what we know from the fossil record accurately records the timing of extinction. This is a working assumption, and probably not an unreasonable one for those groups, like trilobites, with a prolific record. However, until some fossil record is found for the refugia that must have existed in the Llandovery there remains the possibility that extinction was staggered over a longer time period than current opinion allows.

The late Ordovician glacial episode

The glaciogenic rocks at the end of the Ordovician coincide, on the whole, with the extinction events. The evidence can be summarized, briefly. Glaciogenic sediments in North Africa have their counterparts elsewhere in regressive sequences of Hirnantian age that are dominated by clastics and the development of karst surfaces on limestones deposited in inshore Ordovician tropical environments. Few places, other than some deep water graptolitic sites, pass through the glacial interval without its leaving an obvious lithological signature. A eustatic sea-level drop of up to 100 m has been claimed (Brenchley & Newall 1984). The biotas of the Hirnantian are dominated by the so-called *Hirnantia* fauna, named from the characteristic brachiopod. An almost invariable associate is the trilobite *Mucronaspis*. This fauna, widespread in the Hirnantian, extended from higher towards lower latitudes accompanying the glacial episode (Jaanusson 1979; Rong 1984). In detail there is more than one brachiopod community grouped within the Hirnantia fauna as loosely applied in the past (Rong & Harper 1988). The implication would seem to be that this fauna was associated with a latitudinal expansion of the cool water 'province' during the glacial episode, extending into temperate or even subtropical palaeolatitudes, although Rong (1984) seems to prefer the idea that the Hirnantia fauna was eurythermal. The tropical belt did survive, although much compressed latitudinally, in a few sites such as Anticosti Island. There is some evidence of a double glacial pulse (Brenchley 1984), but so far not much evidence to suggest a polyphase glaciation like that of the Pleistocene.

The regressive nature of the Hirnantian must produce a taphonomic bias towards comparatively inshore faunas. The Hirnantia fauna has now been subdivided into a number of benthic community types (Rong & Harper 1988) but the equivalents of deeper-water assemblages typical of the earlier Ordovician are hard to recognize. The critical intervals as far as extinctions are concerned are the Hirnantian and the underlying Rawtheyan stage of the Angle-Welsh area, that equates approximately with the Gamachian interval recognized on the eastern side of North America. The Hirnantian is considered to be a short time interval, 2 Ma, or even less. At species level at least, extinctions were probably scattered through this Rawtheyan–Hirnantian interval; correlation problems make for difficulties in summarizing timing of extinctions on single diagrams for the world. Inevitably this results in grouping of data within intervals, particularly the Hirnantian. With the definition of the base of the Silurian at the *Akidograptus acuminatus* biozone the Hirnantian now includes a short, post-glacial interval rather than coinciding closely with the glacial interval itself.

Patterns of faunal change

The number of Lazarus taxa indicated above impose a limitation on what can be read directly from the rocks through this interval. Groups with a continuous record include graptolites, brachiopods, trilobites and conodonts, and these provide the most informative basis for looking for patterns. Only one superfamily of brachiopods, the Gonambonitacea, becomes extinct within, or at the end of, the Hirnantian. Cocks (1988) comments that the degree of brachiopod extinction 'across the boundary appears to have been far less than previously reported', whereas 'extinctions at the late Hirnantian do not appear to have been greater than at the end Caradoc or end Rawtheyan'. The Hirnantian sees 34 genera make their last appearance compared with some 57 others that continue into the Silurian. Sheehan & Coorough (1989) comment that the most cosmopolitan of brachiopod genera were those that persisted from Ashgill to Silurian. In general the effects of the Ashgill glaciation on brachiopod extinction were spread over the whole of the last Ashgill to earliest Silurian interval (figure 7) rather than concentrated at one specific horizon. Briggs *et al.* (1988) showed that maximum 'last appearances' of genera of trilobites was within, or at the end of, the Rawtheyan or its equivalents. Although Brenchley (1984) claimed that the mortality of most trilobite families coincided with the Rawtheyan extinction event, and with the onset of profound glaciation, a recent study of Owen (1986) showed that, although rare, examples of most of the Ordovician trilobite families persisted into the Hirnantian (Briggs *et al.* 1988) (figures 6 and 8). Their rarity may partly be the result of the facies changes in this interval. The pattern of generic extinction can be compared with a typical series boundary event such as the Tremadoc–Arenig boundary (figure 3) or even with the Cambrian–Ordovician boundary discussed above. The general patterns are remarkably similar, the main point of difference being a low rate of originations early in the Silurian. If this is so, in what feature does the identification of the Ordovician–Silurian boundary coinciding with one of the more important extinction events in the record reside?

As far as the trilobites are concerned the answer has to be the termination of so many major

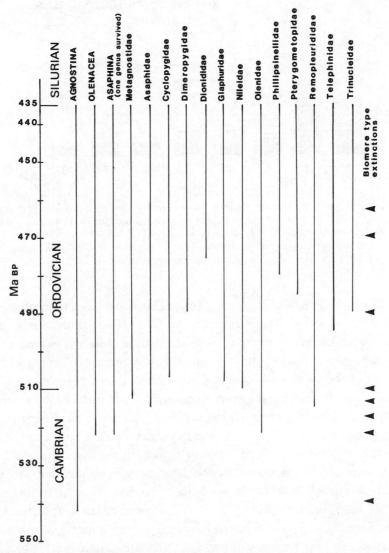

FIGURE 6. Range of higher trilobite taxa that become extinct at the Ordovician–Silurian boundary. Arrows on right mark some of the more important biomere-type events that affect platform taxa, showing how the families affected at the end of the Ordovician have passed through these.

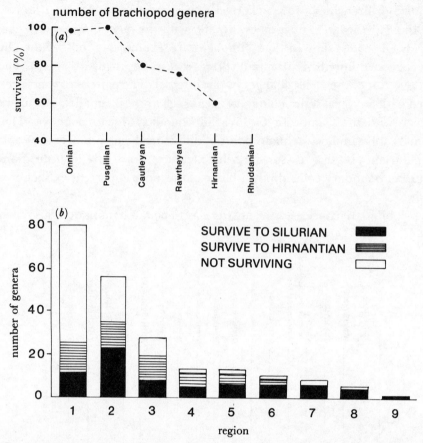

FIGURE 7. Brachiopods through the Ordovician–Silurian boundary, showing continuous high turnover rather than catastrophic extinction. (*a*). Percentage of genera surviving from one stage to the next through the late Ordovician early Silurian period, after Brenchley (1984). (*b*). Brachiopods divided according to their occurrence in one or more biogeographic regions (1–9) by Sheehan & Coorough (1989) showing how those most widely distributed in the Ashgill were also the strongest survivors.

clades after a history of tens of millions of years, and in some cases more than 100 Ma (figure 6). These include groups that passed through the Cambrian–Ordovician boundary, such as Agnostida and Olenacea. The extinction event seems to be different in kind from other events (including biomere-like events, or those connected with regressive–transgressive events) that are characteristic of the lower Palaeozoic, as mentioned above. Because many of these major clades persist into the Hirnantian, the event in question is not that producing generic mortality within or at the end of the Rawtheyan, but one that is likely to have occurred at or towards the end of the glacial episode. The onset of major glaciation did not terminate the clades. It could be argued that previous extinctions coincident with the glaciation rendered a remnant trilobite fauna more vulnerable to extinction later on. Given the taphonomic bias against finding deeper water trilobite faunas through the regressive Hirnantian interval, this kind of argument would be difficult to prove; there are other intervals within the Ordovician (for example at the Tremadoc–Arenig boundary) where deeper faunas are equally elusive. There has been no success in finding widespread iridium anomalies associated with the Ordovician–Silurian boundary event, and so at the moment there are no grounds for suspecting the involvement of extra-terrestrial agents in the extinction.

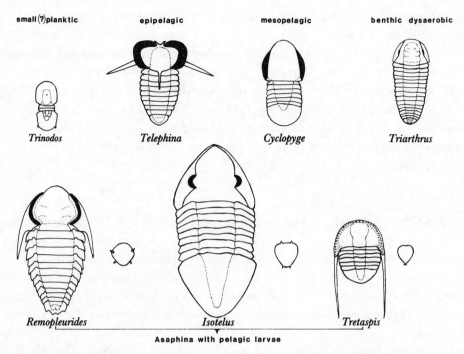

small(?)planktic   epipelagic   mesopelagic   benthic dysaerobic

Trinodos   *Telephina*   *Cyclopyge*   *Triarthrus*

Remopleurides   *Isotelus*   *Tretaspis*

Asaphina with pelagic larvae

FIGURE 8. The morphological range and life habits of trilobites belonging to clades that became extinct at the Ordovician–Silurian boundary. Bottom row shows planktic asaphoid larvae of mostly benthic Asaphina with very disparate adult morphologies.

Trilobite extinctions: oceanic crisis?

Assuming that I am looking for a terrestrial cause for trilobite extinctions, it is informative to see whether there are any common features between the families that became extinct. Considering the life habits of trilobites, the common factors appear to be that nearly all of the trilobites were deeper shelf forms, or had a phase in their life cycle as part of the oceanic plankton (figure 8).

(i) Asaphina

This suborder of trilobites was the largest group in terms of numbers of constituent families extinguished at the Ordovician–Silurian boundary. As reclassified by Fortey & Chatterton (1988) it includes such typical Ordovician families as Trinucleidae, Nileidae, Asaphidae, Cylopygidae and Remopleurididae, which do not survive into the Silurian. One member of the family Raphiophoridae, *Raphiophorus*, is the lone asaphine survivor, surviving until late in the Silurian. Asaphina include a great range of morphology, and presumably a matching range of life habits. The life habits of Cyclopygidae are the best known (Fortey 1985). These generally eschew shallow shelf environments, and were probably mostly mesopelagic. They persisted from the Arenig to the Ashgill with remarkably little morphological change. This may be attributed to the general persistence and stability of the environment they occupied at the edge of Ordovician continents. This stability was upset by the Ordovician–Silurian extinction. Most of the remainder of the group were benthic. Some Asaphina occupied shelf habitats (some Asaphidae) and the group can be found on all the palaeocontinents in the late Ordovician, including relatively high palaeolatitudes around Gondwana and low palaeolatitudes in Laurentia. What can unite such a disparate selection of trilobites to make them vulnerable to extinction? The only conspicuous unifying feature is the possession of a globular and inflated larval stage termed the asaphoid protaspis, (Fortey & Chatterton 1988). Chatterton & Speyer, (1989) maintain that this larval type was specially adapted for a planktonic existence. Some of these larvae grew to several millimetres in length. Their change into benthic adults was accompanied by something akin to a metamorphosis. Unlike the cyclopygids, which spent all of their lives in the water column, the asaphoid larvae settled at some intermediate growth stage. None the less this great group were obligate inhabitants of the oceanic column at one stage in their lives. Hence it is reasonable to conclude that disruption of oceanic structure is implied in their demise.

(ii) *Other groups*

Non-asaphine groups that did not survive the Ordovician include the following (Briggs *et al.* 1988, fig. 9.5).

(i) Agnostidae. The last of a great group ranging from the early Cambrian, agnostids were particularly associated with oceanic settings throughout their history, and may have been pelagic (Robison 1972). Their end might be explained by oceanic disruption.

(ii) Olenidae. As discussed above, olenids are associated with deeper water and dysaerobic palaeoenvironments.

(iii) Telephinidae. This group included activity swimming pelagic animals of frequent occurrence in oceanic biofacies.

(iv) Nileidae. Nileids are mostly found in outer shelf habitats; Dionididae had similar habits and habitats.

(v) Pterygometopidae, Glaphuridae and Dimeropygidae. These groups are the only major trilobite clades that become extinct that are characteristic of more inshore facies. Pterygometopidae may be paraphyletic.

Overall, there is a striking dominance of trilobites that either lived habitually in deeper water facies or had a prolonged planktic larval stage, or pelagic adult stage. Survivors (cheirurids, illaenids, homalonotids, harpids, scutelluids, encrinurids, many proetides, lichids, dalmanitids, odontopleurids and calymenids) are either well known as inhabitants of reef-like calcareous habitats, or were common in inshore clastics around Ordovician Gondwana. It was these habitats and their occupants that seemed to have survived the end Ordovician event. The necessity for the existence of early Silurian refugia has been stated. Although there seems to be no problem about the persistence of clastic facies, which dominate boundary sequences, there is a scarcity of early Llandovery carbonates. They must have persisted somewhere in the equatorial region: the only examples known to me are in Estonia.

Hence it can be concluded that the Ordovician–Silurian event was quite different from the Cambro–Ordovician event (or indeed any of the intra-Ordovician or Cambrian biomere events) as far as the trilobites were concerned. The long-lived clades, the deep-water taxa, or those with specialized planktic larvae, were greatly affected. This is not to deny that the end Ordovician glaciation had an important effect on shelf taxa. Like the brachiopods (Sheehan 1975, 1982), which were less profoundly affected at high taxonomic level than the trilobites (Brenchley 1984; Cocks 1988), shelf trilobites had a high rate of generic extinction, coinciding with the onset of major glaciation, but the major phylogenetic continuity was not broken. The major event was in exterior and oceanic sites. In that the ranges of the last representatives of the affected families extend into the Hirnantian, it is likely that this event occurred at or near the end of that stage (faunas are too rare to say whether extinctions were exactly synchronous). In that pelagic larvae and adults, and benthic adults were all affected it seems likely that a substantial breakdown in oceanic structure was involved. This implies major anoxia.

Graptolites through the Ordovician–Silurian interval

Many graptolitic rocks accumulated in oceanic sites, producing a record of nearly continuous sedimentation through the later Ordovician and into the Silurian. That the passage from Ordovician into Silurian produced a profound change in the graptolite fauna has been recognized for a long time (Elles 1922; Bulman 1958). Recent studies through boundary sequences have provided more detailed narrative accounts of the changes in the graptolite faunas through the critical interval than are available for any other group (Williams 1983, 1987; Koren 1988; Melchin & Mitchell 1988). The question of which taxa survived the extinction events has been clarified by recent studies of the phylogenetics of graptolites (Cooper & Fortey 1986; Mitchell 1987). It is now clear, for example, that biserial graptolites in the Llandovery are closely related to very few of those in the Ashgill, although this is not necessarily reflected in the current taxonomy (Melchin & Mitchell 1988). The main extinction event for the graptolites probably occurred well below the end of the Hirnantian, within or at the end of the *pacificus* biozone (sometimes considered as a subzone of the *anceps* biozone). This horizon marks the end of the dicellograptids, and the bulk of the Ordovician biserials and retiolitimorphs; all of these are good clades. Where graptolitic and non-graptolitic facies occur together, as in China, the end of the *pacificus* biozone or its local equivalent marks the widest spread of the Hirnantia fauna (Mu 1988), and it can be concluded that the graptolite extinction coincides not with the onset of glaciation but with its maximum extent. The interval succeeding the *pacificus* biozone is marked by a generally barren mudstone in the Dob's Linn stratotype, but this contains within it the *extraordinarius* band, the only British occurrence of the biozone of *Climacograptus extraordinarius*, which represents a low point in graptoloid diversity wherever it has been recognized. A few biserial graptoloids are present. It is believed that these comprise the stock that all the spectacular radiation of the Silurian graptoloids was derived from. Although the fauna of the succeeding *persculptus* biozone is also limited in variety, the sister groups of most of the Silurian graptoloids are probably present therein, including the earliest monograptid, *Atavograptus ceryx* (Rickards & Hutt 1970). The base of the succeeding *acuminatus* biozone, with a slightly more diverse fauna, is taken as the conventional base for the Ordovician–Silurian boundary. At Dob's Linn, Scotland, typical graptolitic black shale deposition is resumed in the *persculptus* biozone, having been almost entirely absent during the *extraordinarius* interval. The early Llandovery marks a very widespread extension of organic-rich graptolitic shale over platform areas in general (Cocks & Mackerrow 1973, fig. 1; Thickpenny & Leggett 1988).

This summary suggests that the record of the graptolites is unusually complete through the Ordovician–Silurian boundary interval, probably more complete than that of any other group. Because graptolites were planktonic animals it is clear that, whatever the nature of their extinction crisis, it has to be one that affected the oceanic water column. Skevington (1974, p. 67) suggested that there was a reduction in the overall distribution of graptolite faunas in the late Ordovician that could be implicated in the extinction event: 'in the late Ordovician all graptolite faunas, with rare exceptions, were confined to the tropical zone'. The subsequent description by Legrand (1981, 1988) of extensive graptolitic formations crossing the

[107]

Ordovician–Silurian boundary in the Algerian Sahara, which was close to the glaciated area of the late Ordovician, makes this explanation untenable. Graptolites were capable of living at all palaeolatitudes beyond the end of the Ordovician into the Silurian. So the explanation of their crisis has to encompass a worldwide oceanographic change. It will be recalled that much the same was said above about the trilobite extinctions. To a large extent, the nature of this change depends on the interpretation of graptolite life habits; to what kind of change would they be vulnerable?

Berry *et al.* (1987) have recently proposed the challenging theory that graptolites lived within the early Palaeozoic low-oxygen zone. This may be contrasted with a traditional view that they had a 'superficial drifting mode of life' (Bulman 1970). If they were inhabitants of a low oxygen zone this would make them peculiarly vulnerable to any change that destroyed the stratified structure of the ocean, producing a greater short-term mixing and general oxygenation (Wilde & Berry 1984). Such an event could have accompanied the melting of the Gondwana ice sheet. Could this correspond with the curious facies change in the barren mudstones (hardly a typical black graptolite shale) in the *extraordinarius* interval at Dob's Linn? On the other hand, if the more conventional notion of graptolite life habits were accepted, with some depth stratification accounting for the variety of species, then perhaps most species would be regarded as having had a normal oxygen tolerance, and, barring catastrophes, they would have been vulnerable to anoxia.

Conodonts through the Ordovician–Silurian boundary interval

The conodont record is not as complete as the graptolite record, if only because of a shortage of suitable calcareous successions to record their history. Barnes & Bergström (1988) have summarized this history. The most completely known succession is that on Anticosti Island, Canada. Only 8 out of 32 genera listed in their figure 5 pass between Ordovician and Silurian worldwide, and the change is as profound as that in other groups. Taxonomic pseudoextinction does not appear to present a difficulty. The correlation with graptolitic facies presents something of a problem, but Barnes & Bergström comment that the main extinction event is 'likely to be in the upper part of the *persculptus* zone'. If so, this is slightly later than the graptolite event. An important observation relates to the phylogenetics of the Silurian survivors. Barnes & Bergström that 'where the origin of the Llandovery stocks is known or can be postulated, they appear to be derived, in almost all cases, from stocks that inhabited the tropical waters of the Midcontinent province during the Ordovician'. The Midcontinent province had a long history in the shallow, tropical seas of the Ordovician. This implies that those clades that became extinct were those that tended to inhabit deeper shelf sites in the Ordovician, including North Atlantic province taxa. As with the trilobites it was oceanward taxa that were particularly affected.

Attempt at a synthesis of the Ordovician–Silurian extinctions

I have attempted to gather together the most important facts concerning the groups that have the most complete records across this boundary. A synthetic explanation has to account for all of these facts, regardless of a preferred explanation for one kind of organism.

(i) Timing (figure 7). The timing of major events does not correspond precisely between different groups, even though all are concentrated within the period of the late Ordovician glaciation and its immediate aftermath. The graptolite event apparently preceded the conodont event. We do not know precisely when the deeper water trilobites became extinct because of their rarity in the Hirnantian but, given the similarity in extinction patterns, it is not unreasonable to assume that they were affected by the same event as the conodonts.

(ii) The regressive event and climatic deterioration associated with the glaciation produced extinctions in the shallow platform faunas, especially at generic level, but the extinction of major clades particularly affected deeper-water forms. Brachiopods, which are characteristic of shelf habitats, had a rather protracted history of extinction through the later Ordovician associated with climatic deterioration, and many Hirnantian genera survived into the Llandovery. The factors possibly involved with extinguishing platform taxa have been described by Brenchley (1984), Jablonski (1986) and Sheehan & Coorough (1989), and include: the spread of cold climate; the reduction in area of certain, especially tropical, habitats ('species–area effect'); and the dominance of the clastic Hirnantia–Dalmanitina biofacies and lithofacies. Legislating between these factors is probably impossible.

(iii) Many groups have a Lazarus interval through the Llandovery, or the Hirnantian plus Llandovery (in some cases even longer). Refugia for these taxa must have existed. The wide spread of monotonous Llandovery muds, often anoxic, readily accounts for the taphonomic bias against Llandovery records of such taxa. Taxa within groups such as echinoderms that are liable to exhibit such Lazarus effects may not have died out precisely at the end of the Ordovician.

(iv) An interesting observation is that the major extinction events, as recorded by the termination of clades in the trilobites and conodonts, particularly affected deeper water taxa. In this respect the Ordovician–Silurian event differs from most others in the Lower Palaeozoic, where deeper-water taxonomic continuity is claimed as the norm. Planktic graptolites were also profoundly affected. It can be concluded that major extinction involved an oceanic crisis.

If all four points hold, a scenario to explain them is not easy to find. Saying that the late Ordovician extinction was caused by the glaciation is unsatisfactorily vague, because it does not account for the different timing of events, nor for the different effects on different groups of organisms. Presumably brachiopod extinction is primarily related to their comparatively inshore history through this interval, and hence directly to glacial influence. Diverse Lower Palaeozoic brachiopod faunas are generally more characteristic of inshore environments. One can find rather conflicting accounts of the extent of the brachiopod crisis. For example, Sheehan (1982) stresses a dramatic turnover in brachiopod faunas between late Ordovician and Silurian, whereas Cocks (1988) indicates a greater degree of continuity than had been supposed. Because Sheehan also emphasizes the early history of brachiopod radiation in the Silurian (a different matter from extinction, as I discussed above) it is possible that these differences are a matter of taxonomy. When Sheehan (1982, p. 477) remarks, for example: 'some of these Silurian groups had well-known ancestors in the Ashgill' this implies that the phylogenetic continuity was not broken by the Hirnantian glaciation, merely that subsequent cladogenesis occurred in the Silurian. The group in question is not Silurian at all, in a phylogenetic sense. However, there was clearly a high rate of generic last appearances among brachiopods throughout the Rawtheyan–Hirnantian interval, and this was probably related to a combination of climatic and glacioeustatic causes, as discussed by many authors.

The most interesting question relates to the oceanic change. To summarize briefly: trilobites that became extinct included benthic forms with a long history in exterior sites, Asaphina with distinctive pelagic larval phases, and pelagic trilobites; conodonts (nektobenthic and pelagic animals according to most authorities) of North Atlantic province type that also included oceanic forms: graptolites with planktic habits, well-known for their wide distribution even at specific level. The deepest-water late Ordovician brachiopod community type, the *Foliomena* community (Sheehan 1973), also failed to survive into the Silurian. The graptolite event predated the conodont event, so far as the stratigraphic correlation allows us to tell.

I can only identify one common cause that is likely to have exterminated the range of trilobites affected, and that is a period of complete anoxia in some niches. Some other options may be eliminated as follows. Spread of cool water accompanying the glaciation is an unlikely factor because some of the trilobites were well-adapted to cool waters anyway. For example, the cyclopygids had a long prior history peripheral to Gondwana in boreal palaeolatitudes. The glacioeustatic regression *per se* cannot account for the change because at least some of the trilobites lived in the water column, and many endured beyond the main Hirnantian regression. The fact that both trilobites that lived in the water column as adults (cyclopygids, telephinids), or as juveniles (Asaphina), and those with benthic adults (olenids, most non-cyclopygid Asaphina) were equally vulnerable, indicates that the causative agency had a considerable bathymetric span. A widespread anoxic event seems to be the only mechanism available to do this. The same event accounted for the conodont extinctions. The degree of anoxia would have to be exceptional to account for the disappearance even of some taxa, like olenids, that may have been able to tolerate reduced oxygen tension (see above). Because of the rarity of Hirnantian deep-water trilobites we cannot be precise about the timing, but of the conodonts were affected by the same event, which is probable given the similarity in extinction pattern, then we can do no better than suggesting that it happened in 'an interval in the upper *persculptus* zone' (Barnes & Bergström 1988). This is at, or very close to, the end of the Ordovician glaciation.

The Hirnantia faunas representing the glacial interval underlie *persculptus* graptolite faunas in China, but elsewhere in the world, Britain for example, they extend into at least the lower part of the *persculptus* interval. The sharp lithological change into dark shale occurs either towards the end of this zone or at the base of the succeeding *acuminatus* biozone; at present further refinement is not possible because of the limits of stratigraphic resolution. But it is likely that the faunal change in trilobites and conodonts coincides closely with the post-glacial 'rebound' that was within the *persculptus* or basal *acuminatus*, or both biozones. If this were so it should be possible to relate the anoxic event to the ensuing transgression, by one of the mechanisms outlined by Hallam (this symposium). One among several obvious possibilities was the sudden release of nutrient-rich water stimulating massive plankton blooms, and concomitant consumption of free oxygen. Whatever the cause, an anoxic event should leave some distinctive geochemical signatures. A variety of such signatures has been recognized 'no older than the base of the *persculptus* zone' (Nowlan *et al.* 1988) in northwestern Canada, and this is consistent with more general studies of widespread reducing conditions in the early Llandovery (Thickpenny & Leggett 1988). The anoxic event is likely to have been the cause of the major extinction in the oceanic realm.

Could the same event have been the cause of the graptolite extinctions? On the evidence currently available this does not seem probable, but stratigraphic correlation problems leave room for manoeuvre. If the graptolite event occurred before the conodont event, as indicated on figure 9, then the two extinctions should have been decoupled, because they are separated by the *C. extraordinarius* range zone in several sections, including Dob's Linn (Williams 1988).

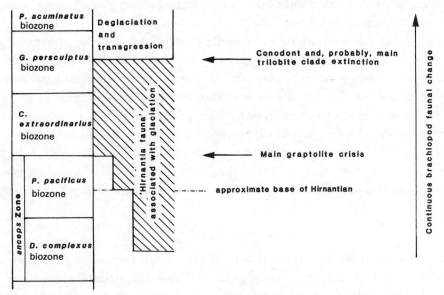

FIGURE 9. Approximate timing of the extinction events affecting higher level taxa of groups with a more or less continuous record through the Ordovician–Silurian boundary interval.

It will be recalled that this interval represented the nadir of graptolite diversity. However, there are parts of the world where the *extraordinarius* biozone is not recognized, for example in China. This pushes the graptolite and other events together in these regions (although there could still be an hiatus). It is possible that the *extraordinarius* interval was very short; the barren mudstone lithology at this interval at Dob's Linn could have accumulated rapidly. Interestingly, a similar barren mudstone interval has been observed to equate with biomere boundaries in the Cambrian (Conway Morris & Rushton 1988). If one accepts the Berry *et al.* (1987) version of graptolite life habits (above), the lethal event for the graptoloids should have been a period of exceptional ventilation, not anoxia, to which one presumes the conodonts and trilobites would have been immune. This view would favour a separate, earlier horizon as the ventilation event; the surviving graptolites would have been the least affected by the anoxic event, and indeed, did go on to proliferate greatly in the Llandovery whereas benthic diversity continued to be rather low. The alternative account of events might be as follows: the apparently different extinction horizons could be explained either by the brevity of the *extraordinarius* interval, perhaps exacerbated by correlation problems, or by the diachronous, shelfward spread of anoxic water from oceanic environments. Gross anoxia caused crises in the graptolite and deeper water groups. Only those species (all apparently diplograptids) that lived in some safe environment could survive; they may have been extreme epiplanktics or shelf taxa. The choice between these hypotheses depends on more refined correlation, independent geochemical evidence, and better ecological understanding of the animals involved. I am reluctant to arbitrate between them here, except to note that the simple explanation is the second, only requiring a single event. However, investigating the problems should afford a fruitful line of further research.

The demise of deeper-shelf trilobites at the end of the Ordovician seems to have had an effect which lasted well into, if not throughout the Silurian. There were no replacements for pelagic trilobites. I can find no convincing Silurian records of the kind of trilobites with reduced or atrophied eyes (atheloptic assemblage) that were characteristic of deep water sites from the mid-Cambrian onwards. None of the trilobite associations from the Wenlock listed by Thomas (1980) seem to be of this kind, and even the lone Ordovician asaphine survivor *Raphiophorus* reappears in an on-shelf basinal facies. By the early Devonian, trilobites with reduced eyes are well-known again, mostly of phacopoid origin. Although the possibility of taphonomic factors being responsible for the Silurian absences cannot be ruled out, it seems as likely that the absence of such morphs indicates a long time-lag in repopulating the full environmental range available to the group. This is particularly interesting in view of hypotheses (see, for example, Jablonski *et al.* (1983)) relating the recruitment of deep water faunas to migrations from shelf environments over a period of time. Could the Ordovician–Silurian boundary have been a major crisis in deep-sea faunas, one that took much of the Silurian to recover from?

6. ANOXIA AND TRANSGRESSION

There is an association between marine transgression and extinction that has been recognized for some time, and Hallam (this symposium) has claimed the pattern as being general. As discussed above, there is some faunal evidence at the Cambrian–Ordovician boundary, and much at the Ordovician–Silurian boundary, for associated anoxic intervals at times of transgression. In detail an increased rate of last appearances of genera appears to be associated with a regressive phase predating the transgression. Relatively minor biomere events through the Cambrian have been associated with anoxia, if not transgression (Palmer 1984), whereas faunal changes at series boundaries in the Ordovician have been associated with regression and subsequent transgression, if not anoxia (Fortey 1984). In general the idea of an association of extinction with anoxia-eustatic cyles is supported, although different authors stress one or the other, and there is a great deal of disagreement as to how these factors might be involved in causing extinction. The Ordovician–Silurian event seems to have been the one wherein the influence on oceanic faunas was most profound. The kind of animals that survived earlier events were vulnerable at this horizon. If anoxia is sought as common cause, it can be concluded that its effectiveness as an agent of extinction must have varied from one event to another.

I thank Dr L. R. M. Cocks, Dr P. D. Taylor and Dr A. W. A. Rushton for reading the manuscript and making helpful suggestions. Dr Cocks and Dr P. J. Brenchley helped me with brachiopod data.

REFERENCES

Barnes, C. R. & Bergström, S. M. 1988 Conodont biostratigraphy of the uppermost Ordovician and lowermost Silurian. In *A global analysis of the Ordovician–Silurian boundary* (ed. L. R. M. Cocks) (*Bull. Br. Mus. nat. Hist.* (Geol.) **43**), 273–284.

Bergström, S. M. 1986 Biostratigraphic integration of Ordovician graptolite and conodont zones – a regional review. *Spec. Pub. geol. Soc. Lond.* **20**, 61–78.

Berry, W. B. N., Wilde, P. & Quinby-Hunt, M. S. 1987 The oceanic non-sulfidic oxygen minimum zone: a habitat for graptolites? *Bull. geol. Soc. Denmark* **35**, 103–114.

Brenchley, P. J. 1984 Late Ordovician extinctions and their relationships to the Gondwana glaciation. In *Fossils and climate* (ed. P. J. Brenchley), pp. 291–316. New York: John Wiley.

Brenchley, P. J. & Cocks, L. R. M. 1982 Ecological associations in a regressive sequence: the latest Ordovician of the Oslo-Asker district, Norway. *Palaeontology* **25**, 783–815.

Brenchley, P. J. & Newell, G. 1984 Late Ordovician environmental changes and their effect on faunas. In *The Ordovician system* (ed. D. L. Bruton), pp. 65–80. Oslo: Universitetsforlaget.

Briggs, D. E. G., Fortey, R. A. & Clarkson, E. N. K. 1988 Extinction and the fossil record of the arthropods. In *Extinction and survival in the fossil record* (ed. G. P. Larwood), pp. 171–209. Oxford: Clarendon Press.

Bulman, O. M. B. 1958 The sequence of graptolite faunas. *Palaeontology* 1, 159–173.

Bulman, O. M. B. 1970 *Treatise on invertebrate paleontology*, vol. V (*Graptolithina* (*revised*)). (163 pages.) Lawrence: University of Kansas Press and Geological Society of America.

Chatterton, B. D. E. & Speyer, S. 1989 Trilobite ontogeny and late Ordovician extinctions. *Paleobiology*. (In the press).

Chen, J.-y., Zou, X.-p., Chen, T.-e. & Qi, D.-l. 1979 Late Cambrian cephalopods of North China – Plectronocerida, Proactinocerida (ord. nov.) and Yanhecerida (ord. nov.). *Acta palaeont. sin.* 18, 1–23.

Chen, J.-y., Qian, Y.-y., Lin, Y.-k., Zhang, J.-m., Wang, Z.-h., Yin, L.-m. & Erdtmann, B.-D. 1985 *Study on Cambrian–Ordovician boundary strata and its biota in Dayangcha. Hunjiang, China*

Cocks, L. R. M. 1988 A global analysis of the Ordovician–Silurian boundary. *Bull. Br. Mus. nat. Hist.* (Geol.) 43, 1–394.

Cocks, L. R. M. 1988a Brachiopods across the Ordovician–Silurian boundary. In *A global analysis of the Ordovician–Silurian boundary* (ed. L. R. M. Cocks) (*Bull. Br. Mus. nat. Hist.* (Geol.) 43), 311–315.

Cocks, L. R. M. & McKerrow, W. S. 1973 Brachiopod distributions and faunal provinces in the Silurian and Lower Devonian. *Spec. Pap. Palaeont.* 12, 291–304.

Conway Morris, S. 1979 The Burgess Shale (Middle Cambrian) fauna. *A. Rev. Ecol. Syst.* 10, 327–49.

Conway Morris, S. & Rushton, A. W. A. 1988 Precambrian to Tremadoc biotas in the Caledonides. In *The Caledonian–Appalachian Orogen* (ed. A. L. Harris & D. J. Fettes), pp. 93–109. *Spec. Pub. geol. Soc. London* 38, 93–109.

Destombes, J. & Henry J.-L. 1987 Trilobites Calmoniidae de l'Ordovicien supérieur du Maroc et les origines de la province Malvino–Cafre. *Lethaia* 20, 129–139.

Elles, G. L. 1922 The graptolite faunas of the British Isles. *Proc. geol. Ass.* 33, 168–200.

Erdtmann, B.-D. 1988 The earliest Ordovician nematophorous graptolites: taxonomy and correlation. *Geol. Mag.* 125, 327–348.

Flower, R. H. 1964 The Nautiloid order Ellesmeroceratida (Cephalopoda). *Mem. Bur. Mines Miner. Resour. N. Mex. Inst. Min. Tech.* 12, 1–234.

Fortey, R. A. 1980 Generic longevity in Lower Ordovician trilobites: relation to environment. *Paleobiology* 6, 24–31.

Fortey, R. A. 1983 Cambrian–Ordovician trilobites from the boundary beds in western Newfoundland and their phylogenetic significance. *Spec. Pap. Palaeont.* 30, 179–211.

Fortey, R. A. 1984 Global Ordovician transgressions and regressions and their biological implications. In *The Ordovician system* (ed. D. L. Bruton), pp. 37–50. Oslo: Universitetsforlaget.

Fortey, R. A. 1985 Pelagic trilobites as an example of deducing the life habits of extinct arthropods. *Trans. R. Soc. Edinb.* 76, 219–230.

Fortey, R. A. & Chatterton, B. D. E. 1988 Classification of the trilobite suborder Asaphina. *Palaeontology* 31, 165–222.

Fortey, R. A. & Cooper, R. A. 1986 A phylogenetic classification of the graptoloids. *Palaeontology* 29, 631–654.

Fortey, R. A., Landing, E. & Skevington, D. 1982 Cambrian–Ordovician boundary sections in the Cow Head Group, western Newfoundland. In *The Cambrian–Ordovician boundary: sections, fossil distributions, and correlations* (ed. M. G. Bassett & W. T. Dean), pp. 95–129. Cardiff: National Museum of Wales.

Fortey, R. A. & Owens, R. M. 1975 Proetida – a new order of trilobites. *Fossils and Strata* 4, 227–239.

Fortey, R. A. & Owens, R. M. 1987 The Arenig series in South Wales. *Bull. Br. Mus. nat. Hist.* (Geol.) 41, 69–343.

Henningsmoen, G. 1957 The trilobite family Olenidae. *Skr. Norske Vid.-Akad. Oslo 1. Mat. Nat. Kl.* 1, 1–303.

Jaanusson, V. 1979 Ordovician. In *Treatise on invertebrate palaeontology*, volume A. Introduction (ed. R. A. Robison & C. Teichert), pp. 136–166. Lawrence: Geological Society of America and University of Kansas Press.

Jablonski, D. 1986 Causes and consequences of mass extinctions: a comparative approach. In *Dynamics of extinction* (ed. D. K. Elliott), pp. 183–227. New York: John Wiley.

Jablonski, D., Sepkoski, J. J., Bottger, D. J. & Sheehan, P. M. 1983 Onshore–offshore patterns in the evolution of Phanerozoic shelf communities. *Science*, Wash. 222, 1112–1125.

James, N. P. & Stevens, R. K. 1987 Stratigraphy and correlation of the Cambro-Ordovician Cow Head Group, western Newfoundland. *Bull. geol. Surv. Canada* 366, 1–143.

Koren, T. N. 1988 Evolutionary crisis of Ashgill graptolites. In *Abstracts of the Vth international symposium on the Ordovician system*, p. 48. Newfoundland: St. John's.

Legrand, P. 1981 Essai sur la paléogéographie del'Ordovicien au Sahara algérien. *Notes mem. Comp. Franc. Petrol* 11, 121–138.

Legrand, P. 1988 The Ordovician–Silurian boundary in the Algerian Sahara. In *A global analysis of the Ordovician–Silurian boundary* (ed. L. R. M. Cocks) (*Bull, Br. Mus. nat. Hist.* (Geol.) 43), 171–176.

Levinton, J. 1988 *Genetics, paleontology and macroevolution.* (656 pages.) Cambridge University Press.

Ludvigsen, R. 1982 Upper Cambrian and lower Ordovician trilobite biostratigraphy of the Rabbitkettle formation, western District of Mackenzie. *Lifesci. Contrib. R. Ontario Mus.* 134, 1–188.

Ludvigsen R. & Westrop, S. R. 1983 Trilobite biofacies of the Cambrian–Ordovician boundary interval in northern North America. *Alcheringa* **7**, 301–319.

Ludvigsen, R., Pratt, B. R. & Westrop, S. R. 1988 The myth of a eustatic sea level drop near the base of the Ibexian series. *Bull. N.Y. St. Mus.* **462**, 65–70.

Melchin, M. J. & Mitchell, C. E. 1988 Late Ordovician extinction among the Graptoloidea. In *Abstracts of the Vth international symposium on the Ordovician system*, p. 58. Newfoundland: St. John's.

Miller, J. F. 1984 Cambrian and earliest Ordovician conodont evolution, biofacies and provincialism. *Spec. Pap. geol. Soc. Am.* **196**, 43–68.

Miller, J. F. 1988 Conodonts as biostratigraphic tools for redefinition and correlation of the Cambrian–Ordovician boundary. *Geol. Mag.* **125**, 349–362.

Mitchell, C. E. 1987 Evolution and phylogenetic classification of the Diplograptacea. *Palaeontology* **30**, 353–406.

Mu, Enzhi 1988 The Ordovician–Silurian boundary in China. In *A global analysis of the Ordovician–Silurian boundary* (ed. L. R. M. Cocks) (*Bull. Br. Mus. nat. Hist.* (Geol.) **43**), 117–132.

Newell, N. D. 1967 Revolutions in the history of life. *Spec. Pap. geol. Soc. Am.* **89**, 63–91.

Norford, B. S. 1988 Introduction to papers on the Cambrian–Ordovician boundary. *Geol. Mag.* **125**, 323–326.

Nowlan, G. S., Goodfellow, W. D. & McCracken, A. D. 1988 Geochemical evidence for sudden biomass reduction and anoxic basins near the Ordovician–Silurian boundary in Northwestern Canada. In *Abstracts of the Vth international symposium on the Ordovician system*, p. 66. Newfoundland: St. John's.

Owen, A. W. 1986 The uppermost Ordovician (Hirnantian) trilobites of Girvan, southeast Scotland with a review of coeval trilobite faunas. *Trans. R. Soc. Edinb.* **77**, 231–239.

Palmer, A. R. 1965 Biomere – a new kind of biostratigraphic unit. *J. Paleont.* **39**, 149–153.

Palmer, A. R. 1979 Biomere boundaries re-examined. *Alcheringa* **3**, 33–41.

Palmer, A. R. 1984 The biomere problem: evolution of an idea. *J. Paleont.* **58**, 599–611.

Patterson, C. & Smith, A. B. 1988 Is the periodicity of extinctions a taxonomic artefact? *Nature, Lond.* **330**, 248–252.

Paul, C. R. C. 1982 The adequacy of the fossil record. In *Problems of phylogenetic reconstruction* (ed. K. A. Joysey & A. E. Friday), (*Spec. Vol. Syst. Ass.* no. 21), pp. 75–118.

Raup, D. M. & Sepkoski, J. J. 1982 Mass extinctions in the marine fossil record. *Science, Wash.* **215**, 1501–1503.

Raup, D. M. & Sepkoski, J. J. 1986 Periodic extinctions of families and genera. *Science, Wash.* **231**, 833–836.

Rickards, R. B. & Hutt, J. E. 1970 The earliest monograptid. *Proc. geol. Soc. Lond.* **1663**, 115–119.

Robison, R. A. 1972 Mode of life of agnostid trilobites. *Int. geol. Congr.* 24 **7**, 33–40.

Rong, J.-y. 1984 Distribution of the Hirnantia fauna and its meaning. In *The Ordovician system* (ed. D. L. Bruton), pp. 101–112. Oslo: Universitetsforlaget.

Rong, J.-y. & Harper, D. A. T. 1988 A global synthesis of the latest Ordovician Hirnantian brachiopod faunas. *Trans. R. Soc. Edinb.* **79**, 383–402.

Ross, R. J. 1951 Stratigraphy of the Garden City formation in northeastern Utah, and its trilobite faunas. *Bull Peabody Mus. nat. Hist.* **6**, 1–161.

Rushton, A. W. A. 1982 The biostratigraphy and correlation of the Merioneth–Tremadoc Series boundary in North Wales. In *The Cambrian–Ordovician boundary: sections, fossil distributions and correlations* (ed. M. G. Bassett & W. T. Dean), pp. 41–59. Cardiff: National Museum of Wales.

Sepkoski, J. J. 1981 A factor analytic description of the fossil record. *Paleobiology* **7**, 36–53.

Sepkoski, J. J. 1984 A kinetic model of Phanerozoic taxonomic diversity. III. Post–Paleozoic families and mass extinctions. *Paleobiology* **10**, 246–267.

Sepkoski, J. J. 1986 Phanerozoic overview of mass extinction. In *Patterns and processes in the history of life* (ed. D. M. Raup & D. Jablonski), pp. 277–295. Berlin: Springer-Verlag.

Sheehan, P. M. 1975 Brachiopod synecology in a time of crisis (late Ordovician–Early Silurian). *Paleobiology* **1**, 205–212.

Sheehan, P. M. 1973 Brachiopods from the Jerrestad Mudstone (early Ashgillian, Ordovician) from a boring in southern Sweden. *Geol. Palaeont.* **7**, 59–76.

Sheehan, P. M. 1982 Brachiopod macroevolution at the Ordovician–Silurian boundary. In *Proc. N. Am. Paleont. Conv.*, **2**, 477–481.

Sheehan, P. M. & Coorough, P. J. 1989 Brachiopod zoogeography across the Ordovician–Silurian extinction event. In *Palaeozoic biogeography* (ed. W. S. McKerrow & C. S. Scotese). (In the press.)

Shergold, J. H. 1988 Review of trilobite biofacies distributions at the Cambrian–Ordovician boundary. *Geol. Mag.* **125**, 363–380.

Skevington, D. 1974 Controls influencing the composition and distribution of Ordovician faunal provinces. *Spec. Pap. Palaeont.* **13**, 59–74.

Sloan, R. E. 1989 A chronology of North American trilobite genera. *Bull. geol. Surv. Canada.* (In the press.)

Smith, A. B. 1988 Patterns of diversification and extinction in early Palaeozoic echinoderms. *Palaeontology* **31**, 799–828.

Stitt, J. H. 1975 Adaptive radiation, trilobite palaeoecology and extinction, Ptychaspid biomere, late Cambrian of Oklahoma. *Fossils Strata* **4**, 381–390.

Stitt, J. H. 1977 Late Cambrian and earliest Ordovician trilobites, Wichita Mountains area, Oklahoma. *Bull. Okla. geol. Surv.* **124**, 1–79.

Taylor, M. E. & Forester, R. M. 1979 Distributional model for marine isopod crustaceans and its bearing on early Palaeozoic, palaeo-zoogeography and continental drift. *Bull. geol. Soc. Am.* **90**, 405–413.

Thickpenny, A. & Leggett, J. K. 1987 Stratigraphic distribution and palaeo-oceanographic significance of European early Palaeozoic organic rich sediments. *Spec. Pub. geol. Soc. Lond.* **26**, 231–248.

Thomas, A. T. 1980 Trilobite associations in the British Wenlock. In *The Calenoides of the British Isles – reviewed* (ed. A. L. Harris, C. H. Holland & B. E. Leake), pp. 447–452. Edinburgh: Scottish Academic Press.

Vail, P. R., Mitchum, R. M., Todd, R. G., Widmier, J. M., Thomson, S., Sangree, J. B., Bubb, J. N. & Hatelid, W. G. 1977 Seismic stratigraphy and global changes of sea level. *Mem. Am. Ass. Petrol. Geol.* **26**, 49–212.

Westrop, S. R. & Ludvigsen, R. 1987 Biogeographic control of trilobite mass extinction at the Upper Cambrian biomere boundary. *Paleobiology* **13**, 84–99.

Wilde, P. & Berry, W. B. N. 1984 Destabilization of the oceanic density structure and its significance to marine extinction events. *Palaeogeog. Palaeoclum. Palaeoecol.* **48**, 143–162.

Wiley, E. O. 1981 *Phylogenetics. The theory and practice of phylogenetic systematics.* New York: John Wiley.

Williams, S. H. 1983 The Ordovician–Silurian boundary graptolite fauna of Dob's Linn, southern Scotland. *Palaeontology* **26**, 605–639.

Williams, S. H. 1988 Dob's Linn – the Ordovician–Silurian boundary stratotype. In *A global analysis of the Ordovician–Silurian boundary* (ed. L. R. M. Cocks) (*Bull. Br. Mus. nat. Hist.* (Geol.) **43**), 17–30.

Winston, D. E. & Nicholls, H. 1967 Late Cambrian and early Ordovician faunas from the Wilberns formation of central Texas. *J. Paleont.* **41**, 66–96.

Ziegler, A. M., Scotese, C. R., McKerrow, W. S., Johnson, M. E. & Bambach, R. K. 1979 Paleozoic palaeogeography. *A. Rev. Earth planet. Sci.* **7**, 473–502.

Phil. Trans. R. Soc. Lond. B **325**, 357–368 (1989)

Printed in Great Britain

The biology of mass extinction: a palaeontological view

By D. Jablonski

Department of the Geophysical Sciences, University of Chicago, 5734 South Ellis Avenue, Chicago, Illinois **60637,** *U.S.A.*

Extinctions are not biologically random: certain taxa or functional/ecological groups are more extinction-prone than others. Analysis of molluscan survivorship patterns for the end-Cretaceous mass extinctions suggests that some traits that tend to confer extinction resistance during times of normal ('background') levels of extinction are ineffectual during mass extinction. For genera, high species-richness and possession of widespread individual species imparted extinction-resistance during background times but not during the mass extinction, when overall distribution of the genus was an important factor. Reanalysis of Hoffman's (1986) data (*Neues Jb. Geol. Paläont. Abh.* **172,** 219) on European bivalves, and preliminary analysis of a new northern European data set, reveals a similar change in survivorship rules, as do data scattered among other taxa and extinction events. Thus taxa and adaptations can be lost not because they were poorly adapted by the standards of the background processes that constitute the bulk of geological time, but because they lacked – or were not linked to – the organismic, species-level or clade-level traits favoured under mass-extinction conditions. Mass extinctions can break the hegemony of species-rich, well-adapted clades and thereby permit radiation of taxa that had previously been minor faunal elements; no net increase in the adaptation of the biota need ensue. Although some large-scale evolutionary trends transcend mass extinctions, post-extinction evolutionary pathways are often channelled in directions not predictable from evolutionary patterns during background times.

Introduction

Extinction has long been recognized as an integral part of the Darwinian equation. Mass extinctions are the most spectacular manifestations of this process, but for many reasons they are particularly problematical as evolutionary factors. Our knowledge of victims and survivors is still painfully sketchy, in terms of taxonomic composition, biological attributes or phylogenetic histories. The details of the timing and magnitudes of mass extinction events are still uncertain, and sampling and preservation can obscure or artificially generate patterns around critical extinction intervals (see Raup, this symposium). Even the definition of mass extinction, and thus the number and distribution of events through the Phanerozoic, is controversial. Nevertheless, the study of extinction has advanced greatly over the past decade, and some generalizations and suggestions for further research are feasible. Here, I focus on the five major mass extinctions of the Phanerozoic fossil record, which are generally agreed upon as significant events in the history of life (see Jablonski (1986a), Flessa *et al.* (1986) and Sepkoski (1986) for overviews), and will say little about driving mechanisms. Instead, I discuss evidence suggesting that mass extinctions remove taxa generally not at risk during times of low ('background') extinction intensity. Taxa or traits that are successful and extinction resistant during most of geological time thus may be lost during mass extinctions, and evolution channelled in directions not predictable from patterns prevalent during background times.

PATTERNS OF EXTINCTION AND SURVIVAL

Neither background nor mass extinction is random. Palaeontologists have long argued, on empirical or theoretical grounds, that certain taxa are relatively extinction resistant (usually to background extinction) because of one set or another of functional, ecological or biogeographical attributes. For example, Alexander (1977) found that Palaeozoic brachiopod genera that cemented to the substratum were geologically longer lived than those attached by a pedicle; Levinton (1974) found that genera of deposit-feeding bivalves were geologically longer lived than suspension-feeders; Hansen (1980) and Jablonski (1986c) found that gastropod species with high larval dispersal capability were geologically longer lived than species with low dispersal. (See Ward & Signor (1983) for an excellent demonstration, in this case for ammonites, that observed differences in extinction rates need not be attributable to differences in morphological complexity, as has sometimes been suggested.) Studies such as these, although valuable, commonly combine background and mass extinctions without checking for differences between the two (exceptions are discussed below). In such instances it is tempting to regard observed patterns as typical of the background times that constitute the bulk of geological record, but explicit analyses are clearly needed.

Recently, some workers have compared mass and background extinctions and found differences not only in numbers of taxa lost, but in kinds of taxa lost. For late Cretaceous molluscs of the North American Coastal Plain, Jablonski (1986b) found that in background times, high species richness (defined as having three or more species in the last 6 Ma of the late Cretaceous of the area) and broad geographical range at the species level enhanced survivorship at the genus level. These observations are not surprising: genera whose species are many or widespread will be more likely to avoid total extinction in the face of most environmental perturbations than will genera with few species or whose species are each restricted to small areas. The relation between geographical range and duration at the species level has been repeatedly documented (see, for example, Jackson 1974; Hansen 1980; Jackson et al. 1985; Hoffman 1986; Jablonski 1986b, c, 1987), and this might be expected to translate into greater survival at the genus level as well (as shown by Jablonski (1986b, figure 2B): overall, genera composed mainly of widespread species have significantly greater median durations than genera composed mainly of restricted species).

Species-level relations between geographical range and geological duration might be an artefact of preservation and sampling (see, for example, Koch (1987); Russell & Lindberg (1988)), but for this particular assemblage of Cretaceous species, a comparison of patterns among taxa with different preservation and collection potential suggests that the relation is not artefactual (Jablonski 1988). If the apparent relation was simply an artefact, then the slope of the regression between range and duration should progressively decrease with increasing robustness and abundance of taxa; that is, a perfect fossil record would yield zero slope, and the worse the record, the steeper the slope. This null expectation was not met for Coastal Plain molluscs.

In contrast to background patterns, neither geographical range of constitutent species nor species richness played a significant role in the survival of genera during the end-Cretaceous mass extinction. The frequency distribution for geographical ranges of species within surviving genera was statistically indistinguishable from that of species within extinct genera: genera were not saved by the magnitudes of individual species ranges (see Jablonski 1986b, figure 2E, F).

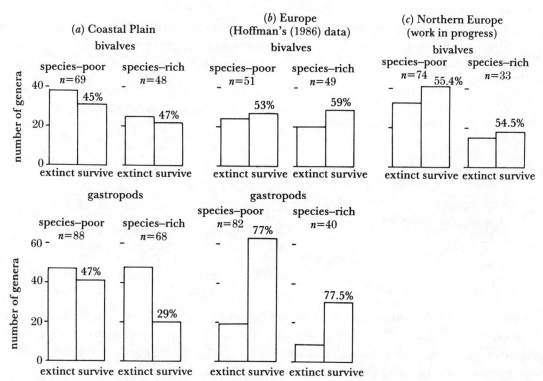

FIGURE 1. Species richness did not play a significant role in survival of genera during the end-Cretaceous mass
extinction, for bivalves and gastropods of the North American Coastal Plain, Europe (re-analysing Hoffman's
(1986) data), and Northern Europe (new analysis, still in progress). For example, 45 % of the 69 species-poor
Coastal Plain bivalve genera, and 47 % of 48 species-poor genera, survived. Note that the only statistically
significant difference between species-rich and species-poor groups, for the Coastal Plain gastropods, is in the
opposite sense to expectations during background extinction.

Further, for bivalves, about 47 % of the species-rich genera and 45 % of the species-poor genera
survived (viewed another way, 40 % of the victims and 42 % of the survivors were species rich)
(see figure 1a and table 1). Even more strikingly, significantly fewer species-rich gastropod
genera survived than did species-poor genera, reversing the survivorship pattern seen for the
preceding 16 Ma (binomial probability = 0.0018).

A synergistic interaction between species richness and geographical range of constituent
species occurs during background times. Thus, for North American molluscs, species-rich
genera composed mainly of widespread species have significantly lower extinction rates than
species-poor genera composed of geographically restricted species, and the other combinations
exhibit intermediate values (Jablonski 1986b). This interaction is also lost during the end-
Cretaceous mass extinction: 59 % of the ordinarily most durable group became extinct, as did
56 % of the ordinarily most vulnerable group (Jablonski 1986b).

Hoffman (1986) attempted a similar analysis for European molluscs, based on the
compilation of Nevesskaya & Solovyev (1981), and failed to find the change in survival
patterns. There are problems with these data: for example, the Soviet authors synonymized
many genera considered to be useful by other workers. For example, species commonly placed
within the scallop genera *Aequipecten*, *Camptonectes*, *Merklinia*, and *Mimachlamys* were all
relegated to *Chlamys* (*Camptonectes perlucidus* Sobetski does appear in the list, but this species is
repeated under *Chlamys*). Nevertheless, I reanalysed these Campanian–Maastrichtian data
with only two modifications: (1) because species richness per genus tends to be a highly skewed

TABLE 1. SPECIES-RICH GENERA WERE EQUALLY REPRESENTED AMONG VICTIMS
AND SURVIVORS OF THE END-CRETACEOUS EXTINCTION

(Note that the statistically significant result for Coastal Plain gastropods is in the opposite sense to expectations during background extinction.)

	species-rich genera		
	victims	survivors	significance (binomial)
Coastal Plain (Jablonski 1986*b*)			
bivalves	40%	42%	n.s. $(0.30 < p < 0.40)$
gastropods	50%	33%	$p < 0.005$
Europe (Hoffman 1986)			
Campanian & Maastrichtian species			
bivalves	45%	52%	n.s. $(0.30 < p < 0.40)$
gastropods	32%	33%	n.s. $(0.40 < p < 0.50)$
Maastrichtian species only			
bivalves	46%	44%	n.s. $(0.30 < p < 0.40)$
gastropods	29%	31%	n.s. $(0.30 < p < 0.40)$
northern Europe			
bivalves	31.25%	30.50%	n.s. $(0.40 < p < 0.50)$

distribution, I avoided comparisons of mean values, which can be deceptive; (2) I excluded 'wastebasket' *sensu lato* (s.l.) genera, broadly defined taxonomic categories with little phylogenetic significance (e.g. *Lucina* s.l., *Cardium* s.l., '*Turbo*' s.l., '*Natica*' s.l. (their quotes)). Following these conventions, I found no significant difference between survivorship of species-rich and species-poor marine genera (see figure 1*b* and table 1). About 53% of the species-poor bivalves and 59% of the species-rich bivalves survived (not significant, binomial probability = 0.30). Among gastropods, 76.8% species-poor genera and 77.5 species-rich genera survived (not significant (n.s.), binomial probability = 0.40). Finally, the frequency distributions of species richnesses of victims against survivors were not significantly different (Kolmogorov–Smirnov tests, $p > 0.50$). The same results obtain if only Maastrichtian species are used: for bivalves (96 genera) 59.3% of the species-poor genera and 59.5% of the species-rich genera survived, and for gastropods (85 genera) 79.7% of the species-poor genera and 80.8% of the species-rich genera. Hoffman's (1986) data for European molluscs, then, actually corroborate the North American results.

Preliminary analysis of my own compilation of northern European bivalve species also supports the North American results rather than Hoffman's conclusions. Of the 107 genera for which I have data so far (figure 1*c* and table 1), 55.4% of the species-poor genera survived, and 54.5% of the species-rich genera survived, again not a significant difference.

Despite the failure of formerly extinction-resistant features such as species richness, the end-Cretaceous extinction was not simply random. For North American molluscs, genera distributed outside the Coastal Plain study area had significantly higher probability of surviving than did genera endemic to that region. For bivalves, 55% of the widespread genera but only 9% of the endemic genera survived; for gastropods 50% of the widespread genera survived but only 11% of the endemic genera survived (see Jablonski (1986*b*) for details and statistical treatments). As noted above, geographical range at the species level played no role in survivorship, but geographical range at the clade level did: a selectivity manifested at a different hierarchical level from those characteristic of background times. Clade geographic range may indeed be a hedge against background extinction as well (this is difficult to demonstrate, because most studies average background and mass extinction situations), but this factor becomes tantamount during the mass extinctions.

The role of endemism cannot yet be assessed in the northern European data, simply because only 5 of 107 bivalve genera analysed so far were endemic to the area in the latest Cretaceous. The scarcity of endemics in north European molluscs (or conversely, the relatively large number of Coastal Plain endemics) may be an artefact of the relatively poor preservation typical of most north European localities relative to those of North America, or it could be a genuine reflection of the nature of the biogeographical barriers defining the respective provinces. Further investigations should clarify these matters, and a quantitative rather than binary biogeographical treatment should be useful.

The vulnerability of endemic genera during mass extinctions has been recorded for other groups and extinction events. Bretsky (1973) found that endemic bivalve genera diversified during background times but suffered disproportionately during the end-Ordovician, late Devonian, end-Permian and end-Triassic events. According to Fortey (1983) and Westrop & Ludvigsen (1987), trilobite extinction events that define the early Palaeozoic biomere boundaries also preferentially removed endemics.

Anstey's (1978, 1986) results for Palaeozoic bryozoans provide some intriguing parallels with end-Cretaceous molluscs. Morphologically simple bryozoan genera (interpreted as ecological generalists, and tending to be species-poor; see also Jablonski (1986b, p. 131)) exhibit fairly steady extinction rates through the Palaeozoic; in contrast, morphologically complex genera (inferred specialists that tend to be species rich) have lower background extinction rates but suffer greater losses during the mass extinctions. As with the end-Cretaceous, these Palaeozoic mass extinctions tended to remove endemics, and Anstey (1986, p. 49) concludes 'The differential loss of long-surviving morphologically complex bryozoans suggests that the terminal Ordovician extinction was both qualitatively and quantitatively different from preceding levels of background extinction'.

Latitudinal patterns in extinction represent another aspect of selectivity that may be evolutionarily important. The tropics are the most diverse regions of the globe, and so they might be expected to lose the greatest number of taxa during a mass extinction, even if tropical losses were no more severe on a percentage basis than those at high latitudes. The sparse data suggest, to some authors at least, that tropical marine organisms suffered disproportionately during mass extinctions (see, for example, Boucot (1983), Sheehan (1985), Jablonski (1986a), Stanley (1988a,b), all on benthic organisms; House (1985) on ammonoids), although Raup & Boyajian (1988) found extinction magnitudes for reef dwellers to be comparable to other invertebrates. Little is known on background origination or extinction rates for tropical marine organisms (see Jablonski & Bottjer (1989a) for review), but several of the authors cited above hold that per-taxon rates are low except in association with mass extinctions, whatever the role of disturbance, environmental mosaics, and other diversity-generating factors over ecological timescales. Tropical distribution, then, may yield extinction resistance during background times yet may be a liability during mass extinctions. Such results – which are in great need of rigorous study – have been taken to implicate global cooling as the cause of mass extinctions (see especially Stanley (1988a, b)). This may be true, but Jablonski (1986d) notes alternative explanations that should also be considered, ranging from potential fragility of the web of interactions among reef and other tropical organisms, so that perturbations similar in magnitude to those at high latitudes could bring more extensive extinctions, to the biogeographical structure of the tropical biota, enriched in endemics and hence more extinction prone in the face of any major perturbation, climatic or otherwise.

This is not to say that natural selection at the individual level was wholly ineffective or random during mass extinction, or that traits can never be of value during both background and mass extinctions. Kitchell *et al.* (1986) attributed the high survivorship of planktonic diatoms across the Cretaceous–Tertiary boundary to the presence a non-planktonic stage in the life cycle, which had already been selected during background times as a response to stressful surface-water conditions. (Unfortunately, they provide no comparative data on background extinction rates in these diatoms.) Sheehan & Hansen (1986) argue that deposit-feeding molluscs show greater end-Cretaceous survivorship than do suspension-feeding molluscs, an echo of Levinton's (1974) more general findings averaging background and mass extinction. Both molluscan and plankton results, however, may reflect instances of pre-adaptations to peculiar end-Cretaceous conditions (inferred disruption and collapse of planktonic food chains, etc.) rather than continuity of selective régimes from background to mass extinction. In any case, results from the end-Cretaceous and other extinction events do suggest a significant loss of effectiveness in at least some of the traits conferring resistance to background extinction. Rather than simply accelerating or emphasizing 'normal' extinction patterns, then, mass extinctions impose a different régime of selectivity that can have unpredictable and lasting effects on the course of evolution.

EVOLUTIONARY CONSEQUENCES

Because the traits that enhance survival during mass extinctions (e.g. broad geographical range at the clade level) need have little correlation with those that enhance survival and diversification during background times, mass extinctions are unlikely to promote the long-term adaptation of the biota. Such 'nonconstructive selectivity' (Raup 1984) can profoundly disrupt both short- and long-term evolutionary processes and remove taxa that are well adapted to background régimes. Taxa or morphologies could be lost not because they are maladaptive as measured under background conditions (which evidently constitute the bulk of evolutionary time), but because they happened to lack the appropriate biogeographic deployment or other traits necessary to weather the mass extinction. Mass extinction patterns will be complicated by the removal of taxa that were already dwindling or 'endangered' (cf. Kauffman 1984), but even expanding clades can be lost if they are endemics or if their centres of distribution lie within vulnerable regions (e.g. the tropical areas discussed above).

One vivid, if inexact, approach to the cost of mass extinctions in terms of lost adaptations involves iterative evolutionary patterns: certain traits are lost during mass extinctions that not only intuitively have adaptive value, but ultimately recur later in the history of the group to great success. Many features, of course, are lost once and for all during mass extinctions, but those that originate again to good effect in another branch of the same clade provide some sense of their evolutionary potential under background conditions. For example, one branch of the naticacean gastropods apparently evolved the ability to drill through the shells of their prey in the late Triassic, but this adaptation disappeared soon thereafter, despite the expansion of available resources that this innovation undoubtedly entailed at the organismic level (Fürsich & Jablonski 1984). More work is needed on late Triassic gastropod systematics, but Fürsich & Jablonski (1984) attribute the loss of the drilling ability to extinction of this innovative lineage in the end-Triassic event. Another naticacean line reoriginated the drilling ability some 120 Ma later, and the group has been expanding in diversity and abundance since that time. Similarly, the ability to bore into hard substrata was first achieved among the mytiloid bivalves late in the Ordovician (Pojeta & Palmer 1976; Wilson & Palmer 1988), and this habit surely opened up new living space and refuges from disturbance and predation. However, this lineage was lost in the end-Ordovician extinction, and bivalve borers do not appear again until the Triassic, some 200 Ma later. Examples probably exist in most higher taxa of invertebrates and vertebrates.

More commonly recognized are lasting effects of evolutionary bottlenecks imposed by mass extinctions. For example, it was the survival of the miocidarid echinoids during the Permo-Triassic extinction that fixed the standard pattern of plating in the echinoid test (Paul 1988); this must have imposed structural constraints on post-Palaeozoic echinoid evolution, but few would argue that it was the plating of the miocidarids that enabled them to survive the end-Permian crisis. Similarly, Aldridge (1988, p. 239) argues that conodont evolution was channelled by an extinction at the end of the early Silurian, which 'eliminated most, if not all, of the innovative ... stocks that might otherwise have gone on to enrich later faunas.' Although plant evolution at high taxonomic levels shows little response to the faunal mass extinctions (see, for example, Boulter *et al.* 1988), Wolfe (1987) argued that the end-Cretaceous event left a 'lasting imprint' on Northern Hemisphere vegetation in terms of distribution and relative abundances of deciduous habit and lower taxa.

Mass extinctions do not simply remove adaptations honed under background régimes, they create opportunities for faunal change by removing dominant taxa and enabling other groups to radiate in the aftermath of the event. As Benton (1987), among others, has pointed out, several radiations that had once been regarded as the triumph of adaptively superior taxa now appear to have been preceded and presumably triggered by the extinction of an earlier group. As Hallam (1987) argues, interactions among major groups seem to involve 'pre-emptive rather than displacive' competition. Removal of a major group sets the stage for diversifications of taxa – often minor components of the pre-extinction biota – that chanced to have traits enhancing survivorship during mass extinctions. Examples of extinction-mediated faunal replacements (reviewed by Benton (1987); see also Jablonski (1986 *b, d*)) include the replacement of therapsids ('mammal-like reptiles') by dinosaurs, dinosaurs by mammals, and archaic carnivorous mammals by the modern order Carnivora. Among marine invertebrates, Sheehan (1982) found that rapid evolution in the 3–5 Ma after the end-Ordovician extinction 'produced most of the important new brachiopod groups that dominated for the following 85 million years', an interval in turn terminated by the late Devonian extinction. The sweeping changes in reef community composition through the Phanerozoic (see Sheehan 1985; Fagerstrom 1987; Stanley 1988 *a, b*), producing a succession of what Copper (1988) terms 'six major evolutionary reef faunas', were also mediated by mass extinctions rather than by progressive competitive exclusion of a dominant group by a new taxon (the scleractinian coral – rudist bivalve interaction in the Cretaceous may be an exception, and deserves further attention; see Scott (1988)).

Some long-term faunal changes transcend individual mass extinction events. Examples include the decline of the trilobites or of Sepkoski's Palaeozoic fauna relative to the post-Palaeozoic fauna (protracted changes that Sepkoski (1984) argues would have unfolded even without mass extinctions), and the 'Mesozoic Revolution' of marine organisms involving escalation of shell-penetrating predators and their increasingly more armoured prey (see Vermeij 1987). However, even here mass extinctions may have played a role. For example, the rather steady pattern of bivalve diversification is broken by diversity drops followed by bursts of origination at the end-Permian and end-Cretaceous extinctions (Hallam & Miller 1988; Miller & Sepkoski 1988). If these bursts represent times of radiation in the absence of dampening interactions with other clades as suggested by Miller & Sepkoski (1988), rather than sampling artefacts around regressive intervals as implied by Bambach & Gilinsky (1988), then mass extinctions become significant at the level of clades and evolutionary novelties. Rebound intervals provide settings of unbridled radiation in which innovations can be captured and new adaptive zones occupied in relative freedom from the pre-emptive competition that typifies clade interactions during background times (see Jablonski & Bottjer (1989a) for a review of theories of clade-level interactions).

Another long-term evolutionary pattern that appears to transcend mass extinctions involves onshore–offshore shifts in marine invertebrate clades (see Bottjer & Jablonski (1988) and Jablonski & Bottjer (1989b) for reviews). Here too, however, extinctions can interrupt trends, and perhaps hasten or delay their progress by millions of years. For example, Frey (1987) and Miller (1988) found that the end-Ordovician extinction eliminated inroads made into offshore shelf environments by suspension-feeding bivalves that had begun diversifying onshore. The bivalves as a group eventually did spread across the shelf, but some of the pathfinding taxa reduced during the end-Ordovician event never recovered.

The rise to dominance of previously unimportant groups therefore does not require an adaptive breakthrough and competitive superiority. As outlined by Jablonski (1986c), a group might have suffered losses along with the rest of its contemporaries during a mass extinction, but undergone a more rapid pre-emptive diversification than the other survivors. On the other hand, new radiations might be seeded from those taxa relatively unscathed by the mass extinction, as appears to have been the case for the placental mammals after the demise of the dinosaurs and the near-extirpation of (North American) marsupials. The most persistent and diverse clades may be those in which major new adaptations and other traits favoured during background times happen to be associated with traits that favour survival during mass extinctions. Such accidental linkages could be extremely important during the rare but far-reaching mass extinction events.

Because mass extinctions are so wide-spaced in evolutionary time, clades are unlikely to remain static in the traits that allowed them to weather those events. Endemism is a liability during mass extinctions, for example, but cosmopolitan survivors will give often rise to new endemics in post-extinction times. At the same time, to the degree that they are heritable, individual-species-, or clade-level traits that confer resistance to background extinction will tend to increase during background times. During those times, loss or retention of traits helpful during mass extinctions will depend on their effects on background survivorship, and their chance linkages with other background-favoured traits. Thus during background times the number of species-rich clades could increase regardless of their geographical ranges relative to species-poor but widespread clades, and the vicissitudes of background extinction could remove those subgroups that represent a clade's best chances of surviving a mass extinction. Conversely, other groups could lose their most vulnerable members in one mass extinction and re-radiate from survivors that happen to retain their extinction resistance during subsequent background times, again altering extinction probabilities from extinction event to extinction event. This could explain, for example, severe extinction of the bivalve order Pholadomyoida during the end-Permian event but not during the end-Triassic event (Hallam & Miller 1988), or the severe end-Permian extinction and relatively mild end-Cretaceous extinction for the echinoderms (Paul 1988).

Threshold or continuum?

The change in rules of extinction and survival from background to mass extinction suggests that mass extinctions play a significant role in shaping the Earth's biota. As Jablonski (1986b, d) notes, however, the change in rules could represent two alternative selective régimes or endmembers of a continuum. Raup (1986) has shown that mass extinction magnitudes are not a separate class unto themselves, but the tail of a curve of extinction intensities when all Phanerozoic stages are surveyed. Threshold effects are of course still possible, but these and other results raise the possibility of a continuum of effects.

McKinney (1988) found echinoid extinction patterns to be similar to the Cretaceous molluscan patterns described above: species-rich clades were more resistant to background extinction, but enjoyed no advantage during mass extinctions. He found, however, a continuum of responses depending on the magnitude of the event. Similarly, Boyajian (1988) found that mass extinctions more severely affected geologically old clades than did background extinctions, and he suggests that clade age is a proxy for such hedges against background extinction as species richness. On examining the underlying age distribution of victims for all stratigraphic stages, he found the number of older clades lost during mass extinction events to be as expected statistically given the large overall extinction. Thus, in one sense, mass extinctions are simply background extinctions writ large. However, when viewed in terms of evolutionary consequences, or in terms of what clades experience over the course of their histories, the change in extinction rules is still a factor. As Raup (1986) shows, most time intervals exhibit low extinction rates, so that, for example, species-rich clades will be buffered against extinction for most of their histories; but those clades will suffer a relatively rapid erasure of that trait's effectiveness as a hedge against extinction when one of the rare, major perturbations arises. The size of the excursion from the small extinction probabilities that typify most of geological time in Raup's (1986) plot may determine how completely the background extinction rules are undermined, but the fact remains that successive time intervals can bring significant shifts in the nature of extinction-resistant traits.

Changes in the rules of extinction and survival, then, need involve no unique forcing mechanism. The failure of species richness and species' geographical ranges to protect clades from mass extinction is probably telling us more about the magnitude and pervasiveness of the perturbation than about which of the many hypothesized causes have come into play. Survival probabilities for the end-Cretaceous event were greatest for clades spread over many provinces, presumably because refugia were few and far between. For tens of millions of years previously, individual species widespread in a single province, or large numbers of species within a single province, were sufficient to enhance clade survivorship, because at any one time perturbations were comfortably below the provincial scale. This changed during the end-Cretaceous mass extinction (and for the other major extinctions, as far as is known), with lasting and unpredictable evolutionary consequences.

Conclusions

In terms of evolutionary effects, mass extinctions tend to remove not only more clades, but different clades from those lost during the times of background extinction. Clades cannot be viewed as static, however: a clade's ability to resist mass extinction can itself change through time. The relative proportion of taxa bearing traits that confer resistance to mass extinction almost certainly changes during the history of major groups. Furthermore, the traits that influence survivorship during mass extinctions vary at relatively low taxonomic levels: multi-province geographical range at the genus level, or extratropical distributions, to name two likely examples. Consequently, analyses of mass extinction selectivity among orders or classes will tend to yield mixed or ambiguous results. Even ammonites, one of the most notoriously volatile invertebrate groups in the fossil record, contain low-diversity, long-lived lineages. Not coincidentally, these low-diversity ammonite clades survive or are the last to disappear at mass extinction boundaries, and tend to be geographically widespread (see Ward & Signor 1983; House 1985).

Whenever traits that confer extinction resistance during mass extinctions (e.g. broad geographical range at the clade level) are poorly correlated with traits that enhance survival and diversification during background times, mass extinctions will disrupt evolutionary patterns and faunal dominance patterns established during background times. By removing or reducing dominant groups, mass extinctions provide opportunities for diversification of taxa that had been minor constituents of the pre-extinction biota. At the same time, mass extinctions can eliminate evolutionary innovations of high selective value if those traits are captured by clades that happen to lack extinction-resistant traits. For many reasons, then, evolution can be channelled in directions not predictable from situations established during background times. A broader range of hypotheses must be considered in analysing the histories of higher taxa and major adaptations, hypotheses that recognize the potential role of extinction processes that may not be congruent during mass extinctions and background times, or across levels in the biological hierarchy.

I thank N. J. Morris and R. J. Cleevely, British Museum (Natural History), for their generous advice and hospitality during my first foray into European Cretaceous bivalves. D. H. Erwin, A. Hoffman, S. M. Kidwell, D. M. Raup and S. Suter provided valuable reviews and discussions. This work was supported by NSF Grants EAR84-17011 and INT86-2045.

REFERENCES

Aldridge, R. J. 1988 Extinction and survival in the Conodonta. In *Extinction and survival in the fossil record* (ed. G. P. Larwood) (Systematics Association special volume no. 34), pp. 231–256. Oxford University Press.

Alexander, R. R. 1977 Generic longevity of articulate brachiopods in relation to the mode of stabilization on the substrate. *Palaeogeogr. Palaeoclimatol. Palaeoecol.* **21**, 209–226.

Anstey, R. L. 1978 Taxonomic survivorship and morphologic complexity in Paleozoic bryozoan genera. *Paleobiology* **4**, 407–418.

Anstey, R. L. 1986 Bryozoan provinces and patterns of generic evolution and extinction in the Late Ordovician of North America. *Lethaia* **19**, 33–51.

Bambach, R. K. & Gilinsky, N. L. 1988 Artifacts in the apparent timing of macroevolutionary 'events'. *Abstr. geol. Soc. Am.* **20**, A104.

Benton, M. J. 1987 Progress and competition in macroevolution. *Biol. Rev.* **62**, 305–338.

Bottjer, D. J. & Jablonski, D. 1988 Paleoenvironmental patterns in the evolution of post-Paleozoic benthic marine invertebrates. *Palaios* **3**, 540–560.

Boulter, M. C., Spicer, R. A. & Thomas, B. A. 1988 Patterns of plant extinction from some palaeobotanical evidence. In *Extinction and survival in the fossil record* (ed. G. P. Larwood) (Systematics Association special volume no. 34), pp. 1–36. Oxford University Press.

Boyajian, G. E. 1988 Mass vs. background extinction: no difference on the basis of taxon age distributions. *Abstr. geol. Soc. Am.* **20**, A105.

Boucot, A. J. 1983 Does evolution occur in an ecological vacuum? II. *J. Paleontol.* **57**, 1–30.

Bretsky, P. W. 1973 Evolutionary patterns in the Paleozoic Bivalvia: documentation and some theoretical considerations. *Bull. geol. Soc. Am.* **84**, 2079–2096.

Copper, P. 1988 Ecological succession in Phanerozoic reef ecosystems: is it real? *Palaios* **3**, 136–151.

Fagerstrom, J. A. 1987 *The evolution of reef communities.* (628 pages.) New York: Wiley.

Flessa, K. W. *et al.* 1986 Causes and consequences of extinction. In *Patterns and processes in the history of life* (ed. D. M. Raup & D. Jablonski), pp. 235–257. Berlin: Springer-Verlag.

Fortey, R. A. 1983 Cambrian–Ordovician trilobites from the boundary beds in western Newfoundland and their phylogenetic significance. *Spec. Pap. Palaeontol.* **30**, 179–211.

Frey, R. C. 1987 The occurrence of pelecypods in early Paleozoic epeiric sea environments, Late Ordovician of the Cincinnati, Ohio area. *Palaios* **2**, 3–24.

Fürsich, F. T. & Jablonski, D. 1984 Late Triassic naticid drillholes: carnivorous gastropods gain a major adaptation but fail to radiate. *Science, Wash.* **224**, 78–80.

Hallam, A. 1987 Radiations and extinctions in relation to environmental change in the marine Lower Jurassic of northwest Europe. *Paleobiology* **13**, 152–168.

Hallam, A. & Miller, A. I. 1988 Extinction and survival in the Bivalvia. In *Extinction and survival in the fossil record* (ed. G. P. Larwood) (Systematics Association special volume no 34), pp. 121–138. Oxford University Press.

Hansen, T. A. 1980 Influence of larval dispersal and geographic distribution on species longevity in neogastropods. *Paleobiology* **6**, 193–207.

Hoffman, A. 1986 Neutral model of Phanerozoic diversification: implications for macroevolution. *Neues Jb. Geol. Paläont. Abh.* **172**, 219–244.

House, M. R. 1985 Correlation of mid-Palaeozoic ammonoid evolutionary events with global sedimentary perturbations. *Nature, Lond.* **313**, 17–22.

Jablonski, D. 1986*a* Causes and consequences of mass extinctions: a comparative approach. In *Dynamics of extinction* (ed. D. K. Elliott), pp. 183–229. New York: Wiley.

Jablonski, D. 1986*b* Background and mass extinctions: the alternation of macroevolutionary regimes. *Science, Wash.* **231**, 129–133.

Jablonski, D. 1986*c* Larval ecology and macroevolution in marine invertebrates. *Bull. mar. Sci.* **39**, 565–587.

Jablonski, D. 1986*d* Evolutionary consequences of mass extinctions. In *Patterns and processes in the history of life* (ed. D. M. Raup & D. Jablonski), pp. 313–329. Berlin: Springer-Verlag.

Jablonski, D. 1987 Heritability at the species level: analysis of geographic ranges of Cretaceous mollusks. *Science, Wash.* **238**, 360–363.

Jablonski, D. 1988 Estimates of species duration [response to Russell & Lindberg]. *Science, Wash.* **240**, 969.

Jablonski, D. & Bottjer, D. J. 1989*a* The ecology of evolutionary innovation: the fossil record. In *Evolutionary innovations* (ed. M. H. Nitecki). University of Chicago Press.

Jablonski, D. & Bottjer, D. J. 1989*b* Onshore-offshore trends in marine invertebrate evolution. In: *Biotic and abiotic factors in evolution* (ed. R. M. Ross & W. D. Allmon). University of Chicago Press.

Jackson, J. B. C. 1974 Biogeographic consequences of eurytopy and stenotopy among marine bivalves and their evolutionary significance. *Am. Nat.* **108**, 541–560.

Jackson, J. B. C., Winston, J. E. & Coates, A. G. 1985 Niche breadth, geographic range, and extinction of Caribbean reef-associated cheilostome Bryozoa and Scleractinia. In *Proceedings of the 5th International Coral Reef Congress*, vol. 4, pp. 151–158. Mourea: Antenne Museum.

Kauffman, E. G. 1984 The fabric of Cretaceous marine extinctions. In *Catastrophes in Earth history* (ed. W. A. Berggren & J. A. Van Couvering), pp. 151–246. Princeton University Press.

Kitchell, J. A., Clark, D. L. & Gombos, A. M. Jr 1986 Biological selectivity of extinction: a link between background and mass extinction. *Palaios* **1**, 504–511.

Koch, C. F. 1987 Prediction of sample size effects on the measured temporal and geographic distribution patterns of species. *Paleobiology* **13**, 100–107.

Levinton, J. S. 1974 Trophic group and evolution in bivalve molluscs. *Palaeontology* **17**, 579–585.

McKinney, M. L. 1988 Extinction selectivity: a key to macroevolutionary processes. *Abstr. geol. Soc. Am.* **20**, A105.

Miller, A. I. 1988 Spatio-temporal transitions in Paleozoic Bivalvia: an analysis of North American fossil assemblages. *Hist. Biol.* **1**, 251–273.

Miller, A. I. & Sepkoski, J. J. Jr 1988 Modeling bivalve diversification: the effect of interaction on a macroevolutionary system. *Paleobiology* **14**, 364–369.

Nevesskaya, L. A. & Solovyev, A. N. (eds) 1981 *Development and change of the mollusks in the Mesozoic-Cenozoic boundary.* (141 pages.) Moscow: Nauka.

Paul, C. R. C. 1988 Extinction and survival in the echinoderms. In *Extinction and survival in the fossil record* (ed. G. P. Larwood) (Systematics Association special volume no 34), pp. 155–170. Oxford University Press.

Pojeta, J. Jr & Palmer, T. J. 1976 The origin of rock boring in mytilacean pelecypods. *Alcheringa* **1**, 167–179.

Raup, D. M. 1984 Evolutionary radiations and extinctions. In *Patterns of change in Earth evolution* (ed. H. D. Holland & A. F. Trendall), pp. 5–14. Berlin: Springer-Verlag.

Raup, D. M. 1986 Biological extinction in Earth history. *Science, Wash.* **231**, 1528–1533.

Raup, D. M. & Boyajian, G. E. 1988 Patterns of generic extinction in the fossil record. *Paleobiology* **14**, 109–125.

Russell, M. P. & Lindberg, D. R. 1988 Estimates of species duration. *Science, Wash.* **240**, 969.

Scott, R. W. 1988 Evolution of Late Jurassic and Early Cretaceous reef biotas. *Palaios* **3**, 184–193.

Sepkoski, J. J. Jr 1984 A kinetic model of Phanerozoic taxonomic diversity. III. Post-Paleozoic families and mass extinctions. *Paleobiology* **10**, 246–267.

Sepkoski, J. J. Jr 1986 Phanerozoic overview of mass extinction. In *Patterns and processes in the history of life* (ed. D. M. Raup & D. Jablonski), pp. 277–295. Berlin: Springer-Verlag.

Sheehan, P. M. 1982 Brachiopod macroevolution at the Ordovician–Silurian boundary. In *Procceedings of the 3rd North American Paleontological Convention*, vol. 2, pp. 477–481.

Sheehan, P. M. 1985 Reefs are not so different – they follow the evolutionary pattern of level-bottom communities. *Geology* **13**, 46–49.

Sheehan, P. M. & Hansen, T. A. 1986 Detritus feeding as a buffer to extinction at the end of the Cretaceous. *Geology* **14**, 868–870.

Stanley, S. M. 1988*a* Paleozoic mass extinctions: shared patterns suggest global cooling as a common cause. *Am. J. Sci.* **288**, 334–352.

Stanley, S. M. 1988*b* Climatic cooling and mass extinction of Paleozoic reef communities. *Palaios* **3**, 228–232.

Vermeij, G. J. 1987 *Evolution and escalation.* (526 pages.) Princeton University Press.

Ward, P. W. & Signor, P. W. III 1983 Evolutionary tempo in Jurassic and Cretaceous ammonoids. *Paleobiology* **9**, 183–198.

Westrop, S. R. & Ludvigsen, R. 1987 Biogeographic control of trilobite mass extinction at an Upper Cambrian 'biomere' boundary. *Paleobiology* **13**, 84–99.

Wilson, M. A. & Palmer, T. J. 1988 Nomenclature of a bivalve boring from the Upper Ordovician of the midwestern United States. *J. Paleontol.* **62**, 306–308.

Wolfe, J. A. 1987 Late Cretaceous–Cenozoic history of deciduousness and the terminal Cretaceous event. *Paleobiology* **13**, 215–226.

Note added in proof – 29 *April* 1989

Becker (1986) argues for differences in background and mass extinction patterns among ammonoids at the late Devonian extinction. In their study of Palaeozoic seastars, Blakes & Guensburg (1988) discuss adaptations related to a predatory mode of life that were evidently lost to the group at the end of the Ordovician, and re-evolved in the Mesozoic some 250 Ma later. Seilacher (1988) concludes his morphogenetic comparison of nautiloids and ammonites: 'There was certainly nothing wrong about ammonite shell construction. Still only the less perfect design of the *Nautilus* shell did survive the catastrophe. Can there be a better example to illustrate that other criteria matter in mass extinctions than in normal selection?'

References

Becker, R. T. 1986 Ammonoid evolution before, during and after the 'Kellwasser-Event' – review and preliminary new results. In *Global bio-events* (ed. O. H. Walliser), pp. 181–188. Berlin: Springer-Verlag.

Blake, D. B. & Guensburg, T. E. 1988 The water vascular system and functional morphology of Paleozoic asteroids. *Lethaia* **21**, 189–206.

Seilacher, A. 1988 Why are nautiloid and ammonite sutures so different? *Neues Jb. Geol. Paläont. Abh.* **177**, 41–69.

Phil. Trans. R. Soc. Lond. B **325**, 369–386 (1989)

Printed in Great Britain

Mass extinctions among tetrapods and the quality of the fossil record

By M. J. Benton†

Department of Geology, The Queen's University of Belfast, Belfast BT7 1NN, U.K.

The fossil record of tetrapods is very patchy because of the problems of preservation, in terrestrial sediments in particular, and because vertebrates are rarely very abundant. However, the fossil record of tetrapods has the advantages that it is easier to establish a phylogenetic taxonomy than for many invertebrate groups, and there is the potential for more detailed ecological analyses.

The relative incompleteness of a fossil record may be assessed readily, and this can be used to test whether drops in overall diversity are related to mass extinctions or to gaps in our knowledge. Absolute incompleteness cannot be assessed directly, but a historical approach may offer clues to future improvements in our knowledge. One of the key problems facing palaeobiologists is paraphyly, the fact that many higher taxa in common use do not contain all of the descendants of the common ancestor. This may be overcome by cladistic analysis and the identification of monophyletic groups.

The diversity of tetrapods increased from the Devonian to the Permian, remained roughly constant during the Mesozoic, and then began to increase in the late Cretaceous, and continued to do so during the Tertiary. The rapid radiation of 'modern' tetrapod groups – frogs, salamanders, lizards, snakes, turtles, crocodilians, birds and mammals – was hardly affected by the celebrated end-Cretaceous extinction event.

Major mass extinctions among tetrapods took place in the early Permian, late Permian, early Triassic, late Triassic, late Cretaceous, early Oligocene and late Miocene. Many of these events appear to coincide with the major mass extinctions among marine invertebrates, but the tetrapod record is largely equivocal with regard to the theory of periodicity of mass extinctions.

1. Introduction

Most studies on mass extinctions so far have focused on the fossil record of marine invertebrates (see, for example, Raup & Sepkoski 1982, 1984, 1986; Jablonski 1986; McKinney 1986; Raup & Boyajian 1988). A smaller number of studies have used the fossil record of vascular plants (Niklas *et al.* 1983) and non-marine vertebrates (Benton 1985*a,b*), and there has been an implicit assumption that these records are poorer. Indeed, the fossil record of marine invertebrates generally has the advantages of abundant specimens, good stratigraphic control, closely spaced samples, uniform preservation quality, broad geographic distributions, and a mature taxonomy.

The aims of this paper are to consider the problems and the advantages of the fossil record of the tetrapods, to outline what has been learnt from it about mass extinctions, and to investigate the problems of incompleteness of the fossil record and paraphyly.

† Present address: Department of Geology, University of Bristol, Queen's Road, Bristol BS8 1RJ, U.K.

2. The nature of the fossil record of tetrapods

(a) The evolution of tetrapods

The first tetrapods, according to most recent classifications (see, for example, Carroll 1987; Panchen & Smithson 1988) are the Ichthyostegalia (Ichthyostegidae, Acanthostegidae), known first from the Famennian Stage (*ca.* 365 Ma BP) of the late Devonian. There are earlier records of tetrapods from the Devonian, based on footprints, but skeletal remains only are considered here. The tetrapods radiated during the Carboniferous into nine or more major lineages of temnospondyl amphibians, which were generally bulky carnivores and piscivores that lived near water, and 'lepospondyl' amphibians, which were often smaller and more varied in their modes of life. Another lineage, the anthracosaurs, led to the amniotes, the clade of tetrapods that moved away from a dependence on the water.

The first amniotes, and thus the first reptiles (see, for example, Carroll 1982; Heaton & Reisz 1986) are the Protorothyrididae (= Romeriidae), known first from the Moscovian Stage (*ca.* 300 Ma BP) of the late Carboniferous. During the remaining 15 Ma or so of the Carboniferous, the early amniotes diversified into a number of additional lineages: the Araeoscelidia and the 'Pelycosauria', the most primitive groups of the Diapsida and Synapsida respectively. These two amniote clades dominated tetrapod evolution from the late Carboniferous to the present day. The Diapsida radiated during the Permian and Triassic, and gave rise to sphenodontians, the ancestors of lizards and snakes, crocodilians and dinosaurs in the Triassic, and birds in the Jurassic. The Synapsida radiated in the Permian and Triassic as the mammal-like reptiles, and gave rise to the mammals towards the end of the Triassic. The mammals remained at low diversity during most of the Mesozoic but began to radiate strongly in the late Cretaceous, well before the end-Cretaceous extinction event.

The temnospondyl amphibians continued, in reduced numbers, through the Permian, Triassic and early to middle Jurassic. The modern amphibians (Lissamphibia) apparently arose in the Permian or early Triassic, and gradually increased in diversity during the Mesozoic, but they never became abundant.

The diversity and importance of the major tetrapod groups are indicated in the phylogenetic tree in figure 1. This is based on recent cladistic analyses, although these leave many doubtful relationships, and familial diversity is indicated.

(b) Size of the data set

In all, there are 858 families of living and extinct non-marine tetrapods, and 58 families of exclusively marine tetrapods (Benton 1987, 1988), giving a total of 916 families. Of these, 352 families are still living (including 157 families of birds and 139 families of mammals) (Table 1).

The total of 916 living and extinct tetrapod families was culled in order to strengthen the data set. Eighty-one of the 916 families have no fossil record (mainly birds and mammals), and they were omitted, reducing the total to 835. A further culling was made to exclude the small number of families that have been based on a single species or genus found in one geological formation ('singletons', a total of 78 families). Indeed, some families have been based on single specimens, and they are best omitted until further finds are made. In effect, a singleton family has zero distribution in time; it arises and disappears in a geological instant, and cannot be

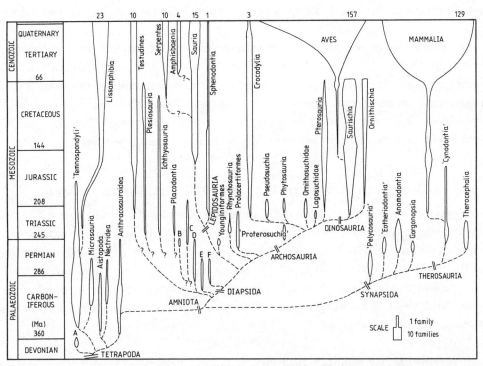

FIGURE 1. Phylogenetic tree of the Tetrapoda, showing relationships, stratigraphic duration and diversity of each group. The major groups are indicated as balloons that show the known stratigraphic range by their height, and the relative numbers of families present by their width (see scale in bottom right-hand corner). Relationships of the groups are indicated by dashed lines on the basis of recent cladistic analyses (see, for example, Benton 1985 c; Gauthier 1986; Heaton & Reisz 1986; Gauthier *et al.* 1988; Kemp 1988; Panchen & Smithson 1988). Abbreviations: A, Ichthyostegalia; B, Pareiasauria; C, Procolophonia; D, Captorhinidae; E, Protorothyrididae; F, Araeoscelidia.

TABLE 1. NUMBERS OF FAMILIES OF TETRAPODS

(Based on data in Benton (1987, 1988).)

	non-marine	exclusively marine	total
Amphibia	102	0	102
Reptilia	240	33	273
Aves	202	0	202
Mammalia	314	25	339
total	858	58	916

sensibly included in calculations of origination rates or extinction rates. The final analysed total of tetrapod families was 754.

(c) Incompleteness

The incompleteness of the fossil record of tetrapods has been described by many authors (see, for example, Pitrat 1973; Bakker 1977; Carroll 1977; Olson 1982; Padian & Clemens 1985; Benton 1985 a, b, 1987). The record of the non-marine tetrapods, which make up the vast bulk of all tetrapods, is particularly poor. Some stratigraphic stages, for example the Aalenian (middle Jurassic), have yielded no identifiable tetrapod fossils at all anywhere in the world, and other stages (e.g. Gzelian (Carboniferous); Toarcian, Bajocian, Callovian, Oxfordian

(Jurassic); Berriasian–Aptian, Cenomanian–Santonian (Cretaceous)) have yielded very few remains.

The incompleteness of the fossil record of terrestrial tetrapods has been characterised in another way by Padian & Clemens (1985, p. 82). Most dinosaur genera are known from only a single stratigraphic stage, which would suggest, in a literal reading of the fossil record, that the dinosaurs experienced total generic mass extinction 24 or 25 times during their history. However, at the family level there is only the one final Cretaceous–Tertiary (K–T) mass extinction event, as dinosaur families generally span more than one stage.

3. ASSESSING THE INCOMPLETENESS OF THE FOSSIL RECORD

Palaeontologists must seek to reduce the incompleteness of the fossil record if its value to macroevolutionary studies is to be enhanced. It is easy to say simply 'collect more specimens', but that gives no idea of how complete our knowledge has become. Do palaeontologists know 80% of the main fossil groups, 50%, 20%, or less than 1%? There are two aspects of the incompleteness of the fossil record, relative incompleteness and absolute incompleteness, and ways must be devised of assessing these. Only then can palaeobiologists find out how well founded their hypotheses are.

(a) Relative incompleteness

It is often said that, for example, birds and bats have very poor fossil records because they live in the air and in trees, where there is relatively little rock deposition. They can only be preserved out of context when a chance accident leads to the burial of a cadaver under the waters of a lake or in the sea. On the other hand, aquatic organisms, and those that live near to water, are much more likely to be preserved as fossils. Can these assumptions of relative preservability be tested quantitatively?

It is possible to estimate the relative completeness of the tetrapod record in a broad way by examining the numbers of families present per stage. The Simple Completeness Metric (SCM) (Paul 1982; Benton 1987, 1988) compares the numbers of families that are known to be present with the numbers that ought to be present. The SCM is based on the fact that tetrapod families span several stratigraphic stages. The family may be represented by fossils throughout its entire duration, or there may be gaps spanning one or more stratigraphic stages where fossils are absent. Jablonski (1986) has termed this the Lazarus Effect, where a taxon apparently disappears, and then reappears higher up in the sequence. The more incomplete the fossil record is for a particular stage, the more Lazarus (hidden) taxa there will be. The SCM ranges from 0% (no fossils at all, e.g. Aalenian) to 100% (all families represented by fossils, e.g. Visean, Ufimian, Scythian). Most other stages have SCM values between 50 and 100%, but values fall below 50% in the early–middle Jurassic (Toarcian–Bajocian), the late Jurassic (Oxfordian), and the late Cretaceous (Turonian–Santonian).

The SCM may also be calculated taxonomically, and it turns out that, among tetrapods, the birds have a 56.9% complete record at the level of the stratigraphic stage, and bats have a surprisingly high value of 75.7% (Benton 1987). The lowest values are for lissamphibians (frogs and salamanders) at 42.0%, and for lepidosaurs (lizards and snakes) at 48.6%. The highest values are for mammal-like reptiles (94.5%), placental mammals (87.0%), and for particular groups of placentals, such as the carnivores (97.1%), artiodactyls (96.6%), perissodactyls (96.2%), and rodents (88.4%).

These figures for relative completeness are calculated at rather coarse levels (families and stratigraphic stages), but they could be estimated at lower levels to make more specific comparisons. They give broad indications of where the gappiness of the fossil record is worst, and where collecting efforts should be concentrated, both stratigraphically and taxonomically. They can also be used to distinguish between episodes of low diversity that may be the result of mass extinction and those that may be the result of collection failure (see below).

(b) Absolute incompleteness

The assessment of the absolute incompleteness of the fossil record is much more difficult and it is probably ultimately impossible. However, one interesting approach to this problem stems from the following idea. Some day in the future, palaeontologists will have a perfect knowledge of the fossil record. In other words, our knowledge can be improved in various ways, and we should seek to identify these and to assess their relative importance.

The elements of our present incomplete knowledge may be categorized, and classified into those that may be improved, and those that lie beyond our control.

(a) Some taxa were never fossilized, and can never be known (e.g. they were soft-bodied, lived in the wrong environment, or were geographically very restricted in distribution). No improvement possible.

(b) Some taxa are unknown because of a lack of study. New localities, new collectors, and new study techniques are required. Can be improved.

(c) Some taxa have been incorrectly identified, assigned to the wrong species, to the wrong genera, or to the wrong families. Restudy of the original material, and careful consideration of the characters may resolve these problems. Can be improved.

(d) Some taxa have been incorrectly dated. Re-examination of the original specimens or localities and revisions of stratigraphic schemes may cause changes. Can be improved.

The last three categories are subject to improvement, but it would be useful to determine which is the most crucial source of error in particular cases. Patterson & Smith's (1987) study showed that 38% of Sepkoski's family records of fish and echinoderms were non-monophyletic, and 15% were incorrectly assigned stratigraphically. Thus, according to their assessment, the major problem was taxonomic (category (c)). However, they did not assess the significance of new collecting (category (b)).

The study of the fossil record of tetrapods by Maxwell & Benton (1987) used a historical analysis to attempt some predictions. The idea was that, if the nature and causes of changes in our knowledge over the past 100 years could be determined, they might provide pointers to potential improvements in the future. Over all, the total number of families had increased this century, so that new finds (category (b)) clearly played a part in improving our knowledge. However, comparisons on a shorter timescale, between Romer's (1966) standard compilation of data, and Benton's (1987) effort, showed that this was less important; both lists gave similar total numbers of families. However, in detail, there has been a great deal of change in 50–70% of family records. For particular extinction events, the changes may be categorized as 20–30% owing to new finds, 20–60% owing to taxonomic revision and 20–30% owing to new evidence on dating. The taxonomic element has been crucial.

These figures are only general indications, but they have shown two important facts. Firstly, new fossil finds are not the sole answer to improving our knowledge, and secondly, changes in our knowledge of the fossil record have not occurred in a regular or systematic way. Detailed analyses of this sort, based on several snapshots of the state of the known fossil record at times in the past decades, may give information on where improvements have occurred and, by extrapolation, where they are likely to occur in the future. The newest idea here is that taxonomic revision may be more important than has been assumed.

4. The problems of paraphyly

A major problem that is emerging in many macroevolutionary studies is paraphyly. It has been suggested (see, for example, Cracraft 1981; Patterson & Smith 1987; Benton 1988) that a large number of the evolutionary groups used in analyses of evolutionary rates are artificial, either in that they include species that evolved from several different ancestors (polyphyletic groups), or that they do not include all of the descendants of a common ancestor (paraphyletic groups). Polyphyletic groups have long been abhored by evolutionists, because they are obviously artificial assemblages of superficially similar organisms. However, paraphyly may be a more significant problem. Patterson & Smith (1987) suggested that 20% of the fish and echinoderm families used in the Raup & Sepkoski (1982, 1984, 1986) studies of diversification and extinction were paraphyletic, the largest category in their categorization of incorrect data. They implied that this, or higher figures, probably pertain for the rest of the Sepkoski data set. Similar values were found in a comparison of 'standard' data sets of tetrapod families, where 20–30% were paraphyletic in comparison with modern cladistically analysed classifications (Maxwell & Benton 1987).

It is important to establish clades, or monophyletic groups, for use in macroevolutionary studies (Cracraft 1981; Benton 1988). Such studies generally focus on supraspecific categories, often families or orders. There is no objective way, of course, to determine the rank of a clade in the taxonomic hierarchy, e.g. whether a particular group is a family or an order. However, such groups should, as far as can be determined, include all of the descendants of a single common ancestor, that is be monophyletic (*sensu* Hennig (1966); that is, holophyletic groups, *sensu* Ashlock (1971)).

A paraphyletic group, such as class Reptilia, is descended from one ancestor, but excludes some of the descendants (here, birds and mammals). The starting point of the clade is a real part of the phylogenetic tree, but the terminations of 'Reptilia' along the lines to mammals and to birds is artificial. 'Reptilia', then, is at least partly a human invention. Graphs of the evolutionary rate of reptiles might show, for example, that they enjoyed rapid rates of origination during Permian and Triassic times, but that these rates dropped off in the Jurassic and Cretaceous. This does not necessarily mean that reptiles were evolving in a sluggish manner, but simply that the new hairy reptiles and feathered reptiles have been arbitrarily excluded from the calculations.

The key to identifying monophyletic groups among Tetrapoda is cladistic analysis, in which patterns of relationship are established on the basis of shared derived characters (synapomorphies). Most tetrapod groups have now been tackled by one or more cladists, and attempts are also being made to analyse the links between these major groups. These latter efforts have generated most controversy (e.g. relationships of sarcopterygian fish and tetrapods, birds and reptiles, early mammals) and this has tended to obscure the fact that a great deal of agreement has become evident in smaller-scale cladograms of particular orders or subclasses. In addition, cladistic analyses of tetrapods have generally not affected the composition of family-level taxa. Even before cladistic methods were widely used, vertebrate systematists defined families on the basis of clear-cut derived characters. It has been in linking the families into orders, then the orders into classes, that character definitions have lost their sharpness, leading to the establishment of artificial taxa on the basis of primitive (plesiomorphous) characters, e.g. Labyrinthodontia, Cotylosauria, Eosuchia, Thecodontia, Prototheria.

Most studies of tetrapod macroevolution have been based on families, and the new classifications have therefore not had as profound an effect as might have been expected. The main changes have arisen in drawing the lower boundaries of families; cladists would tend to exclude 'potential ancestors' from a family unless they display at least one synapomorphy of that family. This has pulled the dates of origin of some families forwards in time. The plesiomorphous taxa are then assigned plesion ranks, possibly equivalent to families. This could potentially give rise to a vast proliferation of new singleton families based on single ill-defined ancestral species. By convention, however, such families are excluded from calculations until a second occurrence is discovered (see above). For example, the family Archaeopterygidae arose and disappeared instantaneously, being represented only by the species *Archaeopteryx lithographica* from rocks of a single age, the Solnhofen Limestone Formation of southern Germany, albeit by several specimens.

The strong rejection of paraphyletic groups by Patterson & Smith (1987) was criticized by Sepkoski (1987). He argued that families are used in palaeobiological studies as convenient proxies for species, which is the level at which macroevolution truly occurs. Use of the family category avoids many of the problems of precise stratigraphic dating and correlation, local effects and gaps in the fossil record that would beset global analyses of species change through time. The Sepkoski (1982) data set of families seems to track species-level patterns of macroevolution even though it includes many paraphyletic families, whereas a culled cladistic data set does not seem to do so. Sepkoski's argument is that the cladistic data set of monophyletic families fails to detect certain mass-extinction events that are well established at the species level, whereas these are clearly indicated in the lists of non-cladistic families. His final argument is that monophyletic families may mask the effects of species-level extinctions, because their temporal shape is biased. He implies that monophyletic groups may actually be inferior to equivalent paraphyletic groups whose upper, arbitrarily established, boundary is determined by some ecological distinction, for example, that corresponds to the loss of many species. For example, one unusual or marginal lineage may prolong a family well beyond a time when all the other included species died out. This notion is testable. It may turn out to have some force, but for vertebrates at least, a rapid survey of non-cladistic and cladistic classifications suggests that the opposite is the case: monophyletic families are generally clearcut packages of species, and their evolutionary fates are closely linked. Paraphyletic families, on the other hand, have few advantages, among vertebrates at least.

5. ADVANTAGES OF USING THE FOSSIL RECORD OF TETRAPODS IN MACROEVOLUTIONARY STUDIES

The fossil record of tetrapods is not as hopeless for studies of mass extinction as has often been assumed. It has a number of advantages over the record of marine invertebrates, and these are noted briefly below.

(a) High probability of identifying clades

Rates of evolution, extinction and origination should be analysed, as noted above, on the basis of monophyletic groups. Tetrapods have proved highly amenable to cladistic analysis, in contrast to fossil invertebrate groups (with the exception of echinoderms and arthropods). This suggests that it may be hard to determine monophyletic families for the bulk of the record of fossil invertebrates. The significance of this problem has been noted above.

(b) Scope for ecological analysis

Many detailed studies of the functional morphology and palaeoecology of single species (autecology) of fossil tetrapods have been carried out, and these often allow detailed reconstructions of their modes of life. Studies have also been made of whole faunas (synecology). This work offers potentially great contributions to detailed palaeobiological interpretation of aspects of extinction events. It may be possible, for example, to compare 'extinction-prone' and 'extinction-resistant' taxa for a broad range of potential ecological correlates: size, diet, position in food chains, locomotory adaptations, reproductive mode, growth rate, habitat preference, geographic distribution and so on. Tetrapods may lend themselves more readily to detailed ecological analysis than many marine invertebrates. This work is facilitated by our knowledge of the ecology of modern terrestrial vertebrates, which is greater than that of modern marine invertebrates.

(c) Maturity of tetrapod systematics

Because *Homo sapiens* is a tetrapod, zoologists have devoted more attention to the species- and generic-level systematics of mammals, birds, reptiles and amphibians than they have to the systematics of brachiopods, annelids, pogonophorans or hyolithids. Our understanding of the relationships and the bounds of living tetrapod species is probably more mature than that of any other group of organisms. This should allow more confident extrapolation of such concepts into the past, and thus better identification of fossil genera and species, better censuses of these taxa and better phylogenetic reconstructions, thereby improving the usefulness of such data for macroevolutionary research.

6. TETRAPOD FAMILY DIVERSITY ANALYSIS

(a) The data

Several authors have recently plotted graphs of the diversity of tetrapod families and orders through time (see, for example, Charig 1973; Pitrat 1973; Bakker 1977; Thomson 1977; Olson 1982; Padian & Clemens 1985; Colbert 1986). However, these graphs have been based largely on data from Romer (1966) and Harland *et al.* (1967), the classic source works. More recent studies (Benton 1985a, b, 1988) have been based on a new compilation of data on families of tetrapods (Benton 1987, 1988). These new compilations differ significantly from those derived from Romer (1966) and Harland *et al.* (1967) in several ways.

(a) New records up to the end of 1986 are included. This has affected the date of origination or extinction of as many as 50% of families.

(b) The latest cladistic classifications have been incorporated, as far as possible, and attempts have been made to test that all families are clades. This has caused significant rearrangements of families of late Palaeozoic and Mesozoic reptiles in particular, by amalgamations and redistributions of genera into monophyletic taxa.

(c) The stratigraphic resolution of family distributions has been improved. As far as possible, the dates of origination and extinction of each family have been determined to the nearest stratigraphic stage, usually by examination of the primary literature. The stage is the smallest practicable division of geological time for this compilation (relevant stage lengths vary from 2 to 19 Ma in length, with a mean duration of 6 Ma). This allows more detailed analysis than simply relying on the Lower, Middle and Upper divisions of geological periods in Romer (1966), Carroll (1987) and elsewhere.

(b) Diversification of tetrapods

The diversity of tetrapods has increased through time, with a particularly rapid acceleration in the rate of increase from the late Cretaceous (Campanian) onwards (figure 2) (Benton 1985a, b). Three major diversity assemblages have been identified (Benton 1985b), which appeared to dominate for a time, and then gave way to another: I (labyrinthodont amphibians, 'anapsids', mammal-like reptiles) dominated from late Devonian to early Triassic times at diversity levels of typically 20–40 families; II (early diapsids, dinosaurs, pterosaurs)

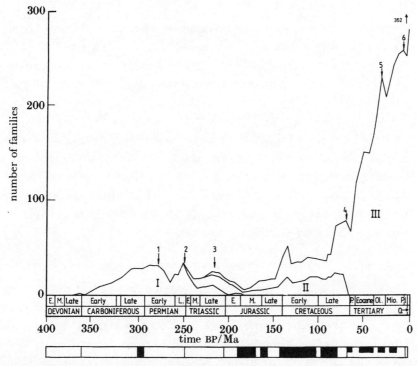

FIGURE 2. Standing diversity with time for families of tetrapods. The upper curve shows total diversity with time, and six apparent mass extinctions are indicated by drops in diversity, numbered 1–6. The relative magnitude of each drop is given in terms of the percentage of families that disappeared. (1) Early Permian (Sakmarian–Artinskian), 58%; (2) late Permian–early Triassic (Tatarian–Scythian), 49%; (3) late Triassic (Carnian–Rhaetian), 22%; (4) late Cretaceous (Maastrichtian), 14%; (5) early Oligocene (Rupelian), 8%; (6) late Miocene (Tortonian–Messinian), 2%. The timescale is that of Palmer (1983). Three assemblages of families succeeded each other through geological time: I, II and III (see text for details). The scm for each stratigraphic stage is indicated in code: values less than 50% (i.e. a poor fossil record) are shown shaded black, values between 50% and 75% are indicated as half black, half blank, and values over 75% are shown fully blank.

dominated during the Mesozoic at diversity levels of 20–50 families; and III (the 'modern' groups: frogs, salamanders, lizards, snakes, turtles, crocodiles, birds, mammals) have dominated from late Cretaceous times to the present day, rising rapidly from overall diversities of 50 to 89 in the Maastrichtian, and then successive peaks of 158 in the early Eocene, 234 in the Late Oligocene, and 279 in the late Miocene.

7. Mass extinctions

(a) Methods

Extinction and origination rates were calculated stage by stage for non-marine tetrapod families based on the new data set. Total extinction (R_e) and total origination (R_s) rates were calculated as the number of families that disappeared or appeared, respectively, during a stratigraphic stage, divided by the estimated duration of that stage (Δt), as $R_e = E/\Delta t$ and $R_s = S/\Delta t$, where E is the number of extinctions and S is the number of originations. Per taxon extinction (r_e) and origination (r_s) rates were calculated by dividing the total rates by the end-of-stage family diversity D (Sepkoski 1978), as $r_e = (1/D)(E/\Delta t)$ and $r_s = (1/D)(S/\Delta t)$. The per taxon rates can be seen as the 'probability of origin' or the 'risk of extinction'. In these calculations, the recent summary geological timescale of Palmer (1983) was used for stage lengths in Ma.

Mass-extinction events are times when large numbers of taxa of diverse taxonomic and ecological position appear to die out in a geological instant (Jablonski 1986). No clear numerical definition of mass extinction has been possible yet, but indications are provided by (a) major drops in overall diversity, and (b) times of unusually high extinction rates.

(b) Diversity drops

There appear to have been six declines in diversity (figure 2, 1–6) that are attributable to mass-extinction events. The other drops on the graph (early Jurassic, end-Jurassic, mid-Cretaceous) probably indicate mainly a change in the quality of the fossil record (Benton 1985a, b), and mass extinctions cannot be assumed here. These three episodes correspond to times when the scm described above gives particularly low values (figure 2). Mass extinctions may lurk within the gaps, but they cannot be assumed.

(c) Extinction and origination rates

The graphs of total rates (figure 3) for tetrapod families show great fluctuations in both origination and extinction rates. There is no clear correlation of high extinction rates with all mass extinction events. Of the highest rates, those in the Artinskian, Tatarian, 'Rhaetian', Maastrichtian, Rupelian, and late Miocene correspond to mass extinctions 2, 3, 4, 5 and 6 (figure 2) respectively. Equally high, or higher, total extinction rates in the Ufimian (late Permian), Tithonian (late Jurassic), Coniacian (late Cretaceous), Thanetian (late Palaeocene), Ypresian (early Eocene), Bartonian–Priabonian (middle–late Eocene), Pliocene and Pleistocene do not match any of the drops in amniote diversity that have been ascribed to mass extinctions.

The total origination rates (figure 3) generally track the total extinction rates quite closely. Both the total extinction and origination rates were found to be dependent on two non-random sources of error, which are noted here.

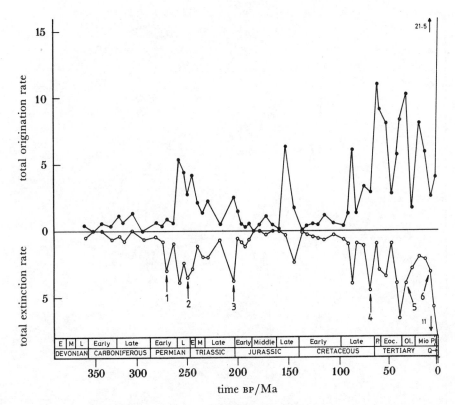

FIGURE 3. Total rates of origination and extinction for families of amniotes, calculated stage by stage for 56 stages between the late Devonian and the Pleistocene. The Miocene was divided into early, middle and late units only, and the Pliocene was treated as a single time unit.

(i) *Variation in the total numbers of taxa available to give rise to new taxa or to go extinct.*

Early parts of both records show very low diversity (1–10 families), whereas the Tertiary portions are two orders of magnitude higher; this must bias the rate values.

(ii) *Lagerstätten effects*

The total origination rates generally track the total extinction rates quite closely; peaks in both rates might have been produced in part by episodes when the fossil record is better than usual, corresponding to particular Fossil-Lagerstätten, such as the Sakamena Group (late Permian), the Solnhofen Limestone (Tithonian), and the Monte Bolca fish beds (Eocene). The improvement in the record boosts the apparent number of family originations and extinctions (Hoffman & Ghiold 1985).

The per taxon rates remove this bias in part. Thus when extinction and origination rates are recalculated relative to the numbers of taxa available (figure 4), the rates do not track each other so closely, although 'Lagerstätten peaks' remain in the Ufimian, Tithonian and Coniacian. There are particularly high per taxon extinction rates at times of mass extinctions corresponding to the Artinskian, Tatarian, and 'Rhaetian' events (1, 2 and 3, figure 2). Per taxon extinction rates are barely elevated at the times of the Maastrichtian, Rupelian, or Late Miocene mass extinctions (4, 5 and 6, figure 2). These mass extinctions correspond to depressed per taxon origination rates (figure 4), as noted by Benton (1985b).

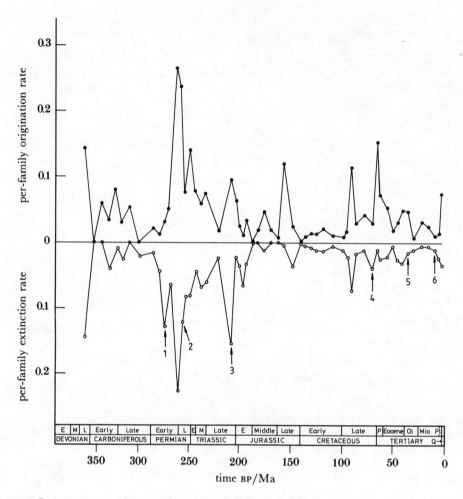

FIGURE 4. Per taxon rates of origination and extinction for families of amniotes. Conventions as in figure 3.

(d) Mass-extinction events

The history of tetrapods has apparently been punctuated by at least six mass-extinction events (figure 2) (Benton 1985 *b*), together with up to seven other possible extinction events. These had widely differing effects, ranging from a 58 % drop in family diversity for the early Permian event to a 2 % drop for the late Miocene event. It has already been argued (Benton 1988) that the fossil record of tetrapods is generally not complete enough to test the hypothesis of periodicity of mass extinctions (Raup & Sepkoski 1984, 1986), but the data from the Triassic record appear to contradict the idea (Benton 1986 *a*, 1988). The possible tetrapod extinctions are described below.

(i) Early Carboniferous (Serpukhovian)

Four families of amphibians died out at the boundary between the early and the late Carboniferous.

'Labyrinthodontia': Proterogyrinidae, Eoherpetontidae;

'Lepospondyli': Adelogyrinidae, Acherontiscidae.

This may correspond to the end-Namurian marine event noted by Sepkoski & Raup (1986,

p. 23), by Saunders & Ramsbottom (1986), and by others. The small number of tetrapod families lost (and each of them is of low diversity) gives lower than normal extinction rates (figures 3, 4). This cannot be regarded as a well-supported tetrapod extinction event.

(ii) *Late Carboniferous (Kasimovian–Gzelian)*

Two families of tetrapods died out.

'Lepospondyli': Ophiderpetontidae, Tuditanidae.

The suggested end-Carboniferous extinction event (Sepkoski & Raup 1986, p. 23) is even less convincing for tetrapods. Only two families of rather rare 'Lepospondyli' disappeared, and the extinction rates were low during both stages (figures 3 and 4).

(iii) *Early Permian (Artinskian)*

Fifteen families died out during this event (1, figure 2):

'Labyrinthodontia': Saurerpetontidae, Trematopsidae, Archeriidae;

'Lepospondyli': Urocordylidae, Hapsidopareiontidae, Ostodolepidae, Lysorophidae;

'Anapsida': Protorothyrididae, Bolosauridae, Mesosauridae;

Diapsida: Araeoscelididae;

Synapsida: Eothyrididae, Edaphosauridae, Ophiacodontidae, Sphenacodontidae.

Six families of 'labyrinthodont' amphibians are known to have survived into the succeeding Kungurian Stage (Trimerorachidae, Eryopidae, Dissorophidae, Archegosauridae, Seymouriidae, Diadectidae), two of 'Lepospondyli' (Keraterpetontidae, Gymnarthridae), and only three of reptiles (Captorhinidae, Caseidae, Varanopidae). This extinction then had its greatest effect on the reptiles, and it marked the sharpest decline in the formerly dominant pelycosaurs (early synapsids). Artinskian extinction rates are high, but not excessively so.

(iv) *Late Permian (Tatarian)*

Twenty-seven families of tetrapods died out at the end of the Permian (2, figure 2).

'Labyrinthodontia': Dvinosauridae, Melanosauridae, Rhinesuchidae, Kotlassidae, Lanthanosuchidae, Chroniosuchidae;

'Anapsida': Captorhinidae, Millerettidae, Pareiasauridae;

Diapsida: Weigeltisauridae, Younginidae, Tangasauridae;

Synapsida: Ictidorhinidae, Gorgonopsidae, Dromasauridae, Endothiodontidae, Cryptodontidae, Aulacocephalodontidae, Dicynodontidae, Pristerodontidae, Cistecephalidae, Diictodontidae, Moschorhinidae, Whaitsiidae, Silphestidae, Procynosuchidae, Dviniidae.

The end-Permian event had a decisive effect on the amphibians, wiping out six families and leaving only three survivors that crossed the Permo-Triassic boundary (Uranocentrodontidae, Benthosuchidae and Brachyopidae). It also caused the end of many major reptilian families, particularly a large number of formerly dominant mammal-like reptiles. The six or seven families that survived into the Triassic rapidly radiated into new forms, but the synapsids had begun to lose their dominance to the diapsids.

The extinction rates for the Tatarian are fairly high (figures 3 and 4), but not as high as those in the Ufimian, the first stage in the late Permian, when there was no drop in tetrapod family diversity.

(v) *Early Triassic (Scythian)*

There was another smaller extinction event about 5 Ma later, at the end of the Scythian Stage, when thirteen tetrapod families died out.

'Labyrinthodontia': Lydekkerinidae, Uranocentrodontidae, Benthosuchidae, Rhytidosteidae, Trematosauridae, Indobrachyopidae;

Diapsida: Proterosuchidae, Euparkeriidae;

Synapsida: Emydopidae, Kingoriidae, Ictidosuchidae, Scaloposauridae, Galesauridae.

The event really affected only the amphibians, as reptilian diversity remained roughly constant as a result of a high origination rate at the time. There was also a small mass extinction of marine invertebrates then (Raup & Sepkoski 1984, 1986).

(vi) *Late Triassic (Carnian–'Rhaetian')*

The three stages of the late Triassic, the Carnian, Norian and 'Rhaetian' (or two, if the 'Rhaetian' is included in the Norian) span 18–25 Ma, depending upon which of the current timescales is employed. Raup & Sepkoski (1984, 1986) have argued that the Late Triassic extinction consisted of a single event, but Benton (1986a, b) has identified at least two phases of extinction in the fossil record of tetrapods (3, figure 2), as well as in that of ammonoids and other groups.

The first, and larger, extinction event occurred at the end of the Carnian Stage. Ten families of tetrapods died out.

Diapsida: Thalattosauridae, Trilophosauridae, Rhynchosauridae, Proterochampsidae;

Synapsida: Kannemeyeriidae, Chiniquodontidae;

'Euryapsida': Nothosauridae, Simosauridae, Cymatosauridae, Henodontidae.

The second, smaller, late Triassic extinction event, at the Triassic–Jurassic boundary 'Rhaetian') was marked by the loss of eight families.

'Labyrinthodontia': Capitosauridae, Plagiosauridae;

'Anapsida': Procolophonidae;

Diapsida: Phytosauridae, Aetosauridae, Rauisuchidae, Ornithosuchidae;

'Euryapsida': Placochelyidae.

These extinctions, although few in number, do seem to have some significance. The last of the thecodontians (four families) disappeared on land, as did the last of the placodonts in the sea. Most of the 'modern' groups of amniotes had appeared during the preceding 12–17 Ma of the Norian Stage: the Testudines, the Crocodylia, and the Mammalia, as well as the Pterosauria, and the Dinosauria in the Carnian. At one time it was thought that the end of the Triassic saw the termination of the formerly abundant 'labyrinthodont' amphibians and the mammal-like reptiles, but these two groups continued in greatly reduced diversity until the Bathonian and Callovian (both middle Jurassic) respectively.

Both the Carnian and the 'Rhaetian' events are associated with peaks in total and per taxon extinction rates of tetrapod families (figures 3 and 4), but the peaks are higher for the latter event.

(vii) *Jurassic–Cretaceous events*

Raup & Sepkoski (1984, 1986) and Sepkoski & Raup (1986) have identified three probable extinction events that affected marine animals between the 'Rhaetian' and the K–T events.

These additional extinction events, with dates of the ends of the stages from Palmer (1983), are

Jurassic: Pliensbachian (193 Ma);

Tithonian (144 Ma);

Cretaceous: Cenomanian (91 Ma).

The data on tetrapod families are particularly weak during parts of this time interval (see above). There are indeed declines in family diversity in the tetrapod data (figure 2) after the Pliensbachian, Tithonian, and Cenomanian, with the decline after the Tithonian standing out best. There are also slight peaks in total (figure 3) and per taxon (figure 4) extinction rates in the Pliensbachian and Cenomanian, with a more marked peak in the Tithonian. At present, the tetrapod data are not good enough to decide either way about the occurrence of these postulated extinction events.

(viii) *Late Cretaceous (Maastrichtian)*

The Cretaceous–Tertiary boundary (K–T) event is surely the best-known mass extinction, and not least for its effects on the reptiles (dinosaurs, pterosaurs and plesiosaurs all died out then). However, in relative terms at least, the percentage loss of families of tetrapods as a whole (4, figure 2) was less than for the two Permian events and the late Triassic events already described. The total extinction rate for the Maastrichtian (figure 3) is higher than any before it, but the per taxon rate for tetrapods (figure 4) is not so impressive, being lower than the 'Rhaetian', Pliensbachian and Coniacian rates, for example. The decline in tetrapod family diversity at the K–T boundary is caused by a slightly elevated extinction rate, and partly by a low origination rate (Benton 1985 b).

Thirty-six families of tetrapods died out at the K–T boundary.

Diapsida: Crocodylia: Uruguaysuchidae, Notosuchidae, Goniopholididae;

Pterosauria: Pteranodontidae, Azhdarchidae;

Dinosauria: Coeluridae, Ornithomimidae, Dromaeosauridae, Saurornithoididae, Oviraptoridae, Elmisauridae, Megalosauridae, Dryptosauridae, Tyrannosauridae, Camarasauridae, Diplodocidae, Titanosauridae, Hypsilophodontidae, Hadrosauridae, Pachycephalosauridae, Nodosauridae, Ankylosauridae, Protoceratopsidae, Ceratopsidae

Sauria: Mosasauridae;

Aves: Baptornithidae, 'Enantiornithes', Lonchodytidae, Torotigidae;

Mammalia (Marsupialia): Pediomyidae, Stagodontidae;

'Euryapsida': Plesiosauria: Elasmosauridae, Cryptocleididae, Polycotylidae.

The K–T event was clearly taxonomically selective: certain major groups became completely extinct during Maastrichtian times: the Pterosauria (two families), the Dinosauria (19 families) and the Plesiosauria (three families). Other groups were less affected: turtles, crocodilians, lizards, snakes, birds and mammals (although two out of three marsupial families died out). Indeed, the mammals continued to radiate without any obvious pause right through the K–T boundary.

(ix) *Late Eocene (Priabonian)*

The major late Eocene extinction event identified by Raup & Sepkoski (1984, 1986), and others, among marine animals is not reflected in the global tetrapod data. Of 183 tetrapod families known at the time, 20 (mainly mammals) died out, corresponding to an elevated extinction rate. There was no diversity drop, however, merely a levelling off (figure 2), as origination rates were also high. This may be a localized, rather than a global, extinction event, corresponding to the 'grande coupure' of French palaeontologists, when numerous species of tetrapods disappeared in Europe.

(x) *Early Oligocene (Rupelian)*

This relatively minor event (5, figure 2) corresponds to a loss of 28 (out of 234) families, mainly of mammals. It has been noted also by Prothero (1985) for North American land mammals, but does not correspond to one of the periodic marine events, even though it was more severe for tetrapods than the late Eocene event. Extinction rates are lower than for the late Eocene event (figure 3), but they are matched by low origination rates, and lead to an overall decline in diversity.

(xi) *Late Miocene (Tortonian–Messinian)*

This event (6, figure 2) also affected the tetrapods, with the loss of 21 families, mainly among mammals, such as certain primates, artiodactyls, notoungulates and cetaceans. It does not match the periodic marine events, where a major extinction occurred earlier, in the Middle Miocene (Raup & Sepkoski 1984, 1986).

(e) Periodicity?

In general, the tetrapod fossil record is not adequate to test Raup & Sepkoski's (1984, 1986) theory of extinction periodicity. Most of the extinctions postulated above (?1, ?2, 4, 6, 7, 8, 9) match marine mass extinctions identified by those authors. However, some (1, 2, 4, 6, 9) do not match very well, and others (3, 5, 10, 11) do not fit the 26-million-year cycles at all. Further, many of the 26-million-year extinctions seem to be absent from the tetrapod data (that is, early Jurassic (Pliensbachian), middle Jurassic (Callovian?), early Cretaceous (Barremian–Aptian?), middle Miocene (Langhian–Serravallian)). Note, however, that Sepkoski & Raup (1986) found only limited evidence in the marine data for the middle Jurassic and early Cretaceous events, which are necessary to fill gaps in the 26-million-year periodicity pattern. Overall, the tetrapod data are suggestive, but by no means conclusive, evidence against periodicity.

I thank the Leverhulme Trust for a Research Fellowship which has funded part of this work. Mrs Libby Mulqueeny drafted the diagrams.

REFERENCES

Ashlock, P. D. 1971 Monophyly and associated terms. *Syst. Zool.* **20**, 63–69.

Bakker, R. T. 1977 Tetrapod mass extinctions – a model of the regulation of speciation rates and immigration by cycles of topographic diversity. In *Patterns of evolution as illustrated by the fossil record* (ed. A. Hallam), pp. 439–468. Amsterdam: Elsevier.

Benton, M. J. 1985 a Mass extinction among non-marine tetrapods. *Nature, Lond.* **316**, 811–814.

Benton, M. J. 1985 b Patterns in the diversification of Mesozoic non-marine tetrapods, and problems in historical diversity analysis. *Spec. Pap. Palaeontol.* **33**, 185–202.

Benton, M. J. 1985 c Classification and phylogeny of the diapsid reptiles. *Zool. J. Linn. Soc.* **84**, 97–164.

Benton, M. J. 1986 a More than one event in the late Triassic mass extinction. *Nature, Lond.* **321**, 857–861.

Benton, M. J. 1986 b The Late Triassic tetrapod extinction events. In *The beginning of the age of dinosaurs* (ed. K. Padian), pp. 303–320. Cambridge University Press.

Benton, M. J. 1987 Mass extinctions among families of non-marine tetrapods: the data. *Mém. Soc. Géol. Fr.* **150**, 21–32.

Benton, M. J. 1988 Mass extinctions and the fossil record of reptiles: paraphyly, patchiness and periodicity. In *Extinction and survival in the fossil record*, Systematics Association special volume no. 34 (ed. G. Larwood), pp. 269–294. London: Academic Press.

Carroll, R. L. 1977 Patterns of amphibian evolution: an extended example of the incompleteness of the fossil record. In *Patterns of evolution as illustrated by the fossil record* (ed. A. Hallam), pp. 405–437. Amsterdam: Elsevier.

Carroll, R. L. 1982 Early evolution of reptiles. *A. Rev. Ecol. Syst.* **13**, 87–109.

Carroll, R. L. 1987 *Vertebrate paleontology and evolution*. San Francisco: W. H. Freeman.

Charig, A. J. 1973 Kurten's theory of ordinal variety and the number of continents. In *Implications of continental drift to the earth sciences* (ed. D. H. Tarling & S. K. Runcorn), vol. 1, pp. 231–245. London: Academic Press.

Colbert, E. H. 1986 Mesozoic tetrapod extinctions: a review. In *Dynamics of extinction* (ed. D. K. Elliott), pp. 49–62. New York: Wiley.

Cracraft, J. 1981 Pattern and process in paleobiology: the role of cladistic analysis in systematic paleontology. *Paleobiology* **7**, 456–468.

Gauthier, J. A. 1986 Saurischian monophyly and the origin of birds. *Mem. Calif. Acad. Sci.* **8**, 1–55.

Gauthier, J. A., Kluge, A. G. & Rowe, T. 1988 The early evolution of the Amniota. In *The phylogeny and classification of the tetrapods, vol. 1 (Amphibians, reptiles, birds) (Systematics Association special volume no. 35A)* (ed. M. J. Benton), pp. 103–155. London: Academic Press.

Harland, J. B. *et al.* 1967 *The fossil record*. London: Geological Society.

Heaton, M. J. & Reisz, R. R. 1986 Phylogenetic relationships of captorhinomorph reptiles. *Can. J. Earth Sci.* **23**, 402–418.

Hennig, W. 1966 *Phylogenetic systematics*. Urbana: University of Illinois Press.

Jablonski, D. 1986 Causes and consequences of mass extinctions: a comparative approach. In *Dynamics of extinction* (ed. D. K. Elliott), pp. 183–229. New York: John Wiley.

Hoffman, A. & Ghiold, J. 1985 Randomness in the pattern of 'mass extinctions' and 'waves of origination', *Geol. Mag.* **122**, 1–4.

Kemp, T. S. 1988 Interrelationships of the Synapsida. In *The phylogeny and classification of the tetrapods, vol. 2 (Mammals) (Systematics Association special volume no. 35B)* (ed. M. J. Benton), pp. 1–21. London: Academic Press.

Maxwell, W. D. & Benton, M. J. 1987 Mass extinctions and data bases: changes in the interpretation of tetrapod mass extinction in the past 20 years. In *Fourth Symposium on Mesozoic Ecosystems, Abstracts* (ed. P. M. Curre & E. M. Koster), pp. 156–160. Drumheller: Tyrrell Museum.

McKinney, M. L. 1986 Taxonomic selectivity and continuous variation in mass and background extinctions of marine taxa. *Nature, Lond.* **325**, 143–145.

Niklas, K. J., Tiffney, B. H. & Knoll, A. H. 1983 Patterns in vascular land plant diversification. *Nature, Lond.* **303**, 614–616.

Olson, E. C. 1982 Extinctions of Permian and Triassic nonmarine vertebrates. *Spec. pap. geol. Soc. Am.* **190**, 501–511.

Padian, K. & Clemens, W. A. 1985 Terrestrial vertebrate diversity: episodes and insights. In *Phanerozoic diversity patterns* (ed. J. W. Valentine), pp. 41–96. Princeton University Press.

Palmer, A. R. 1983 The decade of North American geology 1983 time scale. *Geology* **11**, 503–504.

Panchen, A. L. & Smithson, T. R. 1988 The relationships of the earliest tetrapods. In *The phylogeny and classification of the tetrapods, vol. 1 (Amphibians, reptiles, birds) (Systematics Association special volume no. 35A)* (ed. M. J. Benton), pp. 1–32. London: Academic Press.

Patterson, C. & Smith, A. B. 1987 Is the periodicity of extinctions a taxonomic artefact? *Nature, Lond.* **330**, 248–251.

Paul, C. R. C. 1982 The adequacy of the fossil record. In *Problems of phylogenetic reconstruction* (ed. K. A. Joysey & A. E. Friday), pp. 75–117. London: Academic Press.

Pitrat, C. W. 1973 Vertebrates and the Permo-Triassic extinctions. *Palaeogeogr. Palaeoclim. Palaeoecol.* **14**, 249–264.

Prothero, D. R. 1985 Mid-Oligocene extinction events in North American land mammals. *Science, Wash.* **229**, 550–551.

Raup, D. M. & Boyajian, G. E. 1988 Patterns of generic extinction in the fossil record. *Paleobiology* **14**, 109–125.

Raup, D. M. & Sepkoski, J. J. Jr 1982 Mass extinctions in the marine fossil record. *Science, Wash.* **215**, 1501–1503.

Raup, D. M. & Sepkoski, J. J. Jr 1984 Periodicity of extinctions in the geologic past. *Proc. natn. Acad. Sci., U.S.A.* **81**, 801–805.

Raup, D. M. & Sepkoski, J. J. Jr 1986 Periodic extinctions of families and genera. *Science, Wash.* **231**, 833–836.

Romer, A. S. 1966 *Vertebrate paleontology*, 3rd. edition. University of Chicago Press.

Saunders, W. B. & Ramsbottom, W. H. C. 1986 The mid-Carboniferous eustatic event. *Geology* **14**, 208–212.

Sepkoski, J. J. Jr 1978 A kinetic model of Phanerozoic taxonomic diversity. I. Analysis of marine orders. *Paleobiology* **4**, 223–251.

Sepkoski, J. J. Jr 1982 A compendium of fossil marine families. *Contr. Biol. Geol., Milwaukee Public Mus.* **51**, 1–125.

Sepkoski, J. J. Jr. 1987 [Reply to Patterson & Smith (1987).] *Nature, Lond.* **330**, 251–252.

Sepkoski, J. J. Jr & Raup, D. M. 1986 Periodicity in marine extinction events. In *Dynamics of extinction* (ed. D. K. Elliott), pp. 3–36. New York: Wiley.

Thomson, K. S. 1977 The pattern of diversification among fishes. In *Patterns of evolution as illustrated by the fossil record* (ed. A. Hallam), pp. 377–404. Amsterdam: Elsevier.

Discussion

P. W. KING (*Biology Department, University College London, U.K.*). Is the large increase in number of families of tetrapods since the mid-Cretaceous, contrasting with reasonable stability of that number before then, caused by a genuine increase in the diversity of terrestrial vertebrates, or could it be because of a different amount of variability between species being tolerated within the bounds of one family in different larger taxonomic groups, so that those groups that predominated in pre-Cretaceous times had more diversity in their families than those groups that have flourished since?

Could Dr Benton also briefly describe the pattern of variation of family diversity with time in the teleost fish, as I believe this is in marked contrast to that in terrestrial vertebrates.

M. J. BENTON. Tetrapod families that existed before the late Cretaceous tend to have fewer genera and species than those from late Cretaceous times onwards. Hence there is no evidence that the older families are 'lumped' in comparison with the later ones. These are the facts as they stand, but clearly there are human and preservational factors that could produce this effect. It is hard to see how the magnitude of such factors could be assessed to test the idea that modern tetrapod families might be less diverse and variable than those before late Cretaceous times.

Fishes show rather different patterns of diversification and extinction from the tetrapods. Teleosts began to radiate noticeably in the late Jurassic, and again in the late Cretaceous when overall fish diversity leapt from levels below 40 families to 61–85 families. A second major rise took place in the early and middle Miocene, when overall fish diversity rose to 87–186 families, mainly teleosts. Pleistocene diversity levels (232 families) fall far short of modern fish diversity (459 families), which suggests that many teleost families have poor fossil records. This is not true of tetrapods. Fishes show some of the 'traditional' mass extinctions: late Jurassic (Tithonian), loss of six families, mainly chondrichthyans and teleosts; mid Cretaceous (Cenomanian), loss of ten families of teleosts and 'holosteans'; late Cretaceous (Maastrichtian), loss of 11 families, mainly teleosts. However, the end-Cretaceous decline is small (12.9% drop) and the fishes show no sign of anything approaching a mass extinction in the Tertiary. Further details on fish diversification and mass extinction may be found in Benton (1989).

Reference

Benton, M. J. 1989 Patterns of extinction and evolution in vertebrates. In *Palaeontology and evolution* (ed. D. E. G. Briggs & K. C. Allen), pp. 218–241. London: Belhaven Press.

Phil. Trans. R. Soc. Lond. B **325**, 387–400 (1989)

Printed in Great Britain

The Cretaceous–Tertiary boundary and the last of the dinosaurs

By A. J. Charig

*Department of Palaeontology, British Museum (Natural History), Cromwell Road,
London, SW7 5BD, U.K.*

Disaster theories of the K–T extinctions, more specifically dinosaur extinctions, are presently engendering much controversy. They require (*inter alia*) that those extinctions were sudden and simultaneous worldwide and that they coincided with an allegedly causal disaster at the K–T boundary. This paper reviews the evidence for and against those temporal requirements. The other major requirement is of a biological nature, namely an indication of the manner in which the specified disaster might have extinguished the organisms concerned; yet this causal mechanism, whatever it might have been, apparently had no effect whatever upon other, very similar organisms.

In any particular geographical region, the precise stratigraphic level at which dinosaurs became extinct can be determined only if there is a virtually unbroken succession of potentially dinosaur-bearing continental beds that pass up from the level of the highest dinosaur known to a level well above the K–T boundary. Unfortunately there are surprisingly few regions where such conditions prevail. The problem is further complicated by the difficulties of worldwide stratigraphic correlation and by the fact that specialists in different fields define the position of the K–T boundary on different criteria. Although some alleged discoveries of Palaeocene dinosaurs have long been discredited (the beds were not Palaeocene, or the bones were not dinosaurian), there does seem to be some evidence that dinosaurs died out at different times in different places, sometimes surviving whatever it was that produced the iridium anomaly and sometimes co-existing with Palaeocene palynomorphs and Tertiary-type mammals, or both. In such cases it does not seem unreasonable to postulate a Danian age for the animals in question.

Introduction

It is evident that groups of organisms do not die out in an entirely random manner, unaffected by any external causes. If that were so, it would be expected that the rate of extinction, the proportion of all groups dying out per unit of time, would remain fairly constant. On the contrary, at certain times in the Earth's history the number of groups dying out per unit of time appears to be much greater than at others. Such bursts of extinction, if they seem to have taken place within a sufficiently short interval and to have affected a sufficiently wide diversity of organisms, are called 'mass extinctions'. One such mass extinction seems to have occurred at the end of the Cretaceous, more or less on the K–T boundary.

This variability of the extinction rate leads to several questions.

1. Do such 'mass extinctions' actually occur, or are they illusions resulting from imperfections of the fossil record and from our methods of classification?

2. If there are such things as 'mass extinctions', do they take place at regular intervals?

3. What is the manner of extinction of each particular group? Was it sudden, with all its various member-species disappearing simultaneously everywhere? Or was it gradual, with its

member-species dying out at different times? (The actual extinction of any group, of course, be it a single species or a taxon of higher category, must itself be an instant phenomenon; it takes place at the moment of death of the last surviving individual.)

4. What is the reason for each extinction?

It is obvious that to attempt to answer any of these questions, one needs to know when each extinction occurred, not necessarily in absolute terms, but in terms of a standard strato-chronological scale, uniform worldwide, that is just as useful for making comparisons and correlations. In the case of the third question we need to know the time of extinction of each member-species with some considerable precision.

In recent years, special interest has been attached to what might be termed 'disaster theories' of mass extinctions, in particular the end-Cretaceous extinction and, most especially, the extinction of the dinosaurs as part of that extinction. (I prefer the term 'disaster theory' to the more usual 'catastrophe theory' because the latter term has a special, altogether different meaning in mathematics and could therefore be ambiguous when used in a palaeontological sense.)

Archibald & Clemens (1982) defined a palaeontological catastrophe (i.e. disaster) as 'a single event that set in motion a chain of other events, thereby causing major biological changes and extinctions within at most a few thousand years.' The best known of the many such theories is that originally propounded by Alvarez et al. (1980) (see also Alvarez et al. 1984) in which they suggested that, at the end of the Cretaceous, the Earth was struck by a huge meteorite. They believed that one consequence of this event was the formation of the so-called 'iridium anomaly' in the stratigraphic column; another – so they alleged – was the 'end-Cretaceous mass extinction', dinosaurs included.

I have shown (A. J. Charig, unpublished paper read at the Lyell Meeting of the Geological Society of London and the Palaeontological Association on 'Catastrophes and the history of life', held on 25th February 1987) that, to establish as a credible hypothesis any such theory of the extinction of the dinosaurs, indeed any 'disaster theory', it is necessary to do four separate things.

1. Show that the extinction of the dinosaurs was sudden and simultaneous, worldwide.

2. Show that the time of that extinction coincided exactly with the time of the disaster concerned.

3. Suggest a causal mechanism by which the disaster might reasonably have caused the extinction of the group(s) in question.

4. Explain why that same disaster, that same cause, did not lead to the extinction of other, similar groups of organisms.

This article briefly reviews our knowledge of the timing of the extinction of the dinosaurs in certain parts of the world in terms of the standard strato-chronological scale. Such knowledge is essential to any demonstration of items (1) and (2) above. This includes the question of the relation of the extinction of the dinosaurs to the position of the Cretaceous–Tertiary boundary, in particular the sporadic claims that have been made that dinosaurs survived into the Tertiary.

THE CRETACEOUS–TERTIARY BOUNDARY

In attempting that task, one great difficulty lies in there being many different ways of defining the boundary, none of them entirely satisfactory. Strictly speaking, the various divisions of the stratigraphic column – systems, stages – are defined on the type-sections that stratigraphers have agreed to accept as the global standard for each. Those type-sections, however, are inevitably from different parts of the world, which means that the top of each division is unlikely to coincide exactly with the bottom of the division above; such a succession of disparate type-sections is bound to include many gaps or overlaps between them. We therefore use only *bottoms* in defining the stratigraphical column. Thus, for example, the Permian system runs from the bottom of the Permian type-section (in Russia) to the bottom of the Triassic type-section (in Germany).

A survey of the literature and questioning of my colleagues shows that, in practice, at least five distinct ways of defining the Cretaceous–Tertiary boundary are in common use:

1. *On the fauna and/or flora*

This, the commonest method of stratigraphic correlation, is a somewhat dangerous procedure in so far as it makes one totally unwarranted assumption: that a given species of fossil organism, wherever it appears in the world, is of the same geological age. After all, some species are very long-lived; a species may survive in one part of the world much longer than in another; and its presence or absence may depend very much upon the nature of the local environment. Correlation based upon a single species must therefore be viewed with suspicion. On the other hand, correlation based upon a whole fauna or flora of disparate elements is likely to be more reliable.

2. *On the lithology*

This is even more variable than the palaeontology. It may be employed within a limited area for the recognition of a stratigraphic boundary, but even within such limits it may well be diachronous.

3. *On the geochemistry*

4. *On the 'iridium anomaly' of Alvarez et al.*

Geologists continue to argue as to whether or not this feature is a single, simultaneous, worldwide phenomenon (as well as to whether its origin is terrestrial, i.e. volcanic, or extra-terrestrial).

5. *On magnetic reversals (magnetostratigraphy)*

It is not unreasonable to assume that these at least are simultaneous worldwide. However, it is not always easy to distinguish one reversal from another; to do so may require the use of the adjacent fossils.

Questions put to my colleagues also revealed a remarkable diversity in the methods that they use for dealing with this problem. Those concerned kindly agreed that I might quote them.

C. R. Hill (on Mesozoic plants) states that the terrestrial megaflora passes through the K–T boundary without dramatic change; there is far more change within the Lower Cretaceous and again at the Palaeocene–Eocene boundary. It follows from this that palaeobotanists who study large fossils, wishing to ascertain the position of the K–T boundary, are obliged to use a wide range of fossils from other groups, mostly animal fossils. (On the other hand the calcareous nannoplankton does show a distinct change at that boundary; and, less certainly, some palynologists claim a 'fern spike', a sudden and short-lived abundance of fern spores, at or a little below the K–T boundary.) (Personal communication.)

C. G. Adams (on the Foraminifera) sees a very marked change in planktonic foraminifera and in larger shallow-water benthic foraminifera below the K–T boundary. Whole families disappeared suddenly and simultaneously at the end of the Cretaceous (e.g. the family Globotruncanidae, with about 50 species living at the end of the Maastrichtian). Then, in the Palaeocene, the impoverished fauna was gradually enriched as new forms slowly developed to replace the old ones. Shallow-water foraminifera were affected to a much lesser extent. Adams defines the K–T boundary on the overall marine fauna. (Personal communication.)

E. F. Owen (on Mesozoic brachiopods) reports a situation in his group of animals that is not unlike that of the terrestrial plants. The brachiopod fauna shows no sudden break at the K–T boundary, many genera passing across it with remarkably little change and some persisting even to the present day (e.g. *Crania*, *Terebratulina*). As brachiopods cannot be used in this connexion, and because the lithology is equally useless, fossils from other groups must be employed as indicators of the position of the K–T boundary. Incidentally, observations of the succession in the Netherlands and in Denmark suggest that there was some ecological change at that boundary. (Personal communication.)

M. K. Howarth (on cephalopods) draws the K–T boundary above the last ammonite or belemnite. It is, he says, a less than satisfactory way of drawing a system boundary because it marks the extinction of a group or groups (as evidenced by the fossil record); nevertheless he is forced to use this method for purely practical reasons. The best approach to the defining of a system boundary is, where feasible, to draw it at the first appearance of an incoming fauna. (Personal communication.)

C. Patterson (on Mesozoic fishes) cannot recognize any significant changes in the fossil fishes across the Cretaceous–Tertiary boundary. He is therefore obliged to recognize that boundary upon other elements in the fauna. (Personal communication.)

R. T. J. Moody (on testudinates, *inter alia*) believes that the boundary is marked by a recognizable, though not drastic, change in the turtle fauna; he says that the beginning of the Tertiary era is marked by the appearance of some new cheloniids and of much smaller pelomedusids. With specific reference to the strata of Niger in the southern Sahara, Moody comments that the K–T boundary is not obvious. Its position is best recognized on the lithology of the limestones in the relevant part of the succession, on the presence of the ammonite *Libycoceras* beneath it, and on sharks' teeth; it is hoped that the application of geochemical techniques will prove useful. (Personal communication.) (On the subject of sharks' teeth, Capetta (1987) notes striking changes in the selachian fauna at the K–T boundary, with about 45% of the genera disappearing.)

Also worthy of note are the observations of Whalley (1988). He concluded that evolutionary changes in the Insecta over the K–T boundary provide no evidence of abrupt or catastrophic changes.

Thus there is a considerable diversity of approach to the problem. Incidentally, more than half of the authorities mentioned stated unequivocally that the diverse groups of organisms in which they specialize – megafossil plants, brachiopods, insects, bony fishes – pass across the Cretaceous–Tertiary boundary virtually unchanged. (That fact, though not strictly relevant to the subject of this article, constitutes a remarkable commentary upon the alleged 'universality' of the end-Cretaceous extinction.) Such people are therefore obliged to base their determination of the position of the boundary upon the stratigraphic correlation of a selection of fossils from outside their own area of study.

THE EXTINCTION OF THE DINOSAURS

The title of this article implies that dinosaurs are extinct. Not everyone, however, accepts that that statement is true. Unbelievers fall into two distinct categories:

1. Users of a purely cladistic classification. It is now generally accepted that birds evolved from theropod dinosaurs (Ostrom 1973); therefore, to a cladist, birds *are* theropod dinosaurs. But birds are alive today; therefore, to a cladist, *dinosaurs* are alive today, and cannot be regarded as extinct (figure 1). Cladists have no term for dinosaurs in the usual sense of that word, for they deny that such a group is 'natural', whatever that may mean (e.g. Wiley 1981: 70 *et seq.*).

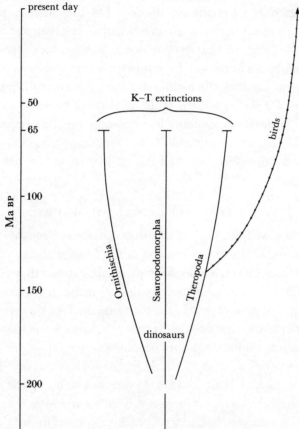

FIGURE 1. The phylogenetic relationship between 'proper' dinosaurs and birds. In a cladistic classification it is obligatory to include the birds *within* the theropod dinosaurs.

This point of view serves only to demonstrate the impracticability of cladistic classification, which classifies organisms according to their phylogeny and not according to their actual characters. Dinosaurs are a clearly defined group of animals, a group that appears to be wholly extinct; the fact that long before they died out there evolved from one of their lineages an entirely different group of animals, the birds, that survived them and persisted right through to the present day is completely irrelevant to any discussion of dinosaur extinction. In any case,

[151]

the lack of a name by which to refer to the dinosaurs proper (i.e. the dinosaurs *sensu stricto*, as usually understood by most people) must surely be a serious hindrance in any discussion that concerns them.

2. Those who believe that dinosaurs *sensu stricto* are still alive today. They draw a parallel with the coelacanths, which, like the dinosaurs, appear to be absent from the fossil record throughout the entire 65 Ma of the Cainozoic, but which were discovered alive and well in the Indian Ocean in 1938 (Smith 1956). This category of people includes a few – a very few – people from the academic world, one of whom (Dr Roy Mackal of Chicago University) has even initiated and led expeditions into the forests of Africa in search of the fabulous 'Mokele-Mbembe' (Agnagna 1983, Bright 1984). Needless to say, his expeditions have proved unsuccessful.

The parallel with the coelacanths, however, is not strictly valid. It is true that the coelacanths show, yet again, the inadequacy of the fossil record; their apparent absence from all Cainozoic rocks serves only to create an illusion. On the other hand, the present-day coelacanths have escaped total extinction until now because they long ago retreated to a highly specialized ecological niche; and for that very reason – because they were animals of modest size, living in the depths of the sea and away from human beings – they escaped our notice until 1938. The dinosaurs present an altogether different case; a group of large terrestrial animals still living on the surface of a well-explored globe populated by some five thousand million people could hardly exist without occasionally leaving some tangible remains or traces of that existence, or without affording some opportunity for a photograph.

For our present purposes, therefore, we shall take it that dinosaurs are truly extinct.

DATING THE LAST OF THE DINOSAURS

We all tend to take it as a well-established fact that dinosaurs *everywhere* died out at the end of the Cretaceous, which, today, means at the end of the Maastrichtian stage. But that is not so; the 'fact' is merely an assumption, a false assumption, for rarely can such dating be logically justified. Fossils, reasonably enough, are generally dated on the strata that contain them; but it is often forgotten that, a century ago or less, the strata themselves may well have been defined by those same fossils. Indeed, the Cretaceous–Tertiary boundary has sometimes been defined as the stratigraphic horizon at which dinosaurs become extinct. Thus, for example, Archibald & Lofgren (1989) cite old coal surveys of the U.S. Geological Survey, in which 'the K–T boundary...was believed to lie about 2–3 cm lower, based upon the highest remains of unreworked dinosaurs'. In those circumstances it is hardly surprising that the dinosaurs seem to vanish simultaneously on that very boundary in various parts of the world! A better example of the circular argument would be hard to find.

We should also remember that the Danian stage, which follows the Maastrichtian and is now regarded by everyone as the basal stage of the Palaeocene (*above* the K–T boundary), was at one time included by many in the Cretaceous rather than in the Tertiary. However, this potentially unwelcome complication may be safely ignored in the present assessment.

Further difficulties are caused by the fact that Upper Cretaceous beds with fragmentary remains of dinosaurs have sometimes been dated to precise stages within the Cretaceous, e.g. to the Campanian or the Maastrichtian, simply on the evidence of the dinosaurs that they contain.

It is, of course, logically impossible to recognize the last dinosaur in any sequence with absolute certainty. The assertion that a given individual is the last dinosaur in that sequence is based upon negative evidence, namely the apparent absence of dinosaur remains in all later beds, and it is therefore capable of falsification by the subsequent discovery of dinosaurs in those later beds.

That apart, the conditions required for a reasonable attempt at such recognition and at relating the position of the last dinosaur to that of the K–T boundary are remarkably stringent. The sequence in a given area needs to consist of a virtually continuous succession of sedimentary rocks of continental origin, potentially dinosaur-bearing, passing up from the level of the last recorded dinosaur for some considerable stratigraphic distance and, if possible, well into the Palaeocene. Only then can we verify, with a reasonable degree of assurance, the absence of later dinosaur remains in younger beds. Disconformities *below* the level of the last recorded dinosaur are, of course, totally irrelevant. An unbroken succession of continental beds, uninterrupted by marine intercalations with well-documented faunas of known age, is very difficult to date and is also less likely than a marine deposit to contain an 'iridium anomaly'. We have here a type of 'Catch 22' situation.

Let us now look at some of the more important parts of the world that have yielded Late Cretaceous dinosaurs.

1. *Western North America, east of the Rockies; especially Montana*

In Montana, detailed studies of continuous stratigraphic sequences across the K–T boundary (Archibald 1982; Archibald & Clemens 1982) have revealed some interesting facts. It had already been claimed many years ago (Sloan & Van Valen 1965) that certain orders of mammals, once thought not to have evolved until the Palaeocene (i.e. the earliest Tertiary; *after* the disappearance of the dinosaurs) had in fact originated in latest Cretaceous times, when they coexisted with dinosaurs and typical Mesozoic mammals. Samples from three different levels in the Hell Creek Formation have now shown that these 'new' mammals increased gradually in both numbers and diversity, whereas the dinosaurs (to be more precise, *Triceratops*) and the older, Mesozoic mammals simultaneously showed a tenfold drop in relative abundance. It has also been noted that the disappearance of the dinosaurs in various regions appears not to be synchronous but to vary in time from one region to another – generally occurring later in more southerly regions with a tropical climate (Van Valen & Sloan 1977). This suggests a gradual climatic change.

On the other hand, doubts were subsequently cast upon these opinions by Smit & van der Kaars (1984). All the 'mixed' faunas of the Hell Creek Formation occur in stream-channel deposits; Smit & van der Kaars believe that the streams were of Palaeocene age, that they cut through the K–T boundary and then re-worked the washed-out remains of dinosaurs and mammals from the Mesozoic rocks, incorporating them with those of the newly dead Palaeocene mammals. However, with the passage of time the beliefs of many of the people concerned (Sloan *et al.* 1986; Smit *et al.* 1987) seem to be approaching a consensus. Stated very briefly, this is that most of the localities concerned are certainly Palaeocene in age, some are certainly Maastrichtian, and others they are not quite sure about. Archibald & Lofgren (1989) remain equivocal.

Meanwhile yet another combination of these workers (Rigby *et al.* 1987) have agreed that six of the dinosaur-bearing localities in the uppermost part of the Hell Creek Formation are definitely to be dated as Palaeocene. They base their view upon the stratigraphic position of the localities and upon their having yielded fossil pollen of unmistakable Palaeocene aspect.

2. *The American Southwest: New Mexico and Texas*

Equally strong support for the idea of Palaeocene dinosaurs, so it seems to me, is afforded by the American Southwest.

The Javelina Formation in the Big Bend region of Texas, according to Stone & Langston (1975), yielded an incomplete, partly articulated skeleton of a large sauropod determined as *?Alamosaurus sanjuanensis*. The matrix in and around the bones, especially the femur, contained many fairly well-preserved palynomorphs (up to 45 species in all); it is not impossible, however, that some are reworked. None of the taxa is restricted to the late Cretaceous. Some appear first in the late Maastrichtian and range up to the present day; pollen of the Chenopodiaceae–Amaranthaceae group, which includes only one Maastrichtian species and is otherwise exclusively Palaeocene to Recent, is well represented. The flora as a whole shows a strong Palaeocene affinity.

Fassett (1982) suggested that the upper part of the Ojo Alamo Sandstone in the San Juan Basin of New Mexico, known to contain abundant dinosaur bones, lay *above* the K–T boundary. The iridium-enriched layer, which should coincide with that boundary, is missing in the San Juan Basin itself; it has been found, however, in the Raton Basin a little to the east, from which its theoretical position in the San Juan Basin has been worked out on the correlation of fossil pollen and spores (figure 2). It is presumed that the central part of the Ojo Alamo had been eroded away before the upper part containing the dinosaur bones was deposited. This means that the dinosaurs must have survived whatever event produced the iridium anomaly.

The most recent mention by the Alvarezes and their fellow-workers of dinosaur extinction in the San Juan Basin (1984) refers only to the magnetic stratigraphy and the question of

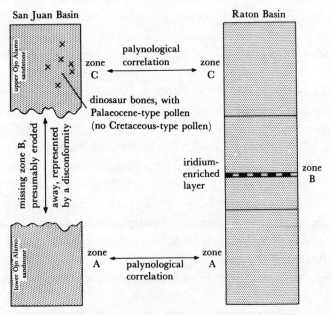

FIGURE 2. Stratigraphic relation of the dinosaur remains in the upper part of the Ojo Alamo Sandstone in the San Juan Basin (New Mexico) to the iridium-enriched layer in the Raton Basin to the east.

whether the extinction occurred in zone 29N or 29R; this can have no conceivable bearing upon the relative positions of the K–T boundary (as determined by the pollen) and the dinosaur extinction. If the pollen correlation used by Fassett is correct (and it was done by Robert H. Tschudy, of the U.S. Geological Survey in Denver – a palynologist of excellent repute), then the impact hypothesis of dinosaur extinction, indeed all disaster theories, fall to the ground.

More recently still Fassett *et al.* (1987) have produced more evidence that tends to confirm Fassett's original findings. Palaeocene-type pollen has been found in the upper part of the Ojo Alamo Sandstone, in one place with a dinosaur bone loose on the surface (which *could* have been reworked) and at another place that appears to be not with, but at the same stratigraphic level as dinosaur material *in situ*. There is no Cretaceous-type pollen at or above the level of the dinosaur bone, and there seems to be no unconformity between the bones and the pollen. This suggests that the Ojo Alamo dinosaurs, if not reworked, are of Palaeocene age, possibly throughout the entire San Juan Basin.

It is surprising that neither of the papers written by Fassett cited above makes any mention of the paper by Stone & Langston.

3. *Mongolia*

Maleev (1954) referred to what he called the 'ankylosaur horizon' at the Bain-Shire locality. According to his stratigraphic table, that horizon contains the genus *Talarurus* and lies above the 'Nemegetu' (= Nemegt); the table equates the upper part of 'ankylosaur horizon' to the Lance and Hell Formations of Montana, placing those two in the Laramie stage and citing the age of all three as Danian. However, even as recently as 1954, Maleev still considered the Danian to be the uppermost stage of the Cretaceous rather than the lowermost of the Palaeocene; he maintained that assignment to the Upper Cretaceous in subsequent papers (for example, in 1955).

Elsewhere the highest deposits of the Mongolian Cretaceous are called the Upper Nemegt Beds. Gradziński *et al.* (1968) declared that their fauna indicated a Campanian or Maastrichtian age. More recently, (1983) Karczewska & Ziembińska–Tworzydo (1983) have stated, on charophytan evidence, that 'the Nemegt Formation is not younger than the Lower Campanian'. Archibald & Clemens (1982) stated that, except for one locality of latest Cretaceous or Palaeocene age, the Gobi Desert succession stops short of the K–T boundary by about 10 Ma. Tatarinov has confirmed (personal communication) that nowhere in Mongolia do the Upper Cretaceous Beds reach the top of the Maastrichtian, that even the highest beds contain dinosaur remains, and that, in consequence, it is impossible to determine the precise level of dinosaur extinction.

4. *Southern France*

Ashraf & Erben (1986), working in the basin of Aix-en-Provence, have located the K–T boundary (defined palynologically) about 60 m *below* the highest of the *in situ* dinosaur eggs.

5. *Transylvania*

The Szentpeterfalva Beds of Transylvania, though not widely exposed, contain a diverse fauna of dinosaurs that were originally described by Nopcsa in a series of papers (1900, 1902,

1904, 1915, 1923, 1928, 1929). There has been much controversy in the past as to the age of the beds. Nopcsa simply believed them to be late Cretaceous; more specifically, they have usually been described as Campanian to Maastrichtian (e.g. Jeletzky 1960), but others have suggested that they might be Danian. There has been no definitive correlation until now. However, the beds and the fauna are presently under study by Grigorescu and Norman & Weishampel, all of whom are quite certain (personal communication) that they are of latest Maastrichtian age. The overlying strata are dated as Oligocene.

6. *Nigeria*

Nopcsa (1925) described a quantity of reptilian material from beds in Sokoto Province, Nigeria, that are regarded by everyone as of Lower Tertiary age (probably Eocene, Landenian). It included several vertebrae that he regarded as dinosaurian – some more specifically as trachodontid.

Swinton (1930, pp. 34–35) showed that these vertebrae were almost certainly crocodilian.

7. *Other regions*

Other parts of the world where allegedly dinosaur-bearing deposits have at one time or another been referred to the Danian stage include Belgium, Spain, Portugal, Austria, India, China, Alaska, Peru and Argentina. But the relevant information published on those regions is either very limited in extent and difficult to assess or is altogether non-existent. They will not be considered here.

Conclusions

Surprisingly enough, the only region where the fossil record extends almost unbroken from the Upper Cretaceous into the Lower Tertiary and appears to document the disappearance of the dinosaurs is western North America (east of the Rockies); we have no direct evidence as to when the dinosaurs disappeared elsewhere, not to within millions of years. It was because of this that the late Thomas J. M. Schopf wrote in 1981 'the problem of the extinction of the dinosaurs boils down to the rather trivial question of what happened to a score of species inhabiting the river and flood-plain habitats adjacent to the North American Western Interior seaway.'

Thus there is no direct evidence to support the idea that the extinction of the dinosaurs was a simultaneous phenomenon, worldwide. It is also true that there is not much evidence to support the opposite contention, that dinosaurs in different places died out at different times; but, without any substantial evidence pointing in either direction, one should logically adhere to the more general solution to the problem, the solution that might reasonably be expected in any comparable case. As to the question of whether or not the dinosaurs survived beyond the Cretaceous–Tertiary boundary, that inevitably hinges upon subjective opinions on the identification of fossil remains found in the field, on whether those fossils were *in situ* or re-worked, on stratigraphic correlation from one area to another, and on what constitutes the K–T boundary. Nevertheless there does seem to be a modest amount of reliable evidence suggesting that in certain regions (notably Montana, New Mexico, Texas, Provence and, *fide* Professor Jaeger's comments (below), India) some dinosaurs did survive into Danian times and beyond the alleged catastrophe, or both.

I record my gratitude to all those colleagues who helped my survey of the various methods used in defining the K–T boundary; to Professor Jean-Jacques Jaeger, of the P. & M. Curie University, who kindly read my manuscript and offered several useful comments and suggestions; and to my wife, Margaret, who drew the diagrams for this paper.

REFERENCES

Agnagna, M. 1983 Results of the first Congolese Mokele-Mbembe expedition. *Cryptozoology* **2**, 103–112.

Alvarez, L. W., Alvarez, W., Asaro, F. & Michel, H. V. 1980 Extraterrestrial cause for the Cretaceous–Tertiary extinction. *Science, Wash.* **208**, 1095–1108.

Alvarez, W., Alvarez, L. W., Asaro, F. & Michel, H. V. 1984 The end of the Cretaceous: sharp boundary or gradual transition? *Science, Wash.* **223**, 1183–1186.

Alvarez, W., Kauffman, E. G., Surlyk, F., Alvarez, L. W., Asaro, F. & Michel, H. V. 1984 Impact theory of mass extinctions and the invertebrate fossil record. *Science, Wash.* **223** (4641), 1135–1141.

Archibald, J. D. 1981 The earliest known Paleocene mammal fauna and its implications for the Cretaceous–Tertiary transition. *Nature, Lond.* **241**, 650–652.

Archibald, J. D. 1982 A study of Mammalia and geology across the Cretaceous–Tertiary boundary in Garfield County, Montana. *Univ. Calif. Publ. geol. Sci.* **122**, i–xvi & 1–286.

Archibald, J. D. & Clemens, W. A. 1982 Late Cretaceous extinctions. *Am. Scient.* **70**, 377–385.

Archibald, J. D. & Lofgren, D. L. 1989 Mammalian zonation near the Cretaceous–Tertiary boundary. (In the press.)

Ashraf, A. R. & Erben, H. K. 1986 Palynologische Untersuchungen an der Kreide/Tertiär-Grenze West-Mediterraner Region. *Palaeontographica* B **200**, 111–163.

Bright, M. 1984 Meet Mokele-Mbembe. *BBC Wildlife*, **December 1984**, 596–601.

Capetta, H. 1987 Extinctions et renouvellements fauniques chez les Sélaciens post-jurassiques. *Mém. Soc. géol. Fr.* (N.S.) **150**, 113–131.

Fassett, J. E. 1982 Dinosaurs in the San Juan Basin, New Mexico, may have survived the event that resulted in creation of an iridium-enriched zone near the Cretaceous/Tertiary boundary. *Spec. Pap. geol. Soc. Am.* **190**, 435–447.

Fassett, J. E., Lucas, S. G. & O'Neill, F. M. 1987 Dinosaurs, pollen and spores, and the age of the Ojo Alamo Sandstone, San Juan Basin, New Mexico. *Spec. Pap. geol. Soc. Am.* **209**, 17–34.

Gradziński, R., Kaźmierczak, J. & Lefeld, J. 1968 Geographical and geological data from the Polish-Mongolian palaeontological expeditions. *Palaeont. pol.* **19**, 33–82.

Jeletzky, J. A. 1960 Youngest marine rocks in western interior of North America and the age of the *Triceratops* Beds; with remarks on comparable dinosaur-bearing beds outside North America. *Geol. Surv. Canada, dept. Mines tech. Surv.* **21**, 25–40.

Karczewska, J. & Ziembińska-Tworzydto, M. 1983 Age of the Upper Cretaceous Nemegt Formation (Mongolia) on charophytan evidence. *Acta palaeont. pol.* **28** (1–2), 137–146.

Maleev, E. A. 1954 Pantsyrnye dinozaury verkhnego mela Mongolii (semeistro Syrmosauridae). *Trudy paleont. Inst.* **48**, 142–170.

Maleev, E. A. 1955 Novye khishchnye dinozaury iz verkhnego mela Mongolii. *Dokl. Akad. Nauk SSSR* **104** (5), 779–782.

Nopcsa, F. 1900 Dinosaurierreste aus Siebenbürgen (Schädel von *Limnosaurus transsylvanicus* nov. gen. et spec.). *Denkschr. Akad. Wiss., Wien* **68**, 555–591.

Nopcsa, F. 1902 Dinosaurierreste aus Siebenbürgen II (Schädelreste von *Mochlodon*). *Denkschr. Akad. Wiss., Wien* **72**, 149–175.

Nopcsa, F. 1904 Dinosaurierreste aus Siebenbürgen III (weitere Schädelreste von *Mochlodon*). *Denkschr. Akad. Wiss., Wien* **74**, 229–263.

Nopcsa, F. 1915 Erdély dinosaurusai. *Magy. allami földt. Intéz. Évk.* **23** (1), 1–24.

Nopcsa, F. 1923 On the geological importance of the primitive reptilian fauna in the uppermost Cretaceous of Hungary; with a description of a new tortoise (*Kallokibotion*). *Q. Jl geol. Soc. Lond.* **79**, 100–116.

Nopcsa, F. 1925 On some reptilian bones from Sokoto. *Occ. Pap. geol. Surv. Nigeria* **2**.

Nopcsa, F. 1928 Dinosaurierreste aus Siebenbürgen IV. Die Wirbelsäule von *Rhabdodon* und *Orthomerus*. *Palaeont. hung.* **1**, 273–304.

Nopcsa, F. 1929 Dinosaurierreste aus Siebenbürgen V. [*Struthiosaurus*] *Geologica hung., ser. palaeont.* **4**, 1–76.

Ostrom, J. H. 1973 The ancestry of birds. *Nature, Lond.* **242**, 136.

Rigby, J. K. Jr. 1987 The last of the North American dinosaurs. In *Dinosaurs past and present* (ed. S. J. Czerkas & E. C. Olson), vol. 2, pp. 118–135. Los Angeles: Natural History Museum of L.A. County; in association with Seattle and London: University of Washington Press.

Rigby, J. K. Jr, Newman, K. R., Smit, J., van der Kaars, S., Sloan, R. E. & Rigby, J. K. 1987 Dinosaurs from the Paleocene part of the Hell Creek Formation, McCone County, Montana. *Palaios* **2**, 296–302.

Schopf, T. J. M. 1981 Cretaceous endings [book review]. *Science, Wash.* **211**, 571–572.

Sloan, R. E., Rigby, J. K. Jr, Van Valen, L. M. & Gabriel, D. 1986 Gradual dinosaur extinction and simultaneous ungulate radiation in the Hell Creek Formation. *Science, Wash.* **232**, (4750), 629–633.

Sloan, R. E. & Van Valen, L. 1965 Cretaceous mammals from Montana. *Science, Wash.* **148** (3667), 220–227.

Smit, J. & van der Kaars, S. 1984 Terminal Cretaceous extinctions in the Hell Creek area, Montana: compatible with catastrophic extinction. *Science, Wash.* **223** (4641), 1177–1179.

Smit, J., van der Kaars, W. A. & Rigby, J. K. Jr. 1987 Stratigraphic aspects of the Cretaceous-Tertiary boundary in the Bug Creek area of eastern Montana, U.S.A. *Mém. Soc. géol. Fr.* (N.S.) **150**, 53–73.

Smith, J. L. B. 1956 *Old fourlegs: the story of the coelacanth*, pp. i–xi, 1–260. London, New York and Toronto: Longmans, Green & Co.

Stone, J. F. & Langston, W. Jr. 1975 Late Maestrichtian? Paleocene palynomorphs associated with the sauropod dinosaur, ?*Alamosaurus sanjuanensis*. *Geol. Soc. Am. Abstr. Prog.* **1975**, 238–239.

Swinton, W. E. 1930 On fossil Reptilia from Sokoto Province. *Bull. geol. Surv. Nigeria* **13**, 1–61.

Van Valen, L. & Sloan, R. E. 1977 Ecology and the extinction of the dinosaurs. *Evol. Theor.* **2**, 37–64.

Whalley, P. 1988 Insect evolution during the extinction of the Dinosauria. *Ent. Gener.* **13**, 119–124.

Wiley, E. O. 1981 *Phylogenetics: the theory and practice of phylogenetic systematics*, pp. i–xv, 1–439. New York, Chichester, Brisbane and Toronto: John Wiley.

Discussion

M. J. BENTON (*Department of Geology, Queen's University of Belfast, U.K.*). 1. The recent proposal that the dinosaurs died out as a result of a temperature-induced change in the sex ratio is an attractive, but unlikely idea. The eggs of modern turtles and crocodilians may hatch as either males or females depending on environmental temperatures early in development. This may have been true for the dinosaurs too – we shall never know – but the late Cretaceous crocodilians and turtles were apparently unaffected by any postulated change in temperature at the time.

2. The statement that paraphyletic groups mask extinction events has to be questioned. Sepkoski (1987) stated that 'the dinosaurs, the symbol of the Cretaceous–Tertiary extinctions, constitute a paraphyletic taxon', and this apparent paradox has been taken up by the present author. However, I would argue that this is mere word-play: the dinosaurs did die out, whether one judges it at the level of species, genera, or families. Indeed, six or seven monophyletic dinosaurian families disappeared at the end of the Cretaceous. This does not alter the suggestion that paraphyletic higher taxa may be of only limited use in macroevolutionary studies.

3. Dr Charig has gone to some length to catalogue the 30 or more findings of supposedly Palaeocene dinosaurs. He has also shown that, on further study, nearly all of these turn out not to be dinosaurs, reworked dinosaurs, or not Palaeocene in age. Because of these disproven cases and the natural desire of palaeontologists to find the 'last' dinosaur, is it not probable that there were no Palaeocene dinosaurs? The search for such beasts, on past performances, is unlikely to be successful.

Reference

Sepkoski, J. J. Jr 1987 [Reply.] *Nature, Lond.* **330**, 251–252.

A. J. CHARIG. 1. If the sex of a dinosaur **was** affected by the ambient temperature early in its development, the critical temperature need not have been the same as for modern reptiles.

2. Dr Benton writes that 'The statement that paraphyletic groups mask extinction events has

to be questioned.' What I said in my lecture (and have now written in my article) was the exact opposite thereof. To paraphrase my wording, it would be the use of **holophyletic** groups (i.e. complete clades, as required in a cladistic classification) that would mask extinction events; thus the dinosaurs, if they are considered to include birds, do **not** become extinct at the K–T boundary. Dr Benton (who has adopted a cladistic approach) argues 'that this is mere word-play: the dinosaurs did die out [because all the constituent] monophyletic dinosaurian families disappeared at the end of the Cretaceous.' That statement, however, is meaningful only if Dr Benton is using the word 'dinosaurs' in a paraphyletic sense, excluding the bird families, in a manner contrary to cladistic thinking. I wholeheartedly disagree with the suggestion in Dr Benton's final sentence.

3. I strongly dispute that I have 'shown that ... nearly all of these findings of supposedly Palaeocene dinosaurs turn out not to be dinosaurs, [to be] reworked dinosaurs, or not Palaeocene in age.' The final sentence of my 'Conclusions' (above) clearly demonstrates otherwise. If I had rejected *nearly* all the findings of supposedly Palaeocene dinosaurs, there would still be no logic in Dr Benton's suggestion that that fact in itself makes it unlikely that there were any Palaeocene dinosaurs at all. As for 'the natural desire of palaeontologists to find the 'last' dinosaur', is it not just as natural for those who are strongly committed to a belief in mass extinctions to be embarrassed by the apparent existence of dinosaurs in the Tertiary?

J.-J. JAEGER (*Laboratoire de Paléontologie des Vertébrés, Université Pierre et Marie Curie, Paris, France*). The Deccan Traps of India have been shown to be another place in the world where the age of dinosaur extinction has been precisely dated. The most recent remains are located in inter-trappean deposits dated at 65.7 ± 2 Ma by ^{39}Ar/^{40}Ar dating and showing reverse magnetic polarity. This reversal, during which almost 80% of the Deccan Traps lavas were deposited, is currently supposed to represent anomaly 29 R.

In the Aix Basin of southern France, several methods failed to yield only precise dating of the numerous dinosaur egg shell localities. In these cases, no iridium anomaly could be located.

A. J. CHARIG. I am very grateful to Professor Jaeger for this important information, previously unknown to me. If the last dinosaur remains in the inter-trappean deposits of India do indeed occur within magnetic reversal zone 29R, then the Indian dinosaurs became extinct at about the same time as those in the San Juan Basin of New Mexico, which means that they too may have survived into Palaeocene times. Considering the abundance of dinosaur eggshell material in the Cretaceous of southern France, the apparent absence in the Aix Basin of an iridium anomaly is particularly unfortunate.

G. B. J. DUSSART (*Christ Church College, Canterbury, Kent, U.K.*). Is there not a systems component to extinctions? One consistent aspect of the evolution of dominant groups appears to have been the success of more homeostatic systems over less homeostatic systems. Thus one might expect the succession of homoiotherms over poikilotherms to be a literally universal occurrence. Topological catastrophe theory tells us, however, that catastrophes are often made up of the coincidental occurrence of one or more events that are not, in isolation, capable of bringing about the catastrophe. Thus dinosaurs might have been under pressure from being behavioural rather than physiological homoiotherms. It would then only take a minor catastrophe such as a bolide impact to give the mammals an opportunity to take over. An alternative promoter of the succession might have been the assumption of egg-eating by mammals at the time. This behaviour could have spread rapidly amongst mammals; as physiological homoiotherms, they might have pursued this mode of life at night, when reptiles were less active and capable of protecting their eggs. Appropriate evidence might be that most of the modern reptiles and birds make elaborate precautions for protecting their eggs. Even though we know that there were dinosaurs which protected their eggs, they might not have done so very successfully, as they are not with us today. A scenario of egg-eating by relatively unspecialized mammals (whose small size might preclude much fossilization) could explain the sudden demise of some of the dinosaurs and the acknowledged lag between the end of the dinosaurs and the ascendancy of the mammals.

To summarise: the main factor of physiological homoiothermy might have operated in concert with subsidiary secondary factors. The second factor might have been a bolide impact in the area currently occupied by North America, egg eating by mammals in other parts of the world, and possibly direct competition between viviparous (but poikilothermic) icthyosaurs and physiologically homoiothermic cetaceans in the seas. Perhaps when considering extinctions, it might be useful to think in terms of causes and triggers. The underlying cause might have been general, but the triggers for extinctions might have been local. The cause produces the contemporaneity of the extinction but the trigger produces the geographic heterogeneity. Similar processes can be seen operating today. Trees are dying in the Harz Mountains of north Germany. The underlying cause is that the trees were planted and grown close to their tolerance thresholds but the trigger is acid rain, which is pushing them into local extinction.

A. J. CHARIG. Dr Dussart's concept of direct competition between ichthyosaurs and cetaceans is, I believe, completely untenable. The last ichthyosaurs known are from the base of the Upper Cretaceous, whereas the first cetaceans are from the top of the Lower Eocene – more than 50 Ma. later!

L. B. HALSTEAD (*Department of Pure and Applied Biology, University of Reading, U.K*). I should like to support Dr Charig in his analysis of the end of the dinosaurs. The work of Sloan *et al.* (1986) and Sullivan (1987) has demonstrated that the dinosaurs were declining in both numbers and diversity for the last 5 Ma of the Cretaceous period. The final 300000 years witnessed a striking acceleration of this decline so that at the end there were only some 12 species distributed among eight genera. It becomes difficult to speak of the final extinction of the dinosaurs as a sudden mass extinction.

References

Sloan, R. E., Rigby, J. K. Jr, Van Valen, L. M. & Gabriel, D. 1986 Gradual dinosaur extinction and simultaneous ungulate radiation in the Hell Creek Formation. *Science, Wash.* **232**, 629–633.
Sullivan, J. M. 1987 *Contr. Sci.* **391**, 1–26.

A. J. CHARIG. I thank Dr Halstead for his support.

Phil. Trans. R. Soc. Lond. B **325**, 401–420 (1989)

Printed in Great Britain

Diversification and extinction patterns among Neogene perimediterranean mammals

By J.-J. Jaeger[1] and J.-L. Hartenberger[2]

[1] *Laboratoire de Paléontologie des Vertébrés, Université Pierre et Marie Curie (Paris VI) and C.N.R.S. URA 720, 4 Place Jussieu, 75252 Paris Cedex 05, France*

[2] *Institut des Sciences de l'Evolution, Université des Sciences et Techniques du Languedoc (Montpellier) and C.N.R.S. URA 327, Place E. Bataillon, 34060 Montpellier, France*

The best mammalian fossil record during the Neogene of Western Europe is that of the rodents, the most successful and diversified mammal order. The study of origination and extinction during the Neogene (24–3 Ma BP) in one of the best-documented areas, Spain and southern France, gives an insight into the dynamics of these communities and indicates the possible nature of the driving forces. Three main periods of time show a high rate of origination: the late Burdigalian (17.5 Ma BP), the early Vallesian (11.5–11 Ma BP) and the early Pliocene (4.2–3.8 Ma BP). Two of these high origination-rate periods are immediately followed by important extinction events during which all cohorts are deeply affected (11.5–11 Ma BP and 4.2–3.8 Ma BP). The most important extinction event seems to occur during the early Vallesian (11.5–11 Ma BP), which probably includes the middle/late Miocene boundary. At the Miocene/Pliocene boundary, and during the early Pliocene, the faunal turnover seems to become faster, inducing a strong decrease of the mean species duration. Whereas the main immigration event, which occurs at 17.5 Ma BP, can be related to other faunal migrations in terms of the closure of the Tethys, as it occurs also in eastern Africa and in southwest Asia, the middle/late Miocene boundary event may have been related to a period of ice growth in the Southern Hemisphere. The extinction event that affects the planktonic foraminifera at 12 Ma BP cannot be chronologically correlated to this southwestern European land-mammal extinction event, because the calibration of the marine fossil record during that time-span has to be precise. Some limited terrestrial faunal exchanges that occur during the Messinian between southwestern Europe and northwestern Africa do not deeply affect the general faunal dynamics. Both allochthonous cohorts of immigrants become rapidly extinct.

Several endemic rodent faunas, indicating insular conditions, have been reported from the southern edge of the western European continent from the middle Miocene up to the Pliocene. All show low taxonomic diversity, strong endemism and short survival. Some of them, like those of the Gargano Islands during the late Miocene, underwent peculiar morphological changes and also speciation. The large number of rodent genera coevolving in the Gargano Islands is indicative of the large surface areas of these islands. The general geographic pattern of southwestern Europe during the Neogene may therefore correspond to a large continental province including Spain and southern France with some kind of fast-modifying archipelago on its southern rim.

1. Introduction

During recent years, a large amount of data have been collected and analysed to relate the main biotic events to physical and geochemical changes. To understand the relations between the physical and biological worlds, two complementary points of view have arisen: some authors think that variation of the physical and chemical parameters of the various environments has immediate and important consequences in the development of life; others, mainly biologists, think that the evolutionary history of a species is largely dependent on other species, which constitute its most influential environment (Van Valen 1973; Stenseth & Maynard-Smith 1984; Hoffman & Kitchell 1984; Benton 1985). However, in this debate, most of the data that have been analysed originated from the marine realm, where physical and geochemical environmental parameters have been extensively studied in recent years. Indeed, most of the palaeobiological data relate to marine invertebrates and to protista. The evolutionary biology of these groups has not been extensively studied, so that the significance of their morphological changes, the species concept in each of these groups and their reproductive biology are not so fully understood as for higher organisms, such as mammals. For example, a mean species duration as estimated in Cainozoic planktonic foraminifera of about 25 Ma BP (Stanley 1979) and of 7.7 Ma BP for the upper half of the Cainozoic (Raup 1987) seems questionable relative to generation time among these groups. On the other hand, the known physical data relative to continental biotas are still incomplete.

In this context, we present here an analysis of the dynamic of mammalian faunas in a time interval of about 21 Ma, from the beginning of the Miocene to the end of the lowermost Pliocene, and in a restricted geographic area, around the Mediterranean.

2. METHODS

The fossil mammals of the perimediterranean area have been extensively studied since the time of G. Cuvier, some 180 years ago. A detailed survey indicates that at least one geographical area has been well documented, the southwestern part of Europe including Spain and southern France. Using the taxonomic results of different authors as a database, we have calculated several measures such as origination and extinction rates, duration of taxa and several classical derived parameters following the methods extensively used by authors such as Simpson (1944), Kurten (1960) and Stanley (1979).

The rodents are the most studied mammals in terms of their taxonomy and are represented, between the earliest Miocene and the late Early Pliocene, by 235 species distributed among 96 genera, giving an average of 2.52 species per genus. Large mammals are also numerous, but unlike the rodents, their taxonomy is still confusing and most of the families have not received comprehensive study in recent years. Therefore, after a preliminary and unsuccessful trial, we focused our study mainly on the rodents. Rodent taxonomy has reached some sort of consensus, having been recently established and verified independently by several research teams. Interest in rodent evolution is largely the result of their abundance in geological formations and of their biochronological value. This seems to be a consequence of their rapid diversification and phyletic evolution. The study of the dynamics of rodent faunas in southwestern Europe between 24 and 3 Ma BP therefore represents the main source of data from which the patterns have been deduced. The results obtained from the interpretation of these data will have to be tested in the future in the light of other faunal provinces and in other mammalian groups. However, the quality of that fossil record is several orders of magnitude higher than that of any other perimediterranean faunal province. Also the intrinsic value of the order Rodentia which, from the middle Paleogene on, represents half the diversity of herbivorous and omnivorous mammals, must also be considered. In this order the species concept used by palaeontologists has been shown to be rather close to the biological one. The reasons seem to be the qualities of high heritability of the dental characters used in taxonomy (Bader 1965), and the high adaptative value of teeth.

As the temporal scale is of major importance for the interpretation of the data, we have chosen the calibrated biochronological zonation proposed by Aguilar (1982), which appeared more detailed and more accurate than the classical land-mammal age zonation of Mein (1975). Also, as the biochronological unit is the marker level, its power of resolution is more accurate than the traditional land-mammal ages (Jaeger & Hartenberger 1975). The correlation between the continental timescale and the marine timescale is discussed by Aguilar & Michaux (1987). The recent timescale proposed by Vai (1988), which has improved the proposal of Berggren et al. (1985), has also been used. Taxonomic information has been collected in numerous recent works. We used mainly the papers of Aguilar (1981), Daams & Freudenthal (1981), Daams et al. (1987), Cuenca Bescos (1988) and Freudenthal (1988). Recent taxonomic publications of additional Spanish and French authors have also been used.

3. Genus duration

The histogram of the distribution of genus durations (figure 1a) has been constructed from extinct genera only. It is roughly similar to that obtained by Gingerich (1977) on several hundred genera of rodents with data collected from Romer (1966). Following a main peak of rather short survival time, there are several accessory peaks, situated respectively around 6–7, 10 and 21 Ma duration. Corresponding peaks are also observable on Gingerich's data. The genera corresponding to the 21 Ma peak belong to panchronic taxa, the living fossils of that period. The two other peaks characterize some genera that have an increased number of species, as shown on figure 3. The study of the distribution of species number relative to genus frequency illustrates a significative difference. There is no correlation between the genus duration and the number of species. The genera with short survival times have, in general, only a few species and their species number shows a logarithmic decrease. The genera with survival times longer than 5 Ma show a very different distribution, with a mode situated around three or four species. In that case, this pattern corresponds to a strategy, suggested by Raup (1987), in which the genera that speciate increase their chance of survival. But this cannot be considered as a general rule, as the longest-surviving genera of our study have only a few species, mainly chronospecies.

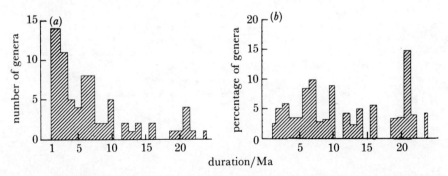

FIGURE 1. Generic duration of southwest European Neogene rodents. Only extinct genera have been plotted. (a) Histogram of longevities of genera of the interval. (b) Histogram of longevities of genera of an instant in time, following the method proposed by Stanley (1979, p. 110).

TABLE 1. RANGE CHART OF GENERA OF RODENTS IN THE NEOGENE OF SOUTHWESTERN EUROPE

(Correlation between marker levels, land-mammal ages and radiometric dating is made following information given in Berggren *et al.* (1985), Aguilar & Michaux (1987) and Vai (1988).)

Mein *et al.* zonation: 0 1 2 3a 3b 4 5 6 7 8 9 10 11 12 13 14a 14b 15a 16a 16b 17

Aguilar *et al.* zonation: A1 A1 A2 A3 A4 A4 A5 B1 B2 C1 C3 C4 C5 C5 D1 D2 D3 D4 D4 E E E F1 F2 F3 G1 G2 G2 G3

marker levels: Coderet, Plaissan, Paulhiac, Caunelles, Lespignan, Laugnac, Estrepouy, Serres de V., Beaulieu, Vx Collonges, Sansan, Grive M, Grenatière, Anwill, Can Llobateres, Montredon, Mollon, Cucuron, Los Mansuetos, Lissieu, La Tour, Alcoy, Caravaca, Celleneuve, Hautimagne, Perpignan, Sète, Seynes, Balaruc, Iles Medas, recent

radiometric correlation: 24 23.5 22.5 22 21 20 19.5 18.5 17.5 17 15.5 14 13 12.5 12 11 10.5 9 8.5 8 7 6 5.5 5 4.5 4 3.5 3 2.5 2

Gliravus
Peridyromys
Microdyromys
Bransatoglis
Glirudinus
Pseudodryomys
Myoglis
Vasseuromys
Altomiramys
Heteromyoxus
Miodyromys
Eomuscardinus
Muscardinus
Armantomys
Paraglirulus
Miomymus
Tempestia
Glirulus
Ramys
Eliomys
Muscardinulus

Palaeosciurus
Heteroxerus
Spermophilinus
Freudenthalia
Aragoxerus
Atlantoxerus
Petinomys
Pliopetes
Pliopetaurista
Miopetaurista
Blackia
Cryptoterus
Forsythia
Pliosciuropterus
Archaeomys
Issiodoromys
Allomys

Plesiominthus
Sminthozapus
Eozapus

Eomys
Eomyops
Rhodanomys
Ritteneria
Ligerimys

TABLE 1 (cont.)

	Cod	Pla	Pau	Cau	Les	Lau	Est	SdV	Bea	VxC	San	GrM	Gre	Anw	CLl	Mon	Mol	Cuc	LMa	Lis	LTo	Alc	Car	Cel	Hau	Per	Sèt	Sey	Bal	IMe	rec
Mein et al. zonation	0	1	2	3a	3b	4	5	6	7	8	9		10	11		12	13		14a	14b	15a		16a	16b	17						
Aguilar et al. zonation	A1	A1	A2	A3	A4	A4	A5	B1	B2	C1	C3	C4	C5	C5	D1	D2	D3	D4	D4	E	E	E	F1	F2	F3	G1	G2	G2	G3		
radiometric correlation	24	23.5	22.5	22	21	20	19.5	18.5	17.5	17	15.5	14	13	12.5	12	11	10.5	9	8.5	8	7	6	5.5	5	4.5	4	3.5	3	2.5	2	
Pseudotheridomys	←•	•	•	•	•	•	•																								
Keramidomys									•	•	•	•	•	•	•	•	•														
Eucricetodon	←•	•	•	•	•	•	•	•		•																					
Pseudocricetodon	←•	•	•	•																											
Adelomyarion	←•	•																													
Melissiodon	←•	•	•	•	•	•	•	•																							
Eumyarion									•	•	•	•	•	•																	
Cricetodon									•	•	•	•	•	•	•																
Megacricetodon									•	•	•	•	•	•	•																
Democricetodon									•	•	•	•	•	•	•	•															
Falbuschia									•	•																					
Lartetomys									•																						
Hispanomys											•	•	•	•	•	•															
Ruscinomys																•	•	•	•	•	•	•				•					
Deperetomys											•	•	•																		
Cricetulodon														•	•	•															
Rotundomys																•															
Kowalskia																•															
Renzimys											•	•	•																		
Cricetus																							•	•	•	•	•	•	•	•	•
Anomalomys									•	•	•	•	•	•	•																
Neocometes									•	•	•	•																			
Prospalax																								•							
Epimeriones																														•	•
Cricetulus																														•	•
Myocricetodon																•															
Progonomys																•	•														
Occitanomys																			•	•	•	•	•	•	•	•	•	•	•	•	•
Parapodemus																			•	•	•	•	•	•	•	•					
Stephanomys																				•	•	•	•	•	•	•	•	•	•	•	•
Apodemus																				•	•	•	•	•	•	•	•	•	•	•	•
Rhagapodemus																		•	•	•	•	•	•	•	•	•					
Valerimys																•	•	•	•	•	•	•	•								
Castillomys																								•	•	•	•	•	•	•	•
Micromys																								•	•	•	•	•	•	•	•
Paraethomys																		•	•	•	•	•	•	•	•	•					
'Gerbillus'																								•							
Trilophomys																							•	•	•	•					
Prosomys																										•					
Promimomys																										•	•	•			
Mimomys																										•	•	•	•	•	•
Rhizospalax	•	•																													
Monosaulax				•	•	•	•	•	•	•	•	•	•																		
Steneofiber									•	•	•	•	•	•																	
Castor																								•	•	•	•	•	•	•	•
Calomyscus																		•													
Dendromus																		•													
Protolophiomys																		•													
Hystrix																										•					

[165]

The commonest life history corresponds to an immigrant genus that becomes rapidly extinct after some unique speciation event.

Several methods allow us to evaluate the duration time of these genera. The arithmetic mean gives a value close to 7 Ma. The Lyellian curve (figure 2) indicates a half-life of 4 Ma, which would give a duration of 8 Ma with a coefficient of two as evaluated by Stanley (1979, p. 119) or a value of 11.5 Ma with a coefficient of 2.89 as demonstrated by Kurtén (1960) and Levinton & Farris (1987). On the other hand, the graphic method suggested by Stanley (1979) of constructing the histogram of genus duration for an instant of time, indicates two different modal values, one around 7 Ma, the other around 21 Ma (figure 1 b). The last value characterizes the panchronic taxa, or living fossils. Gingerich (1977) has found an arithmetic mean of 5.85 Ma for all rodents.

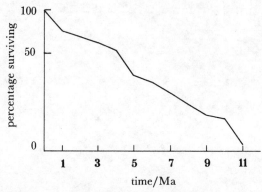

FIGURE 2. Partial (1 to 1 Ma) survivorship curve of genera of rodents of the southwest European Neogene. Only extinct genera have been plotted. Arithmetic mean 7.25; Lyellian mean 8.02 (L).

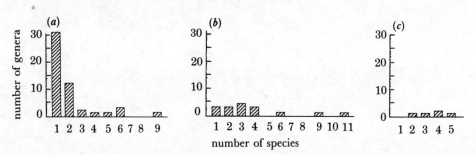

FIGURE 3. Histograms of species number (a) for genera with a duration of 0–5 Ma; (b) for genera with a duration of 5–10 Ma; and (c) for genera with a duration of 10–15 Ma.

Compared with histograms of genus duration for other taxa (Stanley 1979), the Neogene rodents of southwestern Europe show a logarithmic decrease of the genus number relative to their longevity. A similar pattern can also be observed when the duration of species is plotted (figure 4a). This shape may have some palaeobiological meaning, indicating that the palaeontological genus concept is close to the biological genus concept. These biological categories therefore behave in a similar way, but at different rates.

4. Species duration

The histogram of species duration (figure 4a) indicates a distribution between 0.5 and 9 Ma, with a high proportion of species having a short survival time (64 % survive less than 1 Ma).

[166]

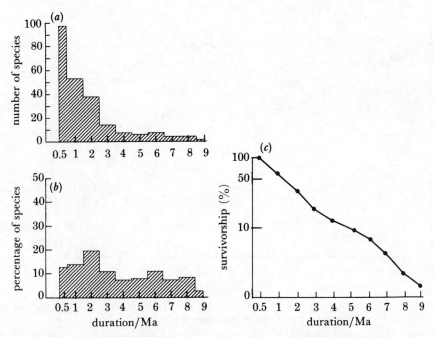

FIGURE 4. Species duration of southwest European neogene rodents. (a) Histogram of longevities of species of the interval; (b) histogram of longevities of species at an instant in time. (c) Survivorship curve of rodent species during the Neogene in southwestern Europe. Half of the species survived 1.25 Ma, corresponding to a Lyellian average of 2.5–3.60 Ma.

The general pattern appears to be quite similar to the histogram for graptolite species duration (figures 9 and 10 in Stanley 1979). The histogram of species duration for an instant of time (figure 4b) (Stanley 1979) indicates a modal duration of species of 2 Ma. On the Lyellian curve, a value of 1.25 Ma corresponds to the 50% ordinate (figure 4c). From different estimates, as for the genus, the mean species duration would have a value of 2.5 Ma as an underestimate or 3.6 Ma. Stanley (1979) indicates values of about 1.6 and 1.4 Ma for Pleistocene mammals, the last value becoming 2 Ma if the quotient of 2.89 is used (Kurtén 1960; Levinton & Farris 1987). For the Miocene rhizomyid rodents of southwest Asia, Flynn (1986) has recently observed an average value of 1.2 Ma. However, this author also gives an estimate of about 2.9 Ma for the smaller-sized species of rhizomyid.

To estimate more precisely the mean duration of species, we used a third method. For each polycohort (figure 8), we have constructed the survivorship curve for the rodent fauna of each zone through time. We measured mean species duration for each 'stratum' from the linear regression of each curve. As indicated by Raup (1987), if the value of the extinction rate is constant through time and from species to species in the cohort, the mean taxon duration must be the reciprocal of the extinction rate. We obtained a series of values between 2.4 and 0.16 Ma. When plotted against the absolute timescale (figure 5) the results suggest a decrease of the species duration during the Neogene.

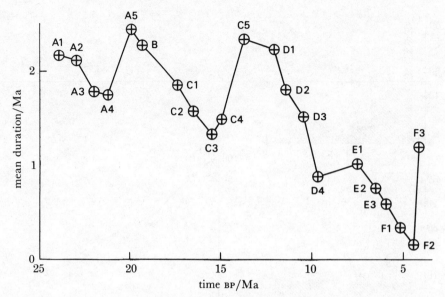

FIGURE 5. Mean average polycohort species duration through time (24–3 Ma BP). A decrease of polycohort species duration through time is observed.

5. ORIGINATION AND EXTINCTION RATES

(a) Origination rates

We have studied the origination rate at the generic level and at the specific level (figures 6 and 7) to appreciate the relative importance of phyletic evolution versus true immigration. Despite the possible origination of new genera through phyletic evolution, this event is much less frequent than the immigration of a new genus or the pseudo-origination of new species through gradual evolution. It should also be remembered that, as this work concerns only one faunal province, most of the immigration and extinction events may be local. Nevertheless, all modes of representation show a similar pattern. The global diversity follows a regular increase from the beginning of the Miocene up to the middle Miocene, followed by a regular decrease. Three major immigration events correspond respectively to the late Burdigalian (C1, 17.5 Ma BP), to the middle/late Miocene boundary (D1, 12 Ma BP) and to the early Pliocene. These immigration events have been recognized for a long time as basic chronostratigraphic events. Additional minor peaks can be observed, their amplitude depending on the mode of representation. Some are the result of immigrations: for example, the lower Vallesian (D1, 12 Ma BP) characterized by the appearance of *Hipparion* and of the earliest murid *Progonomys* (Schaub 1938) and the early Messinian (E1, 6.5 Ma BP) with a limited faunal exchange occurring between northwestern Africa and Spain. Other small peaks may be attributed to minor phases of faunal turnover.

(b) Extinction rate

Some important extinctions follow immediately after the main immigration phases. They can be explained as a simple return to faunal equilibrium following a strong increase of diversity as suggested by Gingerich (1977, 1984) for Pleistocene extinctions. The late Burdigalian extinction event (C1, 17.5 Ma BP) and the one that occurs during the late early Pliocene belong to this category.

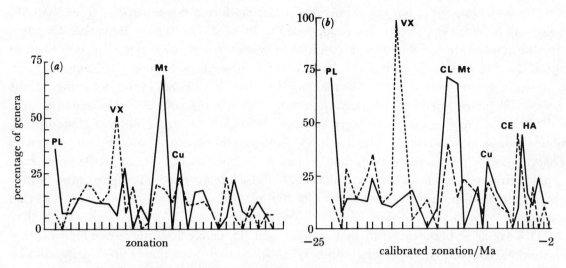

FIGURE 6. Variations of extinction (solid lines) and origination rates (dashed lines) of rodent genera through the Neogene of southwestern Europe. The timescale corresponds to the marker levels (a) and to the absolute timescale in Ma (b).

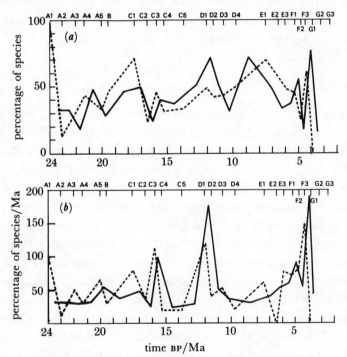

FIGURE 7. Variations of extinction (solid lines) and origination rates (dashed lines) of rodent species through the Neogene of southwestern Europe. (a) Percentage of the number of species represented in each zone. (b) Percentage of the number of species represented in each zone per million years.

The most important extinction event occurs during the early Vallesian (D1, 12–11.5 Ma BP). During this time interval, which most probably includes the middle/late Miocene boundary, the extinction rate reaches much larger values than the origination rate (figures 6 and 7). This period has been already recognized as a period of drastic changes in the composition of southwestern European mammal faunas (Crusafont Pairo 1954). Ecological scenarios have even been suggested for these changes (Weerd & Daams 1978; Daams & Van der Meulen

[169]

1984; Agusti *et al.* 1984) but the extinctions occurring during this timespan (around 0.5 Ma long) had not been suspected to be as important as that. As several large mammals disappear simultaneously (Moyà-Solà 1983), it can now be recognized as the major extinction event in the history of mammals during the Neogene of southwestern Europe. Its amplitude and taxonomic extent makes it very similar to the '*Grande Coupure*' event occurring at the Eocene–Oligocene boundary (Hartenberger 1987, 1988) in the same area. Unfortunately, the calibration of this event is not yet very accurate. It follows the immigration of *Hipparion* and of *Progonomys* into that area. On the basis of numerous observations, the immigration of *Hipparion* in the perimediterranean domain had been dated as 12.5 Ma BP by Berggren *et al.* (1985). However, recent data based on new K/Ar absolute ages and on magnetostratigraphic investigations (Sen 1986) suggest an earlier age of 11.5 or even 11 Ma BP. The Vallesian extinction event can therefore be estimated to date from 11 ± 0.5 Ma BP rather than 12–11.5 Ma BP, as in the timescale that we have used. This more accurate dating indicates that the middle/late Miocene boundary is probably included in the duration of that event. A similar major extinction has been reported among planktonic foraminifera at 12 Ma BP by Hoffman & Kitchell (1984). It has again been discussed by Raup (1987). Whether or not these two events, one occurring in the surface waters of the oceans and the other on land, are slightly diachronous or synchronous, cannot be decided yet as the timescale for the marine middle–late Miocene does not seem to be very accurately calibrated.

6. POLYCOHORT AND COHORT SURVIVORSHIP CURVES

To analyse in greater detail the nature of these extinctions events, we have constructed polycohort (figure 8) and cohort (figure 9) species-survivorship curves similar to those constructed by Hoffman & Kitchell (1984) for Tertiary planktonic foraminifera, by Raup (1978, 1987) for Phanerozoic families of marine organisms and by Hartenberger (1987, 1988) for Palaeogene mammals of western Europe. These curves provide more precise information on the extinction patterns, because of the separate analysis of each faunal stratum. One of the most striking results is the similarity between the polycohort curves, showing all species occurring together at each successive level, and the cohort curves, which describe the extinctions occurring through time in each faunal stratum, each faunal stratum being restricted to the newly appearing species. This may indicate that the new species have an important weight in the determination of the shape of the polycohort curves.

The two strongest extinction events are those occurring during the early Vallesian (11 ± 0.5 Ma BP) and at the end of the early Pliocene (4.2–3.8 Ma BP) respectively. During these two crises, all curves are equally affected and show strong slopes. Other minor extinction events affect only the oldest faunal strata. This is so for the events that occur during C2 (16.5–15.7 Ma BP), the late Miocene (D3, 9 Ma BP) and immediately after the Miocene/Pliocene boundary (figures 5 and 6).

The general structure of these curves warrants some comments. In three zones (B, 19.5–17.5 Ma BP; C4, 15.3–13.8 Ma BP and E3, 6.5–5.2 Ma BP), the slopes of most of the polycohorts are nearly horizontal. The meaning of these patterns of slopes is not yet clear, as two interpretations can be proposed. First, they may correspond to a calibration error with the absolute timescale, the length of each of these zones having been overestimated. In that case, such errors would induce a kind of stepwise pattern indicative of a pseudo-stationary model

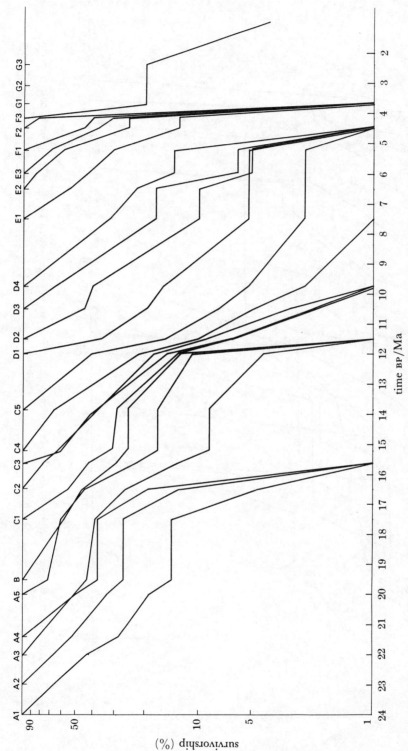

FIGURE 8. Nested polycohort survivorship curves of rodent species of the Neogene of southwestern Europe. The main extinction events occur during the D1–D2 interval (11.5–11 Ma BP) and the F3–G1 interval (4.2–3.8 Ma BP): all the cohorts are affected by these events. Minor events are recorded in C2 (16.5–15.7 Ma BP), in D3 (10.5–9.7 Ma BP) and in F1 (Lower Pliocene).

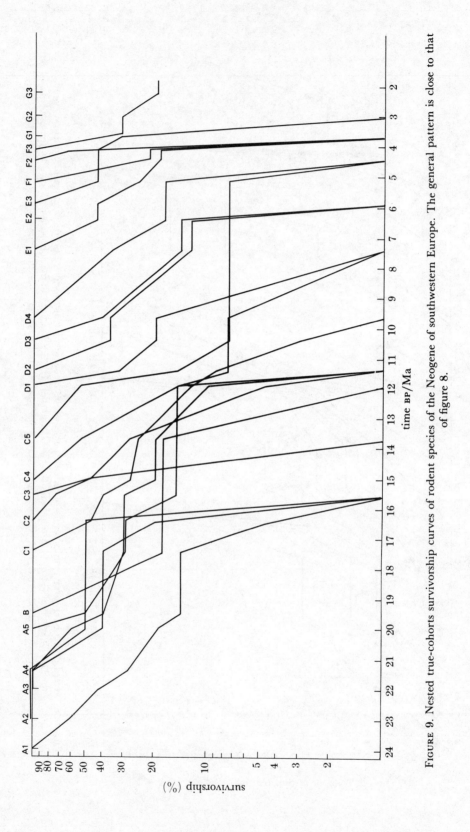

FIGURE 9. Nested true-cohorts survivorship curves of rodent species of the Neogene of southwestern Europe. The general pattern is close to that of figure 8.

(Stenseth & Maynard Smith 1984; Hoffman & Kitchell 1984). On the other hand, if the calibration can be demonstrated to be accurate for these periods, a stationary model, at least for some time intervals, will have to be accepted. However, the general shape of these curves is rather linear, supporting the Red Queen model (Van Valen 1973).

7. Relation with local and global changes

Despite some minor inadequacy in the calibration of our record, the correlations between the extinctions, the originations and the global changes can be discussed. The oceanic oxygen-isotope curve shows continuous increase of mean oceanic temperature from the early Miocene up to 15 Ma BP when a major and rapid decrease occurred. As for planktonic foraminifera (Hoffman & Kitchell 1984), our data do not show any kind of major change in the extinction rate during this drastic temperature drop. The only concomitant trend corresponds to an increase of diversity of rodent communities between 19 and 16 Ma BP followed by a sharp decrease between 16 and 12 Ma BP, which may be related to the temperature decrease indicated by the oceanic oxygen-isotope curve. Pickford (1987), on the basis of the East African mammalian fossil record, tentatively concluded that major shifts had occurred in biogeographic boundaries, inducing faunal exchanges between adjacent provinces, relating to the thermal optimum near 18 Ma BP (NDP1 in Thomas (1985)) and to the temperature decrease that occurs near 15 Ma BP (NDP2 in Thomas (1985)). The first of these changes is also represented in our record and corresponds to a main immigration phase (C1, 17.5 Ma BP) but we have found no indication relative to the second one in terms of origination and extinction rates, the main crisis in our record occurring around 11 Ma BP. Major faunal turnovers among the south Asiatic mammalian communities have been recognized between 20 and 16 Ma BP, at 9.5 Ma BP and at 7.4 Ma BP (Barry et al. 1985). The first of them, related to the closure of Tethys, corresponds to a major land-mammals exchange between Africa and Eurasia, reported all over the Old World.

In a recent oxygen-isotope synthesis, Miller et al. (1987) came to the conclusion that possible ice growth occurred near the middle/late Miocene boundary, for which they suggest an imprecise age of approximately 10–8 Ma BP. They explain the weaknesses of the chronological data as a consequence of the sparse North Atlantic record at these times and of a hiatus or lack of benthic data across this interval in the Indian Ocean record. Also, they relate their hypothesis to the occurrence of a major chronostratigraphic break and offlap events near the middle/late Miocene boundary, which have been suggested by Vail et al. (1977) and Haq et al. (1987), and with the occurrence of canyons incised into continental margins at this time (Farre 1985). Surprisingly however, Miller et al. (1987) indicate that no erosional event has been detected in the middle Miocene when a severe drop in temperature, deduced from oxygen-isotopic composition, occurs.

Some kind of global change seems therefore to be the cause of that main event affecting southwestern European mammalian faunas between 12 and 11 Ma BP, but its precise nature is not yet fully understood, most probably because of an imprecise marine chronological timescale that requires improvement.

The end of the Miocene and the beginning of the Pliocene appear to correspond, following our data, to a period of increased rate of faunal turnover long before the beginning of glacial ages in the North Atlantic region (Shackleton et al. 1984). Several immigrations are rapidly followed by extinctions. The mean species duration among the corresponding polycohorts is therefore especially short. We cannot yet relate these biological events to any other occurrences.

8. Faunal turnover in the adjacent areas

In adjacent areas, the fossil record of rodents is much less well documented than in southwestern Europe. In northwest Africa, the record begins only with the upper part of the middle Miocene (Lavocat 1961; Jaeger 1977). The number of taxa is always fewer than it is at corresponding time levels in southwestern Europe and large gaps still occur in the fossil record (Jaeger 1977). A high level of endemicity is shown if one compares these faunas with those of the adjacent geographic areas. During most of the documented timespan, only one or two genera of rodents, usually represented by different species, are shared with southwestern Europe, with the eastern Mediterranean province or with eastern Africa. This situation, and the low diversity, has led us to consider this faunal province as being isolated from adjacent areas. There is, however, one exception during the late Miocene, when a faunal exchange with southwestern Europe occurred (Brandy & Jaeger 1980; Thomas et al. 1982; Coiffait et al. 1985; Jaeger et al. 1987). Eight taxa, previously considered as endemic to an 'Ibero-Occitan' province, immigrated to North Africa where they became extinct through time (Jaeger et al. 1987).

Several occurrences of North African endemic taxa have also been reported from late Miocene Spanish localities (de Bruijn 1974; Jaeger et al. 1975; Aguilar et al. 1983; Moyà-Solà et al. 1984; Agusti & Galobart 1986; Aguilar & Thaler 1987). The resulting pattern of that faunal exchange is symmetrical to that observed in North Africa, with a progressive extinction of the African taxa. This kind of limited faunal exchange, not as yet clearly recognized on the basis of large mammals (Azzaroli & Guazzone 1980), occurred several times and at several places at the edge of the two different faunal provinces. The most important occurred during the late Miocene between Spain and northwest Africa, but it cannot be considered as being directly attributable to the desiccation of the western Mediterranean sea because it occurs a little earlier. A tectonic cause therefore seems more probable (Jaeger et al. 1987).

Several fossil islands have been identified during the Neogene, through the discovery of more or less endemic fossil mammals, and these relate to the complex tectonic story of the Mediterranean sea. From the early Middle Miocene of Sardinia, de Bruijn & Rümke (1974) have reported a peculiar micromammalian fauna with a mixture of African and European taxa. During the late Tortonian high sea-level stand, several islands occurred in the area that has since become Italy. The best-known is the one from Baccinello (Grosseto) in Tuscany, from which the hominoid *Oreopithecus bambolii* Gervais has been described. Extensive studies have led to an improved calibration of these deposits (Hürzeler 1975; Hürzeler & Engesser 1976) and to a better understanding of the biogeographic affinities of their faunas, which show strong African affinities (Hürzeler 1983; Thomas 1984). Faunas from successive levels have been discovered, indicating a very reduced number of species, an increasing endemicity and a fast rate of phyletic evolution. In the early Pliocene, the isolation ended, as testified by the arrival of a new fauna identical to the other western European early Pliocene faunas (Hürzeler & Engesser 1976). The dating of a volcanic-ash level, located above fossil endemic mammal localities, indicates an age of about 8 Ma BP (Hürzeler & Engesser 1976). The age of the beginning of the isolation, and its duration, are not yet established. From a palaeontological

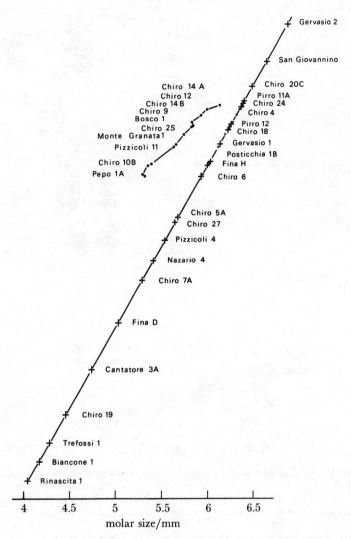

FIGURE 10. Relative position of the Gargano localities, using the increasing complexity of the molar structures and size of the murid *Microtia* as biochronological indexes (from Bastianelli 1989).

point of view, the beginning of the isolation must not have been before 10 Ma BP but a more recent age (8.5–9 Ma BP) cannot be excluded.

An even more spectacular example of island evolution occurring in the Mediterranean domain during the Neogene has been reported by Freudenthal (1971) from the Gargano area, in the province of Foggia, Italy. Numerous fossil mammal remains have been discovered in the '*terra rossa*' of a large number of karstic fissure fillings. The comparison of related taxa indicates that these fissure fillings were deposited during a notable timespan, especially when the amount of morphological changes occurring in various lineages is taken into account. From the available geological and palaeontological evidence, a late Miocene age has been proposed to include the whole sequence of localities (Freudenthal 1985), but an extension into the early Pliocene also cannot be excluded (De Giuli & Torre 1984). The fauna consists of numerous endemic taxa whose phylogenetic relationships are obscured because of their morphologically derived condition, a consequence of a rapid phyletic evolution following the beginning of their

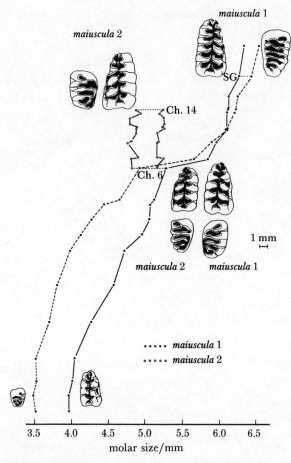

FIGURE 11. Morphological evolution of the largest species of *Microtia*, *M. maiuscula*, from the Gargano. At the level of Chiro 6 (Ch. 6), a second lineage appears, through sympatric speciation or through immigration from an adjacent island. After this event, there is a rapid increase of the morphological complexity in one lineage, whereas the size difference remains stable (from Bastianelli 1989).

isolation. Among them figure a cervid with five horncore-like cranial appendages (Leinders 1984), a giant erinaceid insectivore (Freudenthal 1972; Butler 1980), an unusually large owl (Ballmann 1973, 1976) and numerous endemic rodents (Daams & Freudenthal 1985), which deserve special attention. Until now, despite the fact that all material has not yet been fully investigated, nine rodent genera, distributed among three families (murids, cricetids and glirids), have already been identified with numerous species and chronospecies. Among them is the genus *Microtia*, a murid, members of which show an unusually large size associated with a peculiar molar morphology: the first lower molar shows an increased number of additional lobes in front of the tooth and the third upper molar also shows additional lobes in its distal part. These structures show a marked increasing complexity when the populations from different fissure fillings are compared. As there is no stratigraphic background, the relative age of each locality has been tentatively established by Freudenthal (1976) on the base of the supposed molar-size increase through time among several lineages of murids and cricetids. The alternative hypothesis of an increasing complexity of the molar structure through time as another measure of phyletic evolution, which can be combined with size, suggests a slightly different arrangement in the relative timescale (figure 10) (De Giuli & Torre 1984; Bastianelli

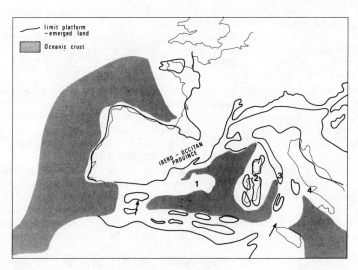

FIGURE 12. Palaeogeographical reconstruction of the Mediterranean during the Late Miocene (Tortonian), with indication of the main localities with endemic faunas. 1, Balearic Islands: 2, Sardinia; 3, Baccinello; 4, Gargano. Arrows indicate the main dispersal routes followed by terrestrial vertebrates.

1989) and allows the identification of several new lineages evolving simultaneously. As shown by Bastianelli (1989) (figure 11), the resulting pattern for the largest species of *Microtia* lineages indicates evidence of either sympatric speciation or immigration from an adjacent island. When compared with the limited number of fossil rodent species in the Pleistocene Mediterranean islands, the number of rodent taxa coevolving in these islands is rather high, suggesting that the area of these islands must have been very large (Diamond 1984). Unfortunately, their exact number and precise extension is largely unknown. Their area may have included a part of the area covered today by the Adriatic sea.

Other insular rodent faunas have also been discovered in the late Miocene of Ibiza (Balearic Islands) (Moyà-Solà *et al.* 1984) and in the early Pliocene of Sardinia (Pecorini *et al.* 1974).

In conclusion, the geography of the southern part of western Europe during Neogene time corresponded to a kind of archipelago, with many islands colonized by taxa originating from Europe or from northern Africa, related to the numerous tectonic, eustatic and climatic changes. In all these insular areas, the faunas underwent a rapid turnover and showed a high extinction rate, as can be expected among insular faunas. None of these numerous endemic taxa survived after the breakdown of the isolation, and the diversity, at least diversity that can be estimated on the base of the fossil record, seems to have been well related to the area of these islands (Diamond 1984). As more data are becoming available, the study of these insular faunas presents a stimulating subject with which to increase our knowledge of the driving forces that control the evolution of these communities. However, most if not all evolutionary novelties seem to have originated in larger faunal provinces where the driving forces seem to be more related to global events.

REFERENCES

Aguilar, J.-P. 1981 *Evolution des Rongeurs Miocènes et paléogéographie de la Méditerranée occidentale.* Ph.D. thesis, University of Montpellier.

Aguilar, J.-P. 1982 Biozonation du Miocène d'Europe occidentale à l'aide des Rongeurs et corrélations avec l'échelle stratigraphique marine. *C. r. Séanc. Acad. Sci., Paris* **294**, 49–54.

34-2

Aguilar, J.-P., Brandy, L. D. & Thaler, L. 1983 Les rongeurs de Salobrena (Sud de l'Espagne) et le problème de la migration messinienne. *Paléobiol. continent.* **14** (2), 3–17.

Aguilar, J.-P. & Michaux, J. 1987 Essai d'estimation du pouvoir séparateur de la méthode biostratigraphique des lignées évolutives chez les Rongeurs néogènes. *Bull. Soc. géol. Fr.* **8** (III), 1113–1124.

Aguilar, J.-P. & Thaler, L. 1987 *Protolophiomys ibericus* nov. gen., nov. sp. (Mammalia, Rodentia) du Miocène supérieur de Salobrena (Sud de l'Espagne). *C. r. Acad. Sci., Paris* **304**, 859–862.

Agusti, J., Moyà-Solà, S. & Gibert, J. 1984 Mammal distribution dynamics in the eastern margin of the Iberian peninsula during the Miocene. *Paléobiol. continent.* **14** (2), 33–46.

Agusti, J. & Galobart, A. 1986 La sucesion de micromamiferos en el complejo carstico de Casablanca (Almenara, Castellon): problematica biogeografica. *Paleontol. Evol.* **20**, 57–62.

Azzaroli, A. & Guazzone, G. 1980 Terrestrial mammals and land connections in the Mediterranean before and during the Messinian. *Palaeogeogr. Palaeoclimat. Palaeoecol.* **29**, 155–167.

Bader, R. S. 1965 Heritability of dental characters in the house mouse. *Evolution* **19**, 378–384.

Ballmann, P. 1973 Fossile Vögel aus dem Neogen der Halbinsel Gargano (Italien). *Scr. geol.* **17**, 1–75.

Ballmann, P. 1976 Fossile Vögel aus dem Neogen der Halbinsel Gargano (Italien), zweiter Teil. *Scr. geol.* **38**, 1–59.

Barry, J. C., Johnson, N. M., Mahmmood Raza, S. & Jacobs, L. L. 1985 Neogene mammalian faunal changes in Southern Asia: correlations with climatic, tectonic and eustatic events. *Geology* **13**, 637–640.

Bastianelli, S. 1989 Les *Microtia* (Rodentia, Muridae) du Gargano: de nouveaux résultats. *C. r. Acad. Sci., Paris.* (In the press.)

Benton, M. J. 1985 The Red Queen put to the test. *Nature, Lond.* **313**, 734–735.

Berggren, W. A., Kent, D. V., Flynn, J. & van Couvering, J. 1985 Cenozoic geochronology. *Bull. Am. Geol. Soc.* **96**, 1407–1418.

Brandy, L. D. & Jaeger, J.-J. 1980 Les échanges de faunes terrestres entre l'Europe et l'Afrique Nord-Occidentale au Messinien. *C. r. Séanc. Acad. Sci., Paris* **291**, 465–468.

de Bruijn, H. 1974 The Ruscinian rodent succession in Southern Spain and its implication for the biostratigraphic correlations of Europe and North Africa. *Senckenbergiana Lethaea* **55**, 435–443.

de Bruijn, H. & Rümke, C. G. 1974 On a peculiar mammalian association from the Miocene of Oschiri (Sardinia) I & II. *Proc. Konink. Nederland. Akad. van Wetensch.* B **77**, 46–79.

Butler, P. 1980 The giant erinaceid insectivore, *Deinogalerix* Freudenthal, from the upper Miocene of Gargano, Italy. *Scr. geol.* **57**, 1–72.

Coiffait, B., Coiffait, P. & Jaeger, J.-J. 1985 Découverte en Afrique du Nord des genres *Stephanomys* et *Castillomys* (Muridae) dans un nouveau gisement de microvertébrés néogènes d'Algérie orientale: Argoub Kemellal. *Proc. Konink. Nederland. Akad. van Wetensch.* B **88**, 167–183.

Crusafont Pairo, M. 1954 La zona pirenaica como filtro-barrera paleobiologico. *Inst. Est. Pirenaicos, Zaragoza,* pp. 317–332.

Cuenca Bescos, G. 1988 Revision de los Sciuridae del Aragoniense y del Rambliense en la fosa de Calatayud-Montalban. *Scr. geol.* **87**, 1–116.

Daams, R. & Freudenthal, M. 1981 Aragonian: the Stage concept versus Neogene Mammal Zones. *Scr. geol.* **62**, 1–17.

Daams, R. & Freudenthal, M. 1985 *Stertomys laticrestatus*, a new glirid (dormice, Rodentia) from the insular fauna of Gargano (Prov. of Foggia, Italy). *Scr. geol.* **77**, 21–27.

Daams, R., Freudenthal, M. & Sierra Alvarez, M. 1987 Ramblian; A new stage for continental deposits of early Miocene age. *Geol. Mijnbouw* **65**, 297–308.

Daams, R. & Van der Meulen, A. J. 1984 Palaeoenvironmental and palaeoclimatic interpretation of micromammal faunal successions in the Upper Oligocene and Miocene of North Central Spain. *Paléobiol. continent.* **14** (2), 241–258.

De Giuli, C. & Torre, D. 1984 Species interrelationships and evolution in the Pliocene endemic faunas of Apricena (Gargano peninsula, Italy). *Geobios, Mém. Spé.* **8**, 379–383.

Diamond, J. A. 1984 'Normal' extinctions of isolated populations. In *Extinctions* (ed. H. M. Nitecki), pp. 191–244. University of Chicago Press.

Farre, J. 1985 *The importance of mass wasting processes on the continental slope.* Ph.D. thesis, University of Columbia.

Flynn, L. J. 1986 Species longevity, stasis and stairsteps in Rhizomyid rodents. *Contributions to Geology, University of Wyoming, special paper* 3, pp. 273–285.

Freudenthal, M. 1971 Neogene vertebrates from the Gargano peninsula, Italy. *Scr. geol.* **3**, 1–10.

Freudenthal, M. 1972 *Deinogalerix koenigswaldi* nov. gen., nov. sp., a giant insectivore from the Neogene of Italy. *Scr. geol.* **14**, 1–19.

Freudenthal, M. 1976 Rodent stratigraphy of some Miocene fissure fillings in Gargano (prov. Foggia, Italy). *Scr. geol.* **37**, 1–18.

Freudenthal, M. 1985 Cricetidae (Rodentia) from the Neogene of Gargano (Prov. of Foggia, Italy). *Scr. geol.* **77**, 29–76.

Freudenthal, M. (ed.) 1988 Biostratigraphy and paleoecology of the Neogene micromammalian faunas from the Calatayud–Teruel Basin (Spain). *Scr. geol.* (Special issue 1.)

Gervais, P. 1872 Sur un singe fossile, d'espèce non encore décrite, qui a été découvert au Monte Bamboli. *C. r. hebd. Séanc. Acad. Sci., Paris* **74**, 1217.

Gingerich, P. D. 1977 Patterns of evolution in the mammalian fossil record. In *Patterns of evolution* (ed. A. Hallam), pp. 469–500. Amsterdam: Elsevier.

Gingerich, P. D. 1984 Pleistocene extinctions in the context of origination-extinction equilibria in cenozoic mammals. In *Quaternary extinctions: a prehistoric revolution* (ed. P. S. Martin & R. G. Klein), pp. 211–222. Tucson: University of Arizona Press.

Haq, B. U., Hardenbol, J. & Vail, P. R. 1987 Chronology of fluctuating sea levels since the Triassic. *Science, Wash.* **235**, 1156–1167.

Hartenberger, J.-L. 1987 Modalités des extinctions et apparitions chez les mammifères du Paléogène d'Europe. *Mém. Soc. géol. Fr.* **150**, 133–143.

Hartenberger, J.-L. 1988 Etudes sur la longévité des genres de Mammifères fossiles du Paléogène d'Europe. *C. r. Acad. Sci., Paris* **306**, 1197–1204.

Hoffman, A. & Kitchell, J. A. 1984 Evolution in a pelagic planktic system: a paleobiologic test of multispecies evolution. *Paleobiology* **10**, 9–33.

Hürzeler, J. 1975 L'âge géologique et les rapports géographiques de la faune de mammifères du lignite de Grosseto. In *Problèmes actuels de paléontologie. Evolution des vertébrés*, pp. 873–876. Paris: Coll. intern. CNRS number 218.

Hürzeler, J. & Engesser, B. 1976 Les faunes de mammifères néogènes du bassin de Baccinello (Grosseto, Italie). *C. r. hebd. Séanc. Acad. Sci., Paris* **283**, 333–336.

Hürzeler, J. 1983 Un alcelaphiné aberrant (Bovidé, Mammalia) des 'lignites de Grosseto' en Toscane. *C. r. Séanc. Acad. Sci., Paris* **296**, 497–503.

Jaeger, J.-J. & Hartenberger, J.-L. 1975 Pour l'utilisation systématique de niveaux-repères en biochronologie mammalienne. In *R.A.S.T.* (ed. Soc. géol. France), p. 201. Paris.

Jaeger, J.-J., Michaux, J. & Thaler, L. 1975 Présence d'un rongeur Muridé nouveau, *Paraethomys miocaenicus* nov. sp. dans le Turolien supérieur du Maroc et d'Espagne. Implications paléogéographiques. *C. r. hebd. Séanc. Acad. Sci., Paris* **280**, 1673–1676.

Jaeger, J.-J. 1977 Les rongeurs du Miocène moyen et supérieur du Maghreb. *Palaeovertebrata, Montpellier* **8**, 1–166.

Jaeger, J.-J., Coiffait, B., Tong, H. & Denys, C. 1987 Rodent extinctions following Messinian faunal exchanges between Western Europe and Northern Africa. *Mém. Soc. géol. Fr.* **150**, 153–158.

Kurtén, B. 1960 Chronology and faunal evolution of the earlier European glaciations. *Soc. scient. Fenn. Comm. Biol.* **21**, 40–62.

Lavocat, R. 1961 Etude systématique de la faune de mammifères et conclusions générales. In *Le gisement à vertébrés miocènes de Beni-Mellal (Maroc). Notes Mém. Serv. Mines Carte géol. Maroc* **155**, 9–11, 29–94, 109–142.

Leinders, J. 1984 Hoplitomerycidae fam. nov. (Ruminantia, Mammalia) from Neogene fissure fillings in Gargano (Italy). Part 1: The cranial osteology of *Hoplitomeryx* gen. nov. and a discussion on the classification of pecoran families. *Scr. geol.* **70**, 1–51.

Levinton, J. S. & Farris, J. S. 1987 On the estimation of taxonomic longevity from Lyellian curves. *Paleobiology* **13**, 479–483.

Mein, P. 1975 Biozonation du Néogène méditerranéen à partir des Mammifères. In *Report on activity of the R.C.M.N.S. working group*, pp. 78–81. Bratislava.

Miller, K. G., Fairbanks, R. G. & Mountain, G. S. 1987 Tertiary oxygen isotope synthesis, sea level history and continental margin erosion. *Palaeoceanography* **2**, 1–19.

Moyà-Solà, S. 1983 Los boselaphini (Bovidae, Mammalia) del Neogeno de la Peninsula Iberica. Ph.D. Thesis, University of Barcelona.

Moyà-Solà, S., Agusti, J. & Pons-Moyà, J. 1984 The Mio-Pliocene insular faunas from the West Mediterranean origin and distribution factors. *Paléobiol. continent.* **14** (2), 347–357.

Pecorini, G., Rage, J.-C. & Thaler, L. 1974 La formation continentale de Capo Mannu, sa faune de vertébrés pliocène et la question du Messinien de Sardaigne. *Rc. Semin. Fac. Sci. Univ. Cagliari*, **13** (Suppl.), 302–319.

Pickford, M. 1987 Concordance entre la paléontologie continentale de l'Est africain et les événements paléo-océanographiques au Néogène. *C. r. Acad. Sci., Paris* **304**, 675–678.

Raup, D. M. 1978 Cohort analysis of generic survivorship. *Paleobiology* **2**, 289–297.

Raup, D. M. 1987 Major features of the fossil record and their implications for evolutionary rate studies. In *Rates of evolution* (ed. Campbell, K. S. W. & Day, M. F.), 1–14.

Romer, A. S. 1966 *Vertebrate paleontology*. University of Chicago Press.

Schaub, S. 1938 Tertiäre und Quartäre Murinae. *Abh. Schweiz. Pal. ges.* **45**, 1–39.

Sen, S. 1986 Contributions à la magnétostratigraphie et à la Paléontologie des formations continentales néogènes du pourtour Méditerranéen. Implications biochronologiques et paléobiologiques. Ph.D. Thesis, University Pierre et Marie Curie, Paris.

Shackleton, N. *et al.* 1984 Oxygen isotope calibration of the onset of ice rafting and history of glaciation in the North Atlantic region. *Nature, Lond.* **307**, 620–623.

Simpson, G. G. 1944 Tempo and mode in evolution. New Haven, Connecticut: Yale University Press.

Stanley, S. M. 1979 *Macroevolution pattern and process*. San Francisco, California: Freeman.

Stenseth, N. C. & Maynard-Smith, J. 1984 Coevolution in ecosystems: Red Queen or stasis. *Evolution* **38**, 870–880.

Thomas, H. 1984 Les origines africaines des Bovidae (Artiodactyla, Mammalia) miocènes des lignites de Grosseto (Toscane, Italie). *Bull. Mus. natn. Hist. nat., Paris* **6**, 81–101.

Thomas, H. 1985 The early and middle Miocene land connection of the Afro-Arabian and Asia: a major event for the Hominoid dispersal. In *Ancestors: the hard evidence* (ed. E. Delson), pp. 42–50. New York: A. R. Liss.

Thomas, H., Bernor, R. & Jaeger, J.-J. 1982 Origines du peuplement mammalien en Afrique du Nord pendand le Miocène terminal. *Géobios* **15** (3), 283–297.

Vai, G. B. 1988 A field guide to the Romagna Apennine geology: the Lamone valley. In *Fossil vertebrates in the Lamone valley* (ed. C. De Giuli & G. B. Vai), pp. 7–37. Faenza: Università di Bologna.

Vail, P. R., Mitchum, R. M. Jr & Thompson, S. 1977 Seismic stratigraphy of global changes of sea-level. Part 4: Seismic stratigraphy. Applications to hydrocarbon exploration. *Mem. Am. Ass. petroleum Geol.* **26**, 83–96.

Van Valen, L. 1973 A new evolutionary law. *Evol. Theory* **1**, 1–30.

van de Weerd, A. & Daams, R. 1978 Quantitative composition of rodent faunas in the Spanish Neogene and palaeoecological implications. *Proc. Konink. Nederland. Akad. van Wetensch.* B **81**, 448–473.

Phil. Trans. R. Soc. Lond. B **325**, 421–435 (1989)

Printed in Great Britain

The case for extraterrestrial causes of extinction

BY D. M. RAUP

Department of Geophysical Sciences, University of Chicago, Chicago, Illinois 60637, *U.S.A.*

The dramatic increase in our knowledge of large-body impacts that have occurred in Earth's history has led to strong arguments for the plausibility of meteorite impact as a cause of extinction. Proof of causation is often hampered, however, by our inability to demonstrate the synchronism of specific impacts and extinctions. A central problem is range truncation: the last reported occurrences of fossil taxa generally underestimate the true times of extinction.

Range truncation, because of gaps in sedimentation, lack of preservation, or lack of discovery, can make sudden extinctions appear gradual and gradual extinctions appear sudden. Also, stepwise extinction may appear as an artefact of range truncation. These effects are demonstrated by experiments performed on data from field collections of Cretaceous ammonities from Zumaya (Spain). The challenge for future research is to develop a new calculus for treating biostratigraphic data so that fossils can provide more accurate assessments of the timing of extinctions.

INTRODUCTION

Ever since Alvarez *et al.* (1980) reported anomalously high iridium concentrations at the Cretaceous–Tertiary (K–T) boundary, intense controversy has surrounded the question of a large-body impact (asteroid or comet) as the primary cause of the terminal Cretaceous mass extinction. Although the K–T problem and iridium anomalies still attract considerable attention, the investigations have broadened to include other geochemical and geophysical evidence of impact and other extinctions in the Phanerozoic. This has stimulated, in turn, new research on the nature of the mass extinctions and on other possible extinction mechanisms, including the effects of sea-level- and climate change and of episodes of intense volcanism.

The question of impact as a possible cause of mass extinction has become extraordinarily complex. Several hundred research papers have been published on the subject since 1980, and commentaries number in the thousands. The literature has thus become unmanageably large, so that a comprehensive review is impossible. More important, the subject involves so many separate scientific disciplines – from palaeontology to astrophysics – that no one individual is competent to judge the merits of all the arguments and counterarguments. Because of this, I refrain from trying to summarize the evidence for the extraterrestrial interpretation of specific extinctions. Instead, the reader is referred to recent reviews by Alvarez (1987) and Hallam (1987); these papers summarize large segments of the basic evidence.

Despite the problems just mentioned, one can easily make a case for the plausibility of large-body impact as a cause of mass extinction.

PLAUSIBILITY ARGUMENTS

Virtually all palaeontologists agree that there have been intervals during the Phanerozoic of unusually high extinction rates. Although the mass extinctions of the late Ordovician, late

Devonian, late Permian, late Triassic and terminal Cretaceous are undoubtedly the largest such events, several of the lesser extinctions are more intense and pervasive than can be explained simply by chance variation in background extinction. Furthermore, these events (or intervals) generally affect organisms over broad geographical areas and encompass many habitats and modes of life. Differences among extinction events exist but are remarkably minor (see Raup & Boyajian (1988) for documentation).

The pervasiveness of the larger extinction events makes it difficult to sustain explanatory mechanisms based on purely biological factors, such as interspecific competition, predation and the like. On the other hand, relatively rapid changes in the physical environment do offer plausible, albeit not necessarily compelling, candidates to explain mass extinctions.

Although an extinction event need not be truly worldwide to produce a significant peak in the Phanerozoic extinction record, a disturbance of substantial geographical extent seems to be required. Among the many ways of producing large-scale environmental perturbation, impacts by asteroids and comets must be considered among the plausible candidates for two reasons: (1) large-body impacts have been common throughout Phanerozoic time; and (2) large impacts have pronounced global effects on atmospheric chemistry, insolation and other relevant environmental factors.

The impact rate over the past 600 million years is reasonably well known. Knowledge of this phenomenon, and its historical record, has increased tremendously in the past three decades with the advent of satellite observations of ancient impact structures, the discovery of geophysical, mineralogical and geochemical indicators of impact, and the increased knowledge of asteroids now in Earth-crossing orbits. Contrary to the conventional wisdom of earlier decades, bombardment of the earth by large objects was not limited to the early Precambrian but rather, has continued at a moderate level to the present day.

Current estimates of the Phanerozoic flux of large impacting objects call for an average of one object of 10 km diameter or larger every 100 million years, and of one object of 1 km or larger every 200000 to one million years (Shoemaker *et al.* 1988)

Although much excellent work on the environmental effects of large-body impact has been done since 1980, many questions remain to be answered. All workers agree, however, that the physical consequences of an impact of one of the larger objects would be severe, many orders of magnitude greater than the effects of a simultaneous detonation of all nuclear bombs and warheads now in existence. Much attention has been given to the debris cloud that would be produced by a large impact and this cloud could indeed be an important element. But recent work on the effects on atmospheric chemistry suggest yet more severe environmental consequences (Prinn & Fegley 1988; Crutzen 1987).

The foregoing discussion is not intended to argue for impact as a better explanation of mass extinction than any other. Its purpose is merely to show that a causal link between impact and extinction is entirely plausible and that this link should be evaluated equally with other explanations. Furthermore, the impact explanation is eminently testable. Although some of the evidence for large-body impact at the K–T boundary is challengeable, such as the putative microtectites and some of the trace element and shocked-quartz data, geology has at its disposal several virtually ironclad indicators of impact. The verification of more than 100 large craters through impact melts, high-pressure minerals (stishovite and coesite) and shatter cones argues for the feasibility of testing proposed extinction–impact pairs (see Grieve 1987). Also, where impacts melts are available, extremely accurate radiometric dating is straightforward.

In the controversies surrounding the search for causes of mass extinction, palaeontologists are continually being asked to provide data on the details of particular extinction events. This has been especially true for the Cretaceous–Tertiary transition, but it applies throughout the Phanerozoic. Common questions include the following.

1. Are the extinctions simultaneous (sudden) or are they spread out over an extended time (gradual)?

2. If the extinctions appear to be gradual overall, is the pattern punctuated by smaller events: the simultaneous extinction of small groups of species?

3. Does a stratigraphic gap between the last occurrence of a group of fossils and a major boundary indicate that the organisms died out before the boundary event?

If mass extinctions are gradual, extending over a considerable time, some of the proposed causes of extinction are not tenable, such as a single comet or asteroid impact. On the other hand, if it can be shown that many species died out in a relatively short time, the idea of a single impact becomes viable, albeit not proven. If mass extinctions are punctuated by 'steps', then proposals of extended comet showers are plausible (Hut *et al.* 1987). Thus stratigraphic data become vitally important in the evaluation of extinction mechanisms.

It has been customary for palaeontologists to accept the first and last occurrences of a species (or larger group) as a literal record of origination and extinction, even though it is widely recognized that the observed ranges of fossils are often truncated by failure of preservation or lack of discovery. As is shown by examples in this paper, range truncation can make a gradual extinction appear sudden or a sudden extinction appear gradual. Until palaeontologists develop the methodology needed to cope with this problem, the contributions of stratigraphic data to our knowledge of mass extinction will be severely limited.

In the sections that follow, I illustrate the problems just mentioned by developing a case study from data on occurrences of late Cretaceous ammonites at Zumaya in northeastern Spain. Most of the phenomena I describe are well known to palaeontologists and stratigraphers, but the formulation may help to focus attention on those situations most in need of new methods of analysis.

ZUMAYA AMMONITES

Figure 1 shows the ranges of 21 ammonite lineages in the 200 metres below the K–T boundary at Zumaya. The data were provided by P. D. Ward (personal communication, April 1988) and are the result of exhaustive collecting and taxonomic work over several years by Ward, in conjunction with J. Wiedmann, J. Mount, and W. J. Kennedy. The data are an update of a similar chart published by Ward *et al.* (1986) and are based on extensive new collections, taxonomic revision, and re-identification. Only material actually in Ward's possession or field identifications made with certainty by Ward are included. Although these studies have been extensive, new collecting will undoubtedly modify details of the record. Future additions and corrections are not relevant to my purpose here, however, because the value of the Zumaya dataset is that it typifies problems common throughout the stratigraphic record.

The 21 lineages in figure 1 are considered by Ward to be separate evolutionary lines, although preservation does not allow specific or subspecific identification in all cases. The

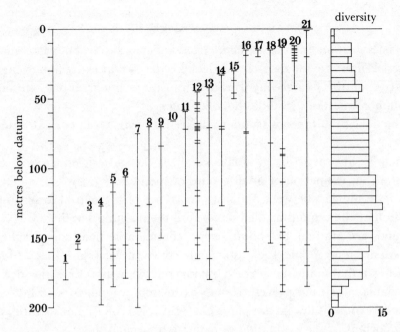

FIGURE 1. Stratigraphic ranges for 21 ammonite lineages at Zumaya, Spain (data provided by P. D. Ward, April 1988). The lineages are defined in table 1. The vertical scale is in metres below the Cretaceous–Teriary boundary at Zumaya. Fossil occurrences are indicated by horizontal tick marks. The histogram on the right shows the changing number of lineages through the section.

TABLE 1. AMMONITE LINEAGES RECOGNIZED BY WARD IN THE 200 m BELOW THE
CRETACEOUS–TERTIARY BOUNDARY AT ZUMAYA, SPAIN

(Lineages are numbered in order of last occurrence (see figure 1 for range chart).)

1. *Baculites anceps*	12. *Glyptoxoceras subcompressum*	
2. *Hauericeras rembda*	13. *Diplomoceras cylindraceum*	
3. *Desmophyllites larteti*	14. *Pachydiscus gollevillensis*	
4. *Pachydiscus epiplectus*	15. *Hoploscaphites constrictus crassus*	
5. *Pseudophyllites indra*	16. *Saghalinites* sp.	
6. *Phylloptychocerus sipho*	17. *Vertebrites kayei*	
7. *Pachydiscus neubergicus*	18. *Phyllopachyceras forbesianum*	
8. *Anagaudryceras* cf. *A. politissimum*	19. *Baculites* sp.	
9. *Hypophylloceras surya*	20. *Anapachydiscus fresvillensis*	
10. *Fresvillia constricta*	21. *Neophylloceras ramosum*	
11. *Pachydiscus jacquoti*		

identifications, as provided by Ward, are given in table 1. For convenience, the lineages are arranged in order of last occurrence and numbered accordingly.

In all, 150 ammonite specimens were identified. Of these, 44 specimens are redundant in the sense that other specimens of the same lineage were found at the same horizon. This leaves 106 occurrences, if an 'occurrence' is defined as one or more specimens of a lineage at a horizon. The 106 occurrences are indicated by the horizontal ticks in figure 1.

The histogram on the right in figure 1 summarizes lineage diversity through the 200 m section. Diversity increases from near zero at the base to a maximum of 15 coexisting lineages at the 125–130 m level. Diversity then declines to zero as the K–T boundary is approached. The low diversity near the base of the section is due to lithofacies: below 200 m, the rocks are dominantly turbidites and contain few ammonites (P. D. Ward, personal communication,

1988). Of greatest interest, of course, is the long upward decline in number of lineages towards the K–T boundary. This decline has been used by several authors as evidence that the ammonites were not suddenly and simultaneously eliminated by the K–T boundary (Ward *et al.* 1986; Wiedmann 1987; but see Ward & MacLeod (1988) for a different view).

Although the extinction pattern at Zumaya is highly relevant to the question of the Cretaceous mass extinction, the dataset also provides a good testing ground for more general aspects of the analysis of stratigraphic range charts. After considering some of these, I return to the question of the gradual extinction of ammonites at Zumaya.

THE HIATUS EFFECT: GRADUAL EXTINCTIONS APPEAR SUDDEN

It is well known that an erosional unconformity, a gap in sedimentation, or a failure of fossil preservation can simulate a sudden extinction. This can be illustrated with the Zumaya range chart by performing a simple experiment: ignore all fossil occurences for an arbitrary thickness of sediment. This is done in figure 2 using an algorithm that eliminates all fossil occurrences between 25 m and 125 m. The remaining ranges are shown as vertical lines connecting the first and last occurrences that survived the experiment (that is, those above 25 m or below 125 m). The lineages have been rearranged to match the new sequence of last occurrences, but the numbering used earlier is maintained so that lineages can be identified. Also, a histogram of last occurrences (apparent extinctions) is shown on the right.

Figure 2 has several interesting features. The ranges of some lineages, such as 1 (on the left) and 21 (on the right) are unaffected by the artificial hiatus because their actual first and last occurrences lie outside the hiatus interval. The ranges of most other lineages are profoundly

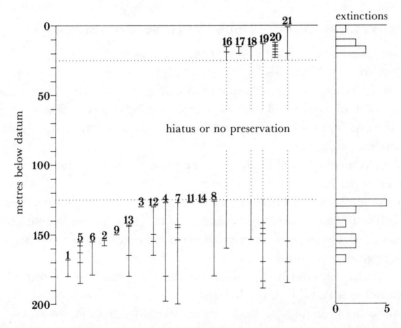

FIGURE 2. Experiment with the effect of imposing a preservational gap in the Zumaya range chart. The pattern is that which would be produced if all fossil occurrences between the 25 and 125 m levels were ignored. Dotted lines indicate Lazarus taxa: those taxa that disappear at the hiatus but reappear above the hiatus. Note that the hiatus produces the appearance of a sudden extinction at the 125 m level even though the full dataset (figure 1) contains no such event.

affected, however, by the loss of information in the 25–125 m interval. For example, the top of lineage 5 drops from 110 m to 155 m because of the loss of the two occurrences in the hiatus interval.

The main effect of the range truncation is to produce a concentration of apparent extinctions close to the base of the hiatus interval (lineages 3, 12, 4, 7, 11, 14 and 8). This is reflected in the histogram of extinctions. Thus we have the appearance of a substantial extinction event at about 125 m even though we know from figure 1 that no such event occurred.

If the hiatus (or interval of no preservation) were not recognized, the pattern in figure 2 could be interpreted as a sudden extinction. In this particular case, the presence of several Lazarus taxa (see Jablonski 1986; McGowran 1986) provides evidence of the hiatus: lineages 16, 18, 19 and 21 are present below the hiatus and reappear above the hiatus.

The pattern produced by the hiatus experiment mimics the real-world situation of brachiopods in the K–T section at Nye Klov (Denmark). In range charts published by Surlyk & Johansen (1984), the brachiopods show a gap in occurrence (documented by Lazarus taxa), immediately above the K–T boundary, and it is impossible to determine whether the termination of species at the K–T boundary is real or merely an artefact of range truncation (see Raup (1987) for discussion). The proposition of sudden and simultaneous extinction is plausible but no more so than the opposing proposition that extinction was gradual throughout the first few metres of the Tertiary.

A substantial gap in the fossil record does not necessarily produce a pattern simulating sudden extinction. In repeated simulations of the type shown in figure 2, spurious extinction events are common but not universal: the outcome in each case depends on the vagaries of the distribution of occurrences above and below the specified hiatus interval.

THE SIGNOR–LIPPS EFFECT: SUDDEN EXTINCTIONS APPEAR GRADUAL

In an important paper, Signor & Lipps (1982) analysed the consequences of limited sampling of the fossil record. In particular, they noted that range truncation often leads to a backward smearing of extinction events and this was later labelled the Signor–Lipps effect (Raup 1986). The work of Signor & Lipps was largely model based and relied primarily on monte-carlo simulations. It is appropriate to explore the phenomenon further by experiments on the Zumaya ammonite dataset.

Figure 3 shows the consequences of imposing a complete and instantaneous extinction of all Zumaya ammonites at the 100 m level. It is assumed, for the experiment, that all ammonites died out at the 100 m level and consequently, all occurrences above this level are ignored. The real occurrences below 100 m are used to construct the range chart as it would appear under these circumstances. The 100 m level was chosen, with reference to the raw data in figure 1, because there was high diversity and no extinction at this level.

The question for this experiment is: what would the preserved record of a sudden annihilation of all lineages look like? The striking result is that the extinctions appear to be gradual, as shown by the histogram of diversity on the right side of figure 3.

The reason that a sudden extinction produced a gradual decline in diversity is obvious from the original data. The stratigraphic tops of all lineages that had originally extended above the extinction horizon (100 m in this case) were truncated back to their last pre-extinction occurrence.

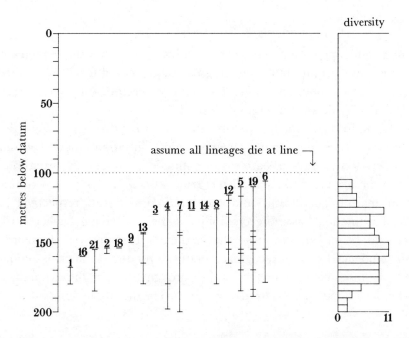

FIGURE 3. Experiment showing the effect of eliminating all fossil occurrences above the 100 m level, simulating the sudden extinction of all lineages at that level. Diversity declines gradually toward the extinction horizon even though the extinction is 'known' to have been sudden. Also, a spurious extinction 'step' appears at the 125 m level.

This experiment shows another kind of artefact that calls for special attention. Note that at about the 125 m level, six lineages terminate almost together (3, 4, 7, 11, 14 and 8). There is nothing in the full dataset to suggest a stepwise extinction and the 'event' is 25 m below the imposed extinction. The step is an indirect result of the 100 m extinction because all higher occurrences of four of the six lineages happen to be above the 100 m level (see figure 1).

It is clear from this experiment that a true extinction in the fossil record may produce a spurious extinction lower in the section. In repeated simulations of this type, with extinctions imposed at different horizons, spurious stepwise extinctions are extremely common.

DISCUSSION

The experiments with the Zumaya dataset have several clear implications, as follows:

(1) a hiatus in fossil preservation often produces the appearance of a sudden extinction where none exists;

(2) extinctions, known to be sudden, often appear to be gradual;

(3) stepwise extinction is often an artefact of the placement of fossil occurrences.

The spurious stepwise extinctions produced by the experiments have further implications. These non-events usually occur at horizons having natural concentrations of fossils rather than at times of true extinction. Given a true extinction event, ranges are truncated back to the next lower horizon of fossil concentration. Only if the fossil concentration happens to coincide with the extinction, is the stepped event an indicator of the extinction. In the general case, a true stepwise extinction event is unlikely to be preserved as such; rather, it is most likely to be smeared backwards in time because of variable range truncation among the involved lineages.

Generality of the Zumaya results

The experimental results presented here are caused by the fact that the stratigraphic ranges at Zumaya are based on discontinuous and irregular sequences of fossil occurrences. To the extent that this condition is common throughout the fossil record, the same patterns will be found regardless of the age or kind of organisms used.

The distortions found in the Zumaya experiments can be avoided only if fossil preservation is virtually continuous. It follows that palaeontological situations more continuous than Zumaya are less likely to distort true patterns of extinction. Conversely, situations with less continuous preservation are more likely to distort true extinctions. The two extremes may be found typically in marine microfossils and terrestrial vertebrates, respectively. Thus an observed pattern of extinction has more credibility if the underlying fossil record is nearly continuous. It should be noted, however, that a nearly complete fossil record does not protect against the tendency for an unrecognized hiatus to produce a spurious extinction event (figure 2). In fact, the more continuous the fossil record, the sharper the extinction produced by a hiatus will be, as long as species are dying out during the hiatus interval.

Towards a new calculus

The foregoing discussion is non-rigorous in the sense that it documents tendencies and typical outcomes: it does not provide the means to assess the likelihood that a given extinction is what it appears to be. In the case of the late Cretaceous disappearance of ammonites at Zumaya, it is important to be able to deduce whether the extinctions at or below the K–T boundary were actually gradual or merely an artefact of the distribution of last occurrences.

The past few years have seen a dramatic increase in the quantity and quality of research on the general problem of evaluating truncated stratigraphic ranges and many of these studies are germane to the Zumaya problem. Especially important is the classic paper by Paul (1982), which presents a preliminary formulation of a calculus for estimating the uncertainty in the placement of first and last occurrences. Follow-up studies include those of Strauss & Sadler (1987) and Springer & Lilje (1988).

Several of the methods used by Paul and others could be applied to the Zumaya case to estimate the probability that a given extinction pattern seen in the range charts is real. Furthermore, the experimental mode used here with the Zumaya data makes it possible to test methodologies by asking whether an imposed extinction event (such as in figure 3) can be reconstructed from the experimental results. But most of the existing statistical methods are directed at problems of time correlation between stratigraphic sections and thus are designed to answer somewhat different questions. For this reason, I will describe a different approach to the problem of assessing extinction patterns. This will involve bootstrapped simulations of what the preserved record should look like under varying conditions of extinction timing.

Simulation of sudden extinction

Starting from the horizon of maximum diversity in the full range chart for Zumaya (15 lineages at 130 m in figure 1), let us assume, as an extreme case, that all lineages actually survived up to the K–T boundary (0 m level) but were killed simultaneously at the boundary. In a preserved record, the ranges of each of these lineages will be truncated backwards in time by an amount typical of the gaps between fossil occurrences at Zumaya. The full range chart (figure 1) contains 85 gaps between fossil occurrences (ticks) and the length of each can be measured. A gap (amount of range truncation) can be assigned to each of the lineages and a picture of probable diversity decline can be constructed. This can then be compared with the diversity decline actually observed to evaluate the efficacy of the model of sudden extinction.

If the occurrences of ammonites at Zumaya were randomly distributed, we could assume an exponential distribution of gap sizes, following Paul, and the ideal case of simultaneous extinction could be modelled by a simple calculation. But the occurrences at Zumaya are not randomly distributed, there being more small gaps than would be predicted by a random model. Therefore, a boostrap technique using the gaps observed in the full Zumaya range chart (figure 1) is preferable.

For each lineage, a gap size is chosen at random (with replacement) from the 85 gaps observed at Zumaya. The gap selected is then used to define a distance (down) to the top of the preserved range of that lineage. When applied to all lineages, this procedure produces one possible pattern of declining diversity. The simulation is repeated many times to build a general picture of diversity decline under the condition of sudden extinction. The result is shown in figure 4: the vertically ruled pattern marked 'sudden' contains 90% of the outcomes of 200

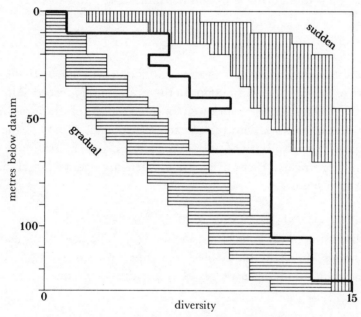

FIGURE 4. Simulation of the pattern of diversity decline expected under two experimental conditions, one representing sudden extinction (vertical ruling) and the other representing gradual extinction (horizontal ruling). The actual diversity decline, from figure 1, is shown as the solid line. The position of the solid line between the simulation bands indicates that ammonite extinction below the K–T boundary at Zumaya was a combination of the sudden and gradaul modes.

simulations. Thus, if all lineages actually lived up to the K–T boundary and all went extinct at that level, the apparent decline in diversity is expected to lie within the indicated band.

Simulation of gradual extinction

The procedure is the same except that a linear decline in actual diversity is postulated. The lineages are assumed to go extinct, one by one, at a uniform rate from the 130 m level to the

K–T boundary. Random picks from the Zumaya gap distribution are then used to further truncate the ranges. The resulting distribution, based on 200 simulations, is shown as the horizontally ruled pattern marked 'gradual' in figure 4.

Evaluation of results

The two alternative models, sudden and gradual extinction, predict patterns of diversity decline that are virtually non-overlapping. The solid line in figure 4 is the track of diversity decline observed in the full range chart (figure 1) and this falls between the two extreme models.

We must conclude that the upper part of the Zumaya section is best explained by a mixture of sudden and gradual extinction. The basic shape of the diversity decline is that of sudden extinction but the position of the curve is to the left of the prediction for sudden extinction and to the right of that for gradual extinction. This suggests that the extinctions were spasmodic, but it does not tell us whether they were concentrated at the K–T boundary.

New field collections made since April 1987, at other localities in the Bay of Biscay Basin demonstrate that the ammonites persisted at high diversity to within a metre of the K–T boundary (Ward & MacLeod 1988). Thus the suddenness of the ammonite extinction is established regionally even though the situation at Zumaya remains uncertain.

CONCLUSION

The case for or against extraterrestrial causes of extinction will not be made firm until we have a clearer picture of the timing of extinctions in the stratigraphic record. Although existing methods of quantitative biostratigraphy work well for some aspects of stratigraphic correlation, new approaches are needed to work with the extinction problem. Although the method used to develop figure 4 has obvious limitations, it does suggest a direction for future research. The method should yield more nearly definitive results when applied to larger datasets: more lineages over longer time spans.

This research was supported by NASA grant NAG 2-237. Also, the study benefited substantially from discussions with colleagues, including J. J. Sepkoski, Jr, David Jablonski, George Boyajian, Charles Marshall, and Mike Foote. Special thanks are due Peter D. Ward for providing his field data on the occurrences of ammonites at Zumaya.

REFERENCES

Alvarez, L. W. 1987 Mass extinctions caused by large bolide impacts. *Physics to-day*, July, pp. 24–33.

Alvarez, L. W., Alvarez, W., Asaro, F. & Michel, H. V. 1980 Extraterrestrial cause for the Cretaceous–Tertiary extinction. *Science, Wash.* **208**, 1095–1108.

Crutzen, P. J. 1987 Acid rain at the K/T boundary. *Nature, Lond.* **330**, 108–109.

Grieve, R. A. F. 1987 Terrestrial impact structures. *A. Rev. Earth planet. Sci.* **15**, 245–270.

Hallam, A. 1987 End-Cretaceous mass extinction event: argument for terrestrial causation. *Science, Wash.* **238**, 1237–1242.

Hut, P., Alvarez, W., Elder, W. P., Hansen, T., Kauffman, E. G., Keller, G., Shoemaker, E. M. & Weissman, P. R. 1987 Comet showers as a cause of mass extinction. *Nature, Lond.* **239**, 118–126.

Jablonski, D. 1986 Causes and consequences of mass extinctions: a comparative approach. In *Dynamics of extinction* (ed. D. K. Elliott), pp. 183–229. New York: John Wiley.

McGowran, B. 1986 Beyond classical stratigraphy. *J. Petrol. Explor. Soc. Aust.*, September 1987, 28–41.

Paul, C. R. C. 1982 The adequacy of the fossil record. In *Problems of phylogenetic reconstruction* (ed. K. A. Joysey & A. E. Friday), pp. 75–117. London: Academic Press.

Prinn, R. G. & Fegley, B. Jr 1988 Biospheric traumas caused by large impacts and predicted relics in the sedimentary record. *Contr. lun. planet. Inst.* **673**, 145.

Raup, D. M. 1986 Biological extinction in earth history. *Science, Wash.* **231**, 1528–1533.

Raup, D. M. 1987 Mass extinction: a commentary. *Palaeontology* **30**, 1–13.

Raup, D. M. & Boyajian, G. E. 1988 Patterns of generic extinction in the fossil record. *Paleobiology* **14**, 109–125.

Shoemaker, E. M., Shoemaker, C. S. & Wolfe, R. F. 1988 Asteroid and comet flux in the neighborhood of the earth. *Contr. lun. planet. Inst.* **673**, 174–176.

Signor, P. W. III & Lipps, J. H. 1982 Sampling bias, gradual extinction patterns, and catastrophes in the fossil record. In *Geological implications of impacts of large asteroids and comets on the Earth* (ed. L. T. Silver & P. H. Schultz), pp. 291–296. Boulder, Colorado: Geological Society of America.

Springer, M. & Lilje, A. 1988 Biostratigraphy and gap analysis: the expected sequence of biostratigraphic events. *J. Geol.* **96**, 228–236.

Strauss, D. & Sadler, P. M. 1987 Confidence intervals for the ends of local taxon ranges. *Tech. Rep. Univ. Calif. (Riverside) Dep. Statist.* **158**, 1–21.

Surlyk, F. & Johansen, M. B. 1984 End-Cretaceous brachiopod extinctions in the Chalk of Denmark. *Science, Wash.* **223**, 1174–1177.

Ward, P. D. & MacLeod, K. 1988 Macrofossil extinction patterns at Bay of Biscay Cretaceous–Tertiary boundary sections. *Contr. lun. planet. Inst.* **673**, 206–207.

Ward, P. D., Wiedmann, J. & Mount, J. F. 1986 Maastrichtian molluscan biostratigraphy and extinction patterns in a Cretaceous/Tertiary boundary section exposed at Zumaya, Spain. *Geology* **14**, 899–903.

Weidmann, J. 1987 The K/T boundary section of Zumaya (Guipuzcoa). In *Field-guide excursion to the K/T boundary at Zumaya and Biarritz* (ed. M. A. Lamolda), pp. 17–23. Leioa, Vizcaya: Jornadas de Paleontologia.

Discussion

P. A. SABINE (19 *Beaufort Road, London, U.K.*). Professor Raup's suggestion that no single individual could present a case for or against an extraterrestrial cause of extinction may be refuted for the Stevns Klint, Denmark, K–T locality by the extraordinarily careful stratigraphical and chemical work done, especially by Hansen and his colleagues of Copenhagen. Recognition of carbon-stained and non-stained white Bryozoa, and ^{13}C measurements by these workers contradicts the hypothesis that the carbon black originated from a giant forest fire started by an impacting bolide (Hansen *et al.* 1987). The carbon-black isotope values are unrelated to meteoritic material but are consistent with a terrestrial origin.

It is highly relevant that the carbon black occurs in the 3.5 m (*ca.* 50 000 years) below and before the boundary clay with its Ir anomaly. The Ir, with large values (at a maximum of 185 ng g^{-1}, *ca.* 45 times the world average of 4 ng g^{-1}), was found to be closely associated with the carbon. It was concluded that the iridium-carrying phase was the carbon black itself. The occurrence recalls the strong affinity of gold and carbonaceous material in the Witwatersrand. Iridium is known (as fluoride) in volcanic gases and a likely explanation is that Ir from such a source was adsorbed on to the carbon black.

There is a general point that this sort of investigation does show the value to be obtained by very careful, precise work when it is based on innovative ideas, and, even more, the necessity to do it. This research provides a powerful comment on Professor Raup's initial suggestion and I ask whether he would accept that in this case the bolide hypothesis has strong evidence against it.

Reference

Hansen, H. J., Rasmussen, K. L., Gwozdz, R. & Kunzendorf, H. 1987 *Bull. geol. Soc. Denm.* **36**, 305–314.

D. M. RAUP. Dr Sabine has suggested that the Hansen group in Copenhagen has 'strong evidence' against the bolide hypothesis of mass extinction, and that this case argues against my suggestion that the problem of mass extinction by extraterrestrial causes is beyond the expertise of any one individual. I will stand with my earlier statement and suggest that the Hansen research is a good illustration of my point.

The Hansen study is important, of course, and is well known in the scientific community. And I agree that it is an excellent example of what can be accomplished with 'very careful, precise work'. Dr Hansen presented his findings and interpretations at the Global Catastrophes in Earth History meeting (Snowbird, Utah) in October 1988, and it was debated fully by the assembled geologists and geochemists, but with no clear resolution. The lack of consensus was, I think, because the subject is extraordinarily complex and because any evaluation of the conclusions requires expert knowledge of Cretaceous palaeontology and biostratigraphy, the biogeochemistry of carbon, the geochemistry of iridium and other trace elements, stable isotopes, diagenetic processes in a complex environment, forest fires and their products, and chemistry of meteorites.

Although I have no doubt that the Hansen team has many (or perhaps all) of the necessary ingredients, any synthesis of the problem made by a single individual must perforce be based largely on hearsay, or to put it another way, on deference to expert colleagues. And this was my major point: unlike many other research problems one deals with on a daily basis, the K–T problem is beyond the resources of any one of us. For the practising palaeontologist or geochemist, this can be both discouraging and exciting. On the one hand, no one is fully competent to evaluate even their own research results, but on the other hand, they are privileged to work in a multidisciplinary milieu one rarely experiences.

P. ASHMOLE (*Department of Zoology, University of Edinburgh, U.K.*). The experiments Professor Raup has described seem to show that one can say very little about the temporal pattern of extinction of a group if the fossils are scarce. I suppose I am right in thinking that the problems would be much less severe if the fossils were, say, an order of magnitude more abundant. Does Professor Raup think that in analysing episodes of extinction one should simply accept that useful data can be obtained only from taxa with abundant fossils in the strata concerned?

D. M. RAUP. Should we ignore scarce fossils when dealing with extinction questions like those at the K–T boundary? This is an interesting and important question. In the Zumaya experiments described in my paper, I showed that it can be impossible to distinguish between sudden and gradual extinction because of limitations in sampling and preservation. Yet the Zumaya case is not atypical of the fossil record: it is better than many (especially terrestrial vertebrate sections) but worse than many others (microfossil sequences, in particular). Should researchers wishing to work with the details of mass extinction convert to micropalaeontology? Or should they simply work harder to find more fossils in sparsely fossiliferous rocks and search for better methods of treating the available data? I don't know.

P. W. KING (*Department of Biology, University College London, U.K.*). It has been suggested that astronomers and physicists should learn more about palaeontology, or even that they should not interfere at all in it. However, I think that it is necessary to teach the palaeontologists a little bit of astronomy.

When the *Apollo* astronauts went to the Moon, they did not just go there for the greater glory of the United States of America: they also performed a lot of useful science. For example, they collected a lot of rocks, which were brought back and, among other things, dated by radioistropic methods. It was found that most of the lunar rock and soil samples were older than almost any rocks found on the Earth, and that for the second half of its life the Moon has been almost totally geologically inactive. However, it has been struck during this time by large bodies, up to asteroidal sizes, which have produced some of the large craters (although most of these are much older). *Apollo 12* astronauts sampled some material from a ray, which comprises ejected material, associated with the crater Copernicus. This is about 100 km in diameter and would have been formed by an object about 10 km in diameter. This material was dated to be about 900 million years old. Copernicus is thus Pre-Cambrian. It is a very well-preserved crater. Nevertheless, there are two craters of similar size on the near-side of the Moon which are, from photographic evidence, even better preserved and less eroded by very small impacts, and are of Phanerozoic age in all probability (I do not know if there are other craters this large and well preserved on the far side of the Moon, but there may well be). If one considers therefore that the Moon has been struck in Phanerozoic times by, say, two objects around 10 km diameter it follows that the Earth, which has a surface area of about 14 times as large, should have been struck by around 28 such asteroidal bodies, of which one or two can reasonably be expected to have been rather larger than the 10 km minimum I am setting. In fact, the greater gravity of the Earth would have a focusing effect and the number of such impacts expected may be slightly over 28. There is no way that the Earth can have avoided being hit.

I assume that such a 10 km diameter body hit the Earth at, for example, the Cretaceous–Tertiary boundary. 'Back of envelope' calculations reveal that, if it is assumed only one third of the bulk of the impacting object is comminuted to fine dust, which enters the atmosphere and remains in suspension long enough to be uniformly deposited worldwide, then each square metre of the surface of the globe will be covered by two kilograms of dust which would make a layer one or two millimetres thick. (I have assumed the remaining asteroidal material would either enter the body of the Earth or escape into outer space. In reality much of the material excavated from the crater by the impact – terrestrial material – would also enter the atmosphere, but it is difficult to calculate how much.) Even the quantity I calculate, which is a very conservative estimate, is sufficient that, when suspended in the upper air as dust, it would cause a great deal of obstruction to the light of the Sun, with, if continued for any length of time, consequent climatic changes. It is not reasonable to believe that an event of this severity would have no measurable effect on life at all.

I am convinced that large impacts are not uncommon in Phanerozoic times. The evidence for a large impact at the K–T boundary (the Iridium and other heavy-metal anomalies, the shock metamorphism of quartz grains, etc.) is, to my mind, overwhelming. I am much less convinced of the likely scale of the environmental effects, although simple physics shows us they must have been severe. Many enthusiasts have proposed a variety of extreme scenarios, and the evidence brought forward in support of them from computer modelling has been confusing and even contradictory. I think that a great deal more work needs to be done in this field, modelling likely effects, climatic and other, which experimental evidence brought to bear from many disciplines of science. The survival, apparently almost unaffected, of many groups across the K–T boundary places a loose upper bound on the scale of the environmental dislocation and may, I think, rule out some of the most extreme models.

In conclusion, I find the evidence of periodicity in mass extinctions to be weak at best, but of the reality of impacting bolides causing environmental dislocation I have no doubt.

The input of physicists to palaeontology has been most fruitful, as it has in other branches of geology, for without methods developed from physics how could absolute dates of rocks be obtained?

D. M. RAUP. I am in complete agreement with the suggestion that our ignornance of the environmental effects of large-body impact is the weak link in the extraterrestrial interpretation of the Cretaceous mass extinction. We have no personal experience with impacts of the size postulated (fortunately!) so the environmental consequences of the impact can only be inferred by extrapolation from small-scale laboratory experiments or by numerical modelling. My understanding is that the uncertainties in both approaches are formidable. And if the physical consequences of a large impact are in doubt, what chance is there of making precise predictions about biological effects and, in turn, confirming or denying these predictions in the fossil record?

Despite this rather dismal view, there is hope from another direction, using the following logic. It is known, as Mr King has pointed out, that the Earth has been hit by many large meteorites during the Phanerozoic. And the geological, geophysical and geochemical techniques for recognizing these events have vastly improved. Within a few years, there should be as good a chronology of Phanerozoic impacts as there now is for extinctions. Armed with these data, it should be straightforward to show whether there is a statistically valid association between large impacts and large extinctions.

A. W. WOLFENDALE, F.R.S. (*Department of Physics, University of Durham, U.K.*). Concerning the title of Professor Raup's lecture 'The case for extraterrestrial causes of extinction', would it be true to say that there has been a gradual extinction of 'the case'? Specifically, does Professor Raup consider that the periodicity in the extinction record is now less secure than hitherto?

D. M. RAUP. Professor Wolfendale has asked whether there has been a gradual extinction of the case for extraterrestrial causes of extinction. From my vantage point, the opposite is true. Eight years ago, there was only the iridium evidence and that was based on only three localities. Furthermore, the iridium data were open to a plethora of alternative interpretations. By now, the K–T iridium anomaly has been confirmed at scores of localities worldwide and many (although not all) of the alternative interpretations have been ruled out by most people working in that subject. In addition, we have very strong evidence from shocked minerals and from osmium isotopes. The missing crater is still a problem to some workers, but there are several good candidates being investigated actively.

Also, the past eight years have seen a vast increase in our knowledge of the details of the terminal Cretaceous extinctions, thanks largely to the field work and laboratory analyses stimulated by the debate. Most (although not all) of the new discoveries have served to shorten the time span of the extinctions and to move it closer to the presumed time of impact.

So, whether or not the extraterrestrial hypothesis is ultimately proven or disproven, I think a gradual strengthening of the case is clear.

Professor Wolfendale asked also about my confidence in the claim of periodicity in the extinction record. This question is entirely separate from that of extraterrestrial causes.

Periodicity, as a description of the extinction record, has been found by several palaeontologists over several decades. Periodicity is involved in the debate over extraterrestrial influences for the sole reason that several astronomers proposed mechanisms in the solar system or galaxy to explain the periodicity, and some of the proposals have used comet or asteroid impact. Thus the connection is not a primary one. Periodicity in the fossil record may turn out to be correct but not driven by extraterrestrial phenomena, or meteorite impact may cause extinction yet have nothing to do with periodicity, or some combination of these may obtain.

The strength of the periodicty proposal itself depends largely on which statistical analyses one finds most convincing. In the past five years, the published papers (and authors) analysing the extinction data are about evenly split on the question. My own view, as a participant, is that the case is extremely strong in the late Mesozoic and Cainozoic but weak or absent in the Palaeozoic.

Phil. Trans. R. Soc. Lond. B **325**, 437–455 (1989)

Printed in Great Britain

The case for sea-level change as a dominant causal factor in mass extinction of marine invertebrates

By A. Hallam

School of Earth Sciences, University of Birmingham, P.O. Box 363, Birmingham B15 2TT, U.K.

A correlation between global marine regressions and mass extinctions has been recognized since the last century and received explicit formulation, in a model involving habitat-area restriction, by Newell in the 1960s. Since that time attempts to apply the species–area relation to the subject have proved somewhat controversial and promoters of other extinction models have called the generality of the regression–extinction relation into question. Here, a strong relation is shown to exist between times of global or regional sea-level change inferred from stratigraphic analysis, and times of high turnover of Phanerozoic marine invertebrates, involving both extinction and radiation; this is valid on a small and large scale. In many cases the most significant factor promoting extinction was apparently not regression but spreads of anoxic bottom water associated with the subsequent transgression. The sea-level–extinction relation cannot be properly understood without an adequate ecological model, and an attempt is made to formulate one in outline.

Introduction

One of the best established results from analysis of the stratigraphic record of the continents is that they have been subjected to a succession of regionally extensive marine transgressions and regressions through the Phanerozoic. That some of these events were global in extent, and hence caused by eustatic changes of sea level, was first proposed by Suess (1906). Eustasy has been a recurrent theme among stratigraphers ever since. The application of the new technique of seismic stratigraphy by Exxon geologists (Vail *et al.* 1977; Haq *et al.* 1987) has been a major stimulus to its study. With respect to the Palaeozoic of North America, Moore (1954) perceived a relation between regressions and extinctions in the marine realm, as had Suess's contemporary, Chamberlin (1909). The idea was pursued further by Newell (1967) who recognized six episodes of Phanerozoic mass extinction, end-Cambrian, end-Ordovician, late Devonian, end-Permian, end-Triassic and end-Cretaceous. Except for the first, all of these have been subsequently accepted by the palaeontological community as genuine mass extinction events, and the end-Cambrian is also recognized as a significant event for certain groups, based on Sepkoski's (1986) analysis of family and generic turnover.

Newell (1967) put forward a qualitative argument that shrinkage of the area of epicontinental sea habitat should have a deleterious effect on neritic organisms and should therefore lead to widespread extinction. Radiation of the survivors would take place during the expansion of habitat area consequent upon a succeeding transgression. A quantitative analysis based on the ecologists' species–area relation was subsequently undertaken by Schopf (1974) and Simberloff (1974) for the biggest extinction event in the Phanerozoic, at the end of the Permian, and was held to provide support for Newell's claim of a strong relation between regression and extinction.

Within the past few years Newell's ideas, and the quantitative application of the taxon–area relation to account for mass extinctions by regression of epicontinental seas, have come under attack from palaeontologists wishing to promote alternative interpretations (McLaren 1983 and 1989; Stanley 1984 and 1986; Hansen 1987). Consequently they require further evaluation in the light of the most up-to-date information. There is also a need to take into account the inferred spread of anoxic bottom waters within epicontinental seas that is commonly associated with the early phase of transgressions; this may well prove to be more significant for mass extinction than regressions. With regard to the physical evidence of eustatic sea-level change, Haq *et al.* (1987) have argued that the correlation of major unconformities is more significant than transgressions and regressions, but Hubbard (1988) has convincingly demonstrated that many such unconformities are of regional, not global extent. It is worth bearing in mind that some episodes of mass extinction may also be regional not global. If these can be shown to correlate with corresponding regional episodes of regression or anoxic spread associated with transgression (Hallam 1986) this will strengthen the case for a relation at least with extensive epeirogeny, if not eustasy. It is always worth bearing in mind that whereas global sea-level curves are abstractions, about which there will probably always be some measure of dispute about because of inevitable uncertainties concerning the frequency, extent and rate of eustatic change, a high degree of consensus can be obtained among stratigraphers about transgressions and regressions on a regional and sometimes a global scale. Transgressions and regressions are tangible phenomena that must have had significant environmental effects on epicontinental environments.

EUSTASY-EXTINCTION CORRELATION AT DIFFERENT SCALES

Notwithstanding the arguments that have been put forward against sea-level change as a dominant control on mass extinctions, it is difficult to disagree with Jablonski's (1986a) assessment that it shows the strongest empirical correlation of all the factors that have been proposed. This correlation is evident at large, intermediate and small scales.

The large-scale correlation can be demonstrated by plotting Newell's (1967) six marine extinction events against a sea-level curve derived from stratigraphic analysis (figure 1). Although there is no published sea-level curve that is not open to dispute, a high measure of consensus exists among stratigraphers that there were significant global regressions at the end of the Ordovician, Permian, Triassic and Cretaceous. It is also noteworthy that the biggest extinction event of all, at the end of the Permian, corresponds, according to the latest estimate, to a very substantial sea-level fall (Holser & Magaritz 1987). Another point to note in figure 1 is that the regressions were quickly followed by significant transgressions. This observation is pertinent to the common association with widespread marine anoxia, to be discussed later.

With regard to an intermediate scale, concerning events at intervals of the order of a few tens of millions of years apart, the ammonoids are a suitable group to study because of their high diversity through a significant portion of Phanerozoic history, and their high rate of generic turnover, implying a high rate of evolution and extinction. This implies that comparatively stenotopic organisms are sensitive to even modest environmental changes. If the taxon–area relation is valid, times of low sea level should correlate with times of low generic diversity because of increased extinction rate; and times of high sea level with high diversity because of increased origination rate, as a consequence of the correlation between sea-level stand and area

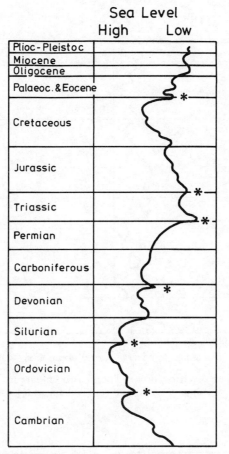

FIGURE 1. The Phanerozoic sea-level curve of Hallam (1984) with asterisks signifying the six marine mass extinction episodes recognized by Newell (1967). Adapted from fig. 6 of Hallam (1984).

of epicontinental sea. Despite the uncertainties involved in palaeogeographic reconstructions, which mean that only approximate results can be obtained, the varying degree of inundation of the continents through time (represented by the sea-level curve) can be established from stratigraphic analysis with less ambiguity than by utilizing unconformity-bound stratigraphic sequences (Vail *et al.* 1977; Haq *et al.* 1987). It has the further advantage of monitoring more directly the factor thought to be responsible for the diversity variations.

Figure 2 presents a plot of ammonoid generic diversity against two sea-level curves. Wiedmann's curve (1986) is based on data from Yanshin (1973) and Sliter (1976) on the changing area of marine inundation of the continents from the Devonian to the Cretaceous inclusive. My curve is based on a best estimate utilising both marine inundation data and data from facies changes up stratigraphic successions; this may account for some of the differences. One point, however, I will readily concede. There was a major fall of sea level in the mid-Carboniferous, as portrayed in the Wiedmann curve, which is clearly apparent from stratigraphic data in both the northern and southern hemispheres (Saunders & Ramsbottom 1986; Veevers & Powell 1987). Furthermore, I have utilized Schopf's (1974) data for the Permian without taking account of intra-Permian events such as an early late-Permian sea-level rise claimed from sequence data by Vail *et al.* (1977).

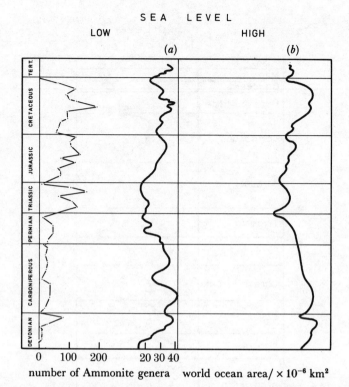

FIGURE 2. Plot of ammonoid generic diversity (broken line), based on data from House (1985), against sea-level curves (a) (Wiedmann 1986) and (b) (Hallam 1984), (solid lines.)

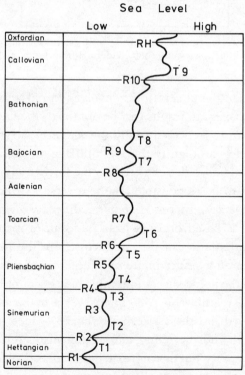

FIGURE 3. Early and Middle Jurassic sea-level curve of Hallam (1988) with succession of transgressive (T) and regressive (R) episodes enumerated. See table 1.

TABLE 1. EARLY AND MIDDLE JURASSIC AMMONITE RADIATIONS AND EXTINCTIONS IN
RELATION TO TRANSGRESSIONS AND REGRESSIONS

(See figure 3.)

transgressive and regressive event	radiations	extinctions
R_{11} end-Callovian	—	kosmoceratids reineckiids tulitids
T_9 early Callovian	kosmoceratids macrocephalitids	—
R_{10} end-Bathonian	—	last stephanoceratids (Cadomitinae) clydoniceratinae
T_8 late Bajocian	perisphinctids	—
R_9 mid-Bajocian	—	sonniniids
T_7 early Bajocian	sonniniids	—
R_8 end-Aalenian	—	most hildoceratids and graphoceratids
R_7 mid-Toarcian	— hildoceratids dactylioceratids	dactylioceratids —
T_6 early Toarcian	(plus migration into boreal realm)	—
R_6 end-Pliensbachian	—	amaltheids
T_5 late Pliensbachian	amaltheids	—
R_5 mid-Pliensbachian	—	liparoceratids
T_4 early Pliensbachian	polymorphitids etc.	—
R_4 end-Sinemurian	—	echioceratids oxynoticeratids
T_3 late Sinemurian	echioceratids oxynoticeratids	—
R_3 mid Sinemurian	—	arietitids
T_2 early Sinemurian	arietitids	—
R_2 end-Hettangian	—	psiloceratids most schlotheimiids
T_1 Hettangian	psiloceratids schlotheimiids	—
R_1 end-Triassic	—	mass extinction

Regardless of their differences there is a discernible relation between sea-level curves and ammonoid diversity (figure 2). Both curves indicate a generally steady rise in sea level through the Devonian, with the time of highest stand in the late Devonian correlating closely with the peak diversity. Thereafter a sharp fall at or close to the end of the period corresponds with a marked drop in diversity to almost zero. Two diversity cycles in the Carboniferous correspond with two major eustatic cycles, with higher diversities being recorded in the early Carboniferous, when sea level stood higher. The sharp diversity fall near the end of the Permian corresponds more closely with my curve than with Wiedmann's, as does the late Triassic diversity maximum followed by a sharp fall to almost zero. Both curves correspond well with the Jurassic diversity pattern, and the mid-Cretaceous diversity peak corresponds better with Wiedmann's curve based on Sliter's (1976) data, suggesting that Cretaceous sea level reached its highest stand at that time. The late Cretaceous sea-level fall shown in both curves matches the gradual diversity decline to zero at the end of the period. It should be noted that the magnitude of sea-level change does not necessarily relate closely to the magnitude of diversity change.

Ammonoids can also be utilized to demonstrate small-scale correlation. Figure 3 is my sea-level curve for the early and mid-Jurassic, with a succession of transgressive and regressive episodes indicated at approximately 5 Ma intervals. These correspond to radiations and extinctions of ammonite families, respectively (table 1). The details of the curve are primarily based on sedimentary cycles in the exceptionally well studied European deposits, but also take into account contemporary transgressions and regressions in other continents. That the sea-level oscillations are more likely to be global than regional in extent is also indicated by the close correlation with ammonite radiation and extinction episodes. Most of the ammonites in question are cosmopolitan in distribution, and if the regressions clearly recorded in Europe were confined to that continent they could have retreated to refuges elsewhere, and returned during the subsequent transgressions, rather in the way that Quaternary organisms have followed shifting climatic belts (Coope 1979).

Bayer & McGhee (1986) confirm the relation outlined above, for the early Middle Jurassic of Central Europe, and claim a hierarchical pattern of change in the vertical sedimentary cyclic sequences corresponding with the pattern of ammonite turnover, more or less at the levels of family, genus and species. Ramsbottom (1981) came to very similar conclusions with respect to the relation of Carboniferous goniatite turnover with eustatic cycles. Other contemporary invertebrate groups such as bivalves have a much slower rate of taxonomic turnover, evidently being less vulnerable to extinction. Consequently they do not exhibit the small-scale relation so well shown by ammonoids.

Regression or anoxia?

The regression model for extinction of organisms in epicontinental seas depends upon the species–area effect being generally applicable. It does indeed appear that there is nearly always a positive correlation among living organisms in a wide range of habitats between species number and area (Connor & McCoy 1979) and study of a variety of island populations confirms the general supposition that the chances of extinction increase with decreasing habitat area (Diamond 1984). Furthermore, Sepkoski (1976) undertook a multiple-regression analysis of Phanerozoic species diversity data that appears to offer support for the species–area relation with respect to ancient epicontinental seas (see also Flessa & Sepkoski (1978)). Nevertheless, several reservations have been expressed about uncritical application of the species–area relation, whatever it may signify, to the fossil record. Thus Wise & Schopf (1981) estimated the changes in diversity of marine species between low and high stands of Quaternary sea level. They found 25 % change in species, 10 % in generic and less than 5 % in family diversity. This agrees with the general observation that large-scale generic or family level changes are insignificant during the Quaternary. They argued accordingly that the substantial family-level drop in the late Permian was more likely to be bound up with a reduction in the number of faunal provinces (Schopf 1979). Similarly, Jablonski (1985) argued that in the modern world most marine faunas would persist in undiminished shallow water regimes around oceanic islands at times of Quaternary sea-level fall of the amount recorded, and thus, that major family-level extinction could not occur.

Stanley (1984, 1986) has observed that one of the biggest Phanerozoic sea-level falls took place, according to Vail *et al.* (1977), in the mid-Oligocene, at a time when no marine mass extinction event is recognizable. Hansen (1987) sees no relation between Palaeogene mollusc extinctions and reductions in shelf area in the Gulf Coast region of North America. More generally, Wyatt's (1987) hypsometric study has demonstrated that the response of continents to sea-level change depends upon their size, and that the global curve of shallow sea area with time that he produced does not vary in a straightforward manner with respect to sea level. Accordingly, predictions of organic diversity change with sea level should take account of their non-linear relation. A comparable point was made by Martin (1981); the species–area curve is steep for small areas but the slope becomes asymptotic as the area increases.

Regarding the Quaternary, one important factor that should not be overlooked is the comparative rapidity of glacioeustatic sea-level change. Bearing in mind that extinction risk increases with time in a stressful environment without further diminution of area (Diamond 1984) it could be that the rapid restoration of less stressful conditions in the Quaternary has served to diminish the extinction rate owing to reduction of habitat area through sea-level fall. The slower and less spectacular sea-level falls of much of the Phanerozoic might have been environmentally more significant because they lasted longer, and quite modest falls in very extensive and extremely shallow-water epicontinental seas with no close modern analogue could have had a correspondingly major effect (Hallam 1981 *a*). For both the Oligocene and Quaternary, however, an important factor might have been the presence of well aerated deep ocean as a consequence of the active circulation of cold, dense bottom water derived from the Antarctic. Thus the deep sea would have provided a refuge that might not have been in existence during the Palaeogene, Mesozoic and much of the Palaeozoic. The notion that the deep sea was more or less anaerobic for much of the geological past goes back to Murray (1895), and most of the modern deep-sea fauna may, for this reason, be no older than Neogene (Briggs 1987). A stagnant deep ocean is proposed for equable periods of earth history by Fischer & Arthur (1977) and mass extinction models invoking this factor have been put forward by Keith (1982) and Wilde & Berry (1984).

A strong empirical correlation exists between the early phases of marine transgression in epicontinental settings and the extensive spread of laminated organic-rich shales (black shales in traditional parlance), which are with good reason held to signify deposition in more or less anoxic water (Hallam & Bradshaw 1979; Hallam 1987 *a*; Thickpenny & Leggett 1987). Because such transgressions normally follow quickly after major regressions, it is not always clear from analysis of extinction events what the critical causal factor was, although in both cases there would have been a reduction in benthic and nektobenthic habitable area. I briefly review the relevant evidence for several of the 29 Phanerozoic mass extinction events recognized by Sepkoski (1986) on the basis of his analysis of fossil marine genera.

Cambrian

Because of their high diversity and rate of turnover, trilobites provide the most promising material for study, and Sepkoski (1986) notes a number of extinction events, most apparently minor (Briggs *et al.* 1988). One of the more notable of these is at the end of the early Cambrian. An extensive, circum-Iapetus, regression event at this time also coincides with the extinction of the archaeocyathids (Palmer & James 1980). The most important event, at least for platform or cratonic trilobites as opposed to those that lived in deeper offshore waters, was at the end of the Cambrian. Thus at this time approximately half of North American trilobite families became extinct (Westrop & Ludvigsen 1987). To account for the extinctions these authors apply a simple biogeographic model following Schopf (1979), involving a reduction in the number of faunal provinces as a result of transgression at the beginning of the Ordovician. On the other hand Briggs *et al.* (1988) perceive a compound event, with extinction rate increasing with end-Cambrian regression prior to the extinctions associated with transgression at the beginning of the Ordovician. Westrop & Ludvigsen consider that facies data on the North American craton argue against invoking an anoxic event. Berry & Wilde (1978) and Thickpenny & Leggett (1987) single out the late Cambrian as a time of very extensive black shale deposition, so the possibility of an anoxic event cannot be excluded. Furthermore, deeper-water trilobites survive the end Cambrian event (Fortey 1983). Of these, at least the olenids and agnostids were probably adapted to cope with low oxygen concentrations.

Ordovician–Silurian

The only event of major importance in this period was at the end of the Ordovician, when one of the five most significant marine mass extinctions of the Phanerozoic took place. This has been generally associated with a regression caused by a glacioeustatic fall of sea level that was itself associated with the extension of the Saharan ice sheet. Briggs *et al.* (1988) claim to observe a compound event associated with the trilobites. A significant reduction in generic diversity is coincident with the stratigraphically well-documented regression, but the final extinction coincides with the subsequent earliest Silurian transgression that can be interpreted as glacioeustatic rebound following disappearance of the Saharan ice sheet. The earliest Silurian is marked by an exceptionally wide distribution of black shales and is a prime candidate for an oceanic anoxic event (Thickpenny & Leggett 1987).

Devonian

With regard to the ammonoids, House (1985) recognizes no fewer than eight extinction events from the early Devonian to the beginning of the Carboniferous that show a strong correlation with the widespread establishment of anoxic conditions associated with global transgressive pulses. Of these the most important, termed by House the 'Kellwasser event', took place at the Frasnian–Famennian boundary, and is one of the five biggest marine extinction events in the Phanerozoic. It is most notable for the dramatic and short-term disappearance of reef ecosystems (McLaren 1983). Whereas Johnson *et al.* (1985) agree with House that the mass extinction at this time is marked by global transgression associated with a widespread anoxic event, Goodfellow *et al.* (1989) argue for an association of the anoxic event with regression at the start of the Famennian. On the basis of conodont biofacies changes Sandberg *et al.* (1988) claim a short-term eustatic rise followed by a fall at the end of the Frasnian.

Carboniferous

According to House (1985) the correlation between widespread anoxia and ammonoid extinction is especially clear for his so-called 'Hangenberg event' at the Devonian–Carboniferous boundary. M. R. House (personal communication) also recognizes a regressive event preceding it, a pattern he compares with that at the Frasnian–Famennian boundary. Sepkoski (1986) recognizes a major mid-Carboniferous event, at the Visean–Namurian boundary, with a relatively low, although significant, extinction rate against a low rate of background extinction. The global diversity reduction at this time among ammonoids has already been noted (figure 2). As regards North America, Saunders & Swan (1984) correlate the severe goniatite diversity reduction with the major regression recorded by the widespread unconformity of the Mississippian–Pennsylvanian boundary. It should, however, be noted that, in exceptionally complete sections in Oklahoma and Texas, this boundary is marked by black shales enriched in phosphate, chalcophile and platinum group elements (Orth *et al.* 1986).

Permian

The largest extinction event of all, which significantly affected nearly all marine groups, correlates with an equally dramatic global regression. According to the most recent estimate, the fall of sea level might have been as much as 280 m, at a rate of 60 m per Ma. It was immediately followed at the start of the Triassic by an even more rapid sea-level rise (Holser & Magaritz 1987). No widespread black shales have been reported for the earliest Triassic, but account should be taken of the fact that stratigraphically continuous sections across the Permian–Triassic boundary are very rare as a consequence of the end-Permian regression. The sections in the Lower Yangtze region near Nanjing, China, are the most complete in the world (Sheng *et al.* 1984); I have observed that the oldest Triassic limestones, though not evidently organic-rich, are finely laminated and almost entirely benthos-free, both features characteristics of anoxic deposits. The only benthic megafossils present are a few thin horizons rich in the bivalve *Claraia*; this is considered to be an opportunistic genus well adapted to dysaerobic conditions (P. Wignall, personal communication). Similar deposits with restricted, probably dysaerobic, faunas, are recorded from basal Triassic deposits in other parts of the world, such as the Caucasus, Kashmir and the southern Alps (Logan & Hills 1973).

Triassic

The only significant Triassic event recognized by Sepkoski (1986) is at the end of the period, when conodonts completely disappeared and ammonites almost completely disappeared. Many other groups were severely affected, including reef communities (Fagerstrom 1987). There is a strong correlation with widespread and substantial regression, but it is also noteworthy that the oldest Jurassic strata in western Europe include extensively distributed laminated black shales and organic-rich limestones signifying anoxic conditions; these correlate with a global transgression (Hallam 1981 *b*). Personal observations in Western Nevada and northern Chile indicate barren, laminated black shales and limestones at the Triassic–Jurassic boundary, anoxic conditions having been established shortly before the end of the Triassic. I saw no evidence of end-Triassic regression. The same is true for northern Peru (Prinz 1985).

Jurassic

Two Jurassic extinction peaks are revealed by Sepkoski's generic data, Pliensbachian–Toarcian and Tithonian. The former can be pinpointed more accurately as an early Toarcian extinction event in Europe affecting the majority of the benthos and nektobenthos at species level, but not the plankton. In global terms, it correlates precisely with a transgression and an oceanic anoxic event as indicated by both facies and carbon isotope data (Hallam 1987 *b*; Jenkyns 1988). This event is significant for more general interpretation because it is not immediately preceded by a globally extensive late Pliensbachian regression (though such a regression is recognizable in western Europe) that points to bottom-water anoxia as the only factor that need be invoked (Hallam 1987 *b*). Whereas Jenkyns (1988) wants the anoxic event and concomitant extinctions to be global, evidence from Andean South America does not appear to accord with this (Hallam 1986); this matter clearly requires further investigation. The Tithonian event is also regional rather than global in extent. Where it is recognizable, in Europe, it correlates with evidence of widespread regression, but not anoxia during the subsequent transgression. In South America there is no evidence of regression or anoxia across the Jurassic–Cretaceous boundary (Hallam 1986).

Cretaceous

Two extinction events are clearly recognizable for this period, at the Cenomanian–Turonian boundary and at the end of the Maastrichtian. The Cenomanian–Turonian boundary event, which severely affected both the benthos and plankton (Elder 1987; Jarvis *et al.* 1988), correlates precisely with a major sea-level rise, perhaps to its highest point in the Cretaceous, and to an oceanic anoxic event clearly recorded by sedimentary facies and carbon isotope data (Schlanger *et al.* 1987; Arthur *et al.* 1987). Jarvis *et al.* (1988) recognize a series of extinction steps successively affecting the microbenthos, inferred deeper-water planktonic foraminifera and intermediate water groups. At the height of the event, the benthic fauna was extremely impoverished and the planktonic fauna consisted only of the shallowest water groups. These successive changes are correlated with a rapid rise of the oxygen minimum zone in the latest Cenomanian, and its progressive penetration into epicontinental sea environments. As with the Toarcian anoxic event, there is no evidence that this mid-Cretaceous event was preceded by a global fall of sea level.

With regard to the much discussed end-Cretaceous event, there is no need here to do more than point out that there is a strong correlation with significant global regression, which is likely to be implicated in many of the extinctions, if not with the plankton (Hallam 1987 *c*) and that locally the basal Tertiary strata are organic-rich shales indicative of anoxic conditions. This is the case, for instance, in the Fish Clay of Denmark. Surlyk & Johanssen (1984) invoke bottom-water anoxia as one of the factors responsible for brachiopod extinctions in that country. It is worth noting that the highest iridium anomalies yet recorded from the K–T boundary come from organic-rich deposits at Stevns Klint, Denmark; Caravaca, Spain and Woodside Creek, New Zealand. A black clay, free of bioturbation also marks the base of the Tertiary at El Kef, Tunisia (Keller 1988 *b*).

Tertiary

The most important extinction event is across the Eocene–Oligocene boundary but it has become evident from close study of the excellent marine record that the extinctions were spread out over several million years, though whether they were gradual (Corliss *et al.* 1984) or stepwise (Hut *et al.* 1987) remains debatable. Though temperature decline is commonly invoked as the prime causal factor, Dockery (1986) finds a good correlation between major benthic extinction events and marine regressions in the Gulf of Mexico coastal plain (see also Dockery & Hansen 1987).

As I have already stated, Stanley's (1984, 1986) argument that the spectacular mid-Oligocene sea-level fall recorded by Vail *et al.* (1977) does not correlate with an extinction event can be countered to some extent by invoking the concept of a well aerated deep ocean. A further point is that the approximately 400 m fall inferred by Vail *et al.* (1977) is a vast overestimate. The latest best estimate based on a variety of data is between 30 and 90 m (Miller *et al.* 1987).

In summary, of the 13 extinction events reviewed above, which include all the most important Phanerozoic events, regression seems to be definitely correlated in 10 cases and more doubtfully in 1, and anoxia definitely associated with transgression in 6 cases and more doubtfully in 4 (table 2). Only for the late Eocene has no correlation been yet established with either regression or anoxia, though following Dockery's (1986) work this remains a possibility.

TABLE 2. CORRELATION OF MARINE REGRESSION AND ANOXIC EPISODES WITH MAJOR PHANEROZOIC EXTINCTION EVENTS

major extinction event	regression	anoxia
late Eocene	?	–
end-Cretaceous	+	?
mid-Cretaceous (Cenomamian–Turonian)	–	+
end-Jurassic	+	–
early Jurassic (Toarcian)	–	+
end-Triassic	+	+
end-Permian	+	?
mid-Carboniferous (Visean–Namurian)	+	?
end-Devonian	+	+
late Devonian (Frasnian–Famennian)	+	+
end-Ordovician	+	+
end-Cambrian	+	?
end-Lower Cambrian	+	–

DISCUSSION

Whereas many factors have been responsible for extinctions of particular species, there are only a few that can plausibly be invoked to account for more or less simultaneous mass extinctions of different types of marine organisms on a global scale. Although this can be considered an advantage in attempting to understand causes it should be noted at the outset that there is an interaction between environmental changes involving climate, volcanicity and sea level, so that determining the critical factor may not prove to be straightforward. Furthermore, it is important to distinguish between proximate and ultimate causes, particularly when evaluating the relative plausibility of extraterrestrial, as opposed to terrestrial, controlling factors.

Bolide impact

To have any kind of plausibility as a cause for extinctions as opposed to more conventional explanations based on changes confined to our own planet, bolide impact theories must satisfy at least two conditions. The extinctions must be demonstrated, by means of the best available stratigraphic evidence, to have been on a spectacular scale and to have taken place over geologically negligible periods of time (McLaren 1983). Secondly, there should be independent physical and chemical evidence, such as tektites, shocked quartz and platinum-group metal anomalies. After nearly a decade of intensive research such evidence remains weak to non-existent for all major extinction events except for that at the Cretaceous–Tertiary boundary (Donovan 1987; Orth & Attrep 1988; Orth 1989). Late Eocene deposits contain two regionally extensive tektite horizons, but neither correlate with a mass extinction (Glass 1988; Keller *et al.* 1987). Even for the Cretaceous–Tertiary extinction, some argue that the evidence can be explained with a purely terrestrial scenario (Hallam 1987*c*). There have undoubtedly been many bolide impacts during the Phanerozoic but the only persuasive case for an association with mass extinctions is the K–T event, which may therefore be unique in this respect. Even assuming the evidence for impact to be conclusive, it is still by no means clear how many of the end-Cretaceous extinctions relate to it, and no really plausible, exclusively impact-based 'killing scenario' has yet been proposed. The finest time resolution allowable from stratigraphic analysis is rarely less than several thousand years and usually much more. For the end-Cretaceous extinctions it has been argued that even for the more spectacular events, such as those affecting calcareous plankton, timescales of at least thousands of years must be invoked (Keller 1988*a*; Brinkhuis & Zachariasse 1988), though this view has been disputed by Smit *et al.* (1988). To propose a series of stepwise extinctions involving multiple impact by comets (Hut *et al.* 1987), is a scientifically weak response to the problem in the absence of independent evidence, requiring acceptance of a dubious astronomical hypothesis (Hallam 1987*c*). Furthermore, a recent analysis by Raup (this symposium) has thrown into doubt the possibility of distinguishing stepwise extinctions from more gradual or catastrophic events.

Climate

Climate can have an indirect affect on sea-level fluctuations at times of substantial polar icecaps as a consequence of glacioeustasy. A more direct influence for climate on Phanerozoic extinctions has been argued by Stanley (1984, 1986, 1988). His general hypothesis, that episodes of cooling, rather than of falling sea-level, have promoted mass extinctions in the marine realm, appears to be well-supported for the western North Atlantic regime in Plio-Pleistocene times. It is debatable whether it can be applied to extinctions earlier in the Tertiary. Whereas the increase in extinction rate across the Eocene–Oligocene boundary appears to correlate well with independent evidence of fall in oceanic temperatures (as determined by oxygen isotope analysis of foraminiferal shells), Shackleton (1986) points out that the main climatic change so determined, followed, rather than preceded, the biostratigraphic events that are generally taken as defining the Eocene–Oligocene boundary; this excludes a causal relation between the climatic and faunal changes. Dockery (1986) has perceived, for Gulf Coast molluscs, a close relation between extinctions and regressive events. For pre-Tertiary extinctions, evidence for the climatic cooling hypothesis is decidedly weak, if not non-existent. Thus the biggest extinction event of all, at the end of the Permian, coincides with a time of climatic amelioration following the disappearance of the last remnants of the Gondwana ice sheet (Veevers & Powell 1987.) Stanley (1988) holds a contrary opinion. Extinctions at the end of the early Carboniferous and Ordovician could be related to growth of ice sheets, but the key factor in these cases could be the regressions not the climatic cooling (Brenchley 1984; Powell & Veevers 1987). Stanley's principal argument in favour of his climatic control hypothesis is that organisms with a tropical distribution have been more subject to extinction than higher latitude organisms. This is indecisive because tropical organisms are generally stenotopic and therefore relatively vulnerable to a variety of environmental changes (cf. Jablonski 1986*b*).

Volcanism

The transmission of large quantities of volcanic dust and aerosols into the atmosphere and stratosphere is known to cause lowering of air temperatures and may also give rise to substantial acidic rainfall, with environmentally deleterious consequences (Rampino *et al.* 1988). Volcanism on a massive scale over an extended period of time has been proposed as an alternative to bolide impact to account for the dramatic extinction event at the end of the Cretaceous (Officer *et al.* 1987) but it is unlikely to have been more than a contributory factor to the general pattern of extinctions (Hallam 1987 *c*). There is at least one good reason why volcanism is unlikely to have been a major causal factor in Phanerozoic marine extinctions; the correlation between major extinction and volcanic events is rather poor. Although the eruption of the Siberian Traps may well coincide quite closely with the end-Permian extinctions (Holser & Magaritz 1987), just as the Deccan Traps eruption apparently coincides with the end-Cretaceous extinctions, the peak of Karoo volcanic activity was early Jurassic not end-Triassic (Fitch & Miller 1984; Aldiss *et al.* 1984). For the other major extinction events there are no correlating major eruptive events. The Paraná (early Cretaceous), North Atlantic (Palaeocene) and Columbia River flood basalts (Miocene) compare in size with the Deccan eruptions but there are no correlative major extinction events. Where a correlation does exist, as at the end of the Palaeozoic and Mesozoic, both volcanism and sea-level changes may be expressions of significant events in the mantle (Loper *et al.* 1988).

Sea-level changes

As has been outlined above, a strong correlation exists between both major and minor extinction episodes in the marine realm and inferred changes of sea level, with the likeliest cause of extinction being bound up with reduction of shallow marine habitat area as a consequence either of regression or the spread of anoxic waters during the ensuing transgression. At times of significant regression the continental climates would be expected to exhibit greater seasonal extremes of temperature, a phenomenon that could well have been a major causal factor in the significant increase in extinction rates that took place among land vertebrates at the end of the Permian, Triassic and Cretaceous (Bakker 1977). It is noteworthy that an important mid-Oligocene extinction event among land mammals (Prothero 1985) appears to coincide with a major regressive event.

In seeking an adequate extinction model for Phanerozoic shallow-marine invertebrates it is inappropriate to lay too much emphasis on the comparatively abnormal circumstances of the Quaternary, with shallow ocean being represented by pericontinental seas. One should take full account of the likelihood that the environmental settings of ancient epicontinental seas were different in important respects from anything today (Hallam 1981 *a*). In such seas of extreme shallowness over extensive stretches, even a modest change of sea level could have had significant environmental consequences. Johnson (1974) stressed the importance of organic adaptations to changed circumstances in understanding the likely causes of extinction of neritic invertebrates. During episodes of sustained enlargement of epicontinental seas, organisms become progressively more stenotopic and an equilibrium is established. They are in effect 'perched', subject to the continued existence of their environment. Extinctions occur to an extent proportional to the speed of regression, degree of stenotopy attained, or a combination of the two.

The ecological model proposed by Hallam (1978) to account for Jurassic ammonite extinctions bears several resemblances to that of Johnson. Evidence of progressively increasing stenotopy is recorded by data on phyletic size increase, a *K*-selected trend. Times of relatively low sea level signify times of restriction and deterioration of neritic habitat, such as increases in the variability of temperature and salinity of extremely shallow water. In familiar ecological parlance, the increased environmental stress favours *r*-selected organisms. This is recorded by the small size of new ammonite taxa that have evolved rapidly from their larger more stenotopic ancestors, often as a result of heterochronous changes (figure 4); thus speciation, involving the origin of new taxa by reproductive isolation at times of restricted connections between epicontinental seas, as well as extinction, is promoted by regression. Comparable models to those of Johnson & Hallam have been put forward for a variety of Palaeozoic and Mesozoic marine invertebrates (see Johnson & Colville 1982; Rollins *et al.* 1979; Klapper & Johnson 1980; Roberts 1981; Ramsbottom 1981; Bayer & McGhee 1985).

None of these models takes into account anoxic events associated with marine transgression as a possible cause of mass extinction, but we have seen that such phenomena may in some cases have been more significant than regression. McLaren (1983) has argued that a sudden spread

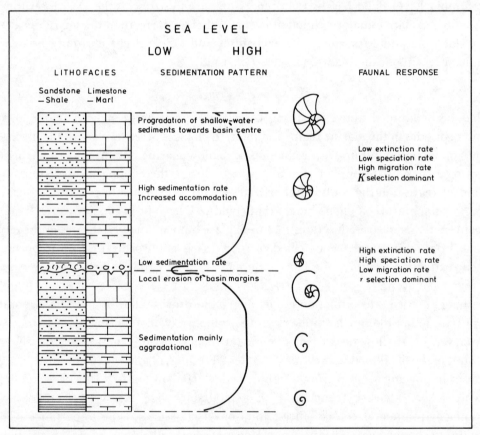

FIGURE 4. Relation of sea-level changes to major sedimentary cyclicity in the European Jurassic, and concomitant faunal response. The sea level oscillations are probably symmetrical and the curve appears cuspate because of the relatively condensed character of the 'transgressive' deposits. The changes schematically represented for the ammonite sequence are (*a*) phyletic size increase (a *K*–selected trend) and (*b*) paedomorphosis, with the ribbed juvenile character of the ancestral taxon extending into the adult of the descendent, following progenesis at a time of high environmental stress (an *r*–selected adaptation).

of anoxic bottom waters at the beginning of the Famennian could have been a consequence of oceanic overturn provoked by bolide impact, but the timescale implied seems much too short. The widespread black shales at extinction horizons normally have a thickness suggesting deposition over periods of time ranging from at least thousands to hundreds of thousands of years. Furthermore, an impact would be expected to disturb oceanic stratification and thereby act against the preservation of anoxic layers. On the other hand the initial spread of anoxic bottom water could have been geologically very rapid, with extinction horizons in the stratigraphic sequence being knife-sharp (Hallam 1987 b). This overcomes an objection raised by McLaren (1983, 1989) that regressions were too slow to have caused mass extinctions as dramatically sudden as, for instance, those at the Frasnian–Famennian boundary. It should be evident from this that a combination of regression and subsequent transgression with concomitant spread of anoxic water provides an exceptionally powerful means of substantially increasing extinction rate among epicontinental marine organisms. That not all sea-level falls (e.g. in the early Devonian) correlate with mass extinction episodes, and that there is no simple relationship between the amount of sea-level change and the extent of extinction, probably indicates that the key factor is reduction of habitat area; this will depend on such factors as continental configuration and topography.

Much uncertainty surrounds the frequency, rate and extent of the underlying causes of sea level change (Hallam 1984). A mechanism involving volume change of ocean ridges seems inadequate to account for the frequently short-term events that many stratigraphers recognize, but the invocation by Goodwin & Anderson (1985) of numerous glacioeustatic events throughout the Phanerozoic to account for small-scale sedimentary cycles seem implausible without evidence of substantial polar icecaps for most of this time, and the evidence from flora and fauna of extended periods of equability. Although there is now evidence of some high latitude ice during equable periods such as the Jurassic and Cretaceous (Frakes & Francis 1988) this does not establish the likelihood of icecaps large enough to affect sea-level in any significant way. Furthermore, anoxic events would be difficult to explain if such icecaps had persisted through the Phanerozoic (Fischer & Arthur 1977). The most important reason for the association of black shales and transgressive episodes is that the early stages of transgression over the continents are characterized by broad stretches of poorly oxygenated shallow water, with restricted circulation with the open ocean, that provide a short transit for organic matter from productive surface water to the bottom sediments. Consequently there is less oxidation and greater retention of organic matter (Hallam 1981 a; Arthur & Jenkyns 1981). The vertical and lateral extent of anoxic waters during a transgressive event would depend on a variety of factors, such as bottom topography, climate and rate and amount of sea-level rise. Some anoxic (perhaps more safely called hypoxic) events in epicontinental seas might have been generated by upwards expansion of the oceanic oxygen-minimum zone and others by phenomena intrinsic to the seas themselves. It may yet prove unnecessary to invoke the proposal of the lack of deep ocean refugia at times of habitat restriction in such seas.

Correlation between mass extinctions and magnetic field reversal patterns at the end of the Palaeozoic and Mesozoic, together with changes in sea level, volcanicity and climate, may be best accounted for by a model involving mantle–core interactions, such as that put forward by Loper et al. (1988). Major epeirogenic movements of the continents at these times could have had a more spectacular influence on sea level than, for example, changes in seafloor spreading rates. The extent to which some such model could provide an adequate causal mechanism for a greater number of mass extinction events remains uncertain, but at any rate it is likely that the extinction events indicate episodic rather than periodic control (Lutz 1987; Quinn 1987).

I am indebted to Dave Jablonski for his critical reading of the manuscript, and to Art Boucot, Richard Fortey, Michael House and Paul Wignall for helpful comments.

REFERENCES

Aldiss, D. T., Benson, J. M. & Rundle, C. C. 1984 Early Jurassic pillow lavas and palynomorphs in the Karoo of eastern Botswana. *Nature, Lond.* **310**, 302–304.

Arthur, M. A. & Jenkyns, H. C. 1981 Phosphorites and paleoceanography. In *Proceedings of the 26th international geological congress, geology of oceans symposium, Paris 1980, (Oceanologica Acta)*, pp. 83–96.

Arthur, M. A., Schlanger, S. O. & Jenkyns, H. C. 1987 The Cenomanian–Turonian oceanic anoxic event, II. Palaeoceanographic controls on organic-matter production and preservation. In *Marine petroleum source rocks* (ed. J. Brooks & A. J. Fleet) (*Geol. Soc. spec. Publ. no. 26*), pp. 401–420. Oxford: Blackwell Scientific.

Bakker, R. T. 1977 Tetrapod mass extinctions – a model of the regulation of speciation rates and immigration by cycles of topographic diversity. In *Patterns of evolution as illustrated by the fossil record* (ed. A. Hallam), pp. 439–468. Amsterdam: Elsevier.

Bayer, U. & McGhee, G. R. 1985 Evolution in marginal epicontinental basins: the role of phylogenetic and ecological factors. In *Sedimentary and evolutionary cycles* (ed. U. Bayer & A. Seilacher), pp. 164–220. Berlin: Springer-Verlag.

Berry, W. B. N. & Wilde, P. 1978 Progressive ventilation of the oceans – an explanation for the distribution of the Lower Paleozoic black shales. *Am. J. Sci.* **278**, 257–275.

Brenchley, P. J. 1984 Late Ordovician extinctions and their relationship to the Gondwana glaciation. In *Fossils and climate* (ed. P. J. Brenchley), pp. 291–316. Chichester: John Wiley.

Briggs, D. E. G., Fortey, R. A. & Clarkson, E. N. K. 1988 Extinction and the fossil record of the arthropods. In *Extinction and survival in the fossil record* (ed. G. P. Larwood) (*System. Ass. Spec. Vol. no. 34*), pp. 171–210. Oxford: Clarendon Press.

Briggs, J. C. 1987 *Biogeography and plate tectonics*. Amsterdam: Elsevier.

Brinkhuis, H. & Zachariasse, X. 1988 Dinoflagellate cysts, sea-level changes and planktonic foraminifers across the Cretaceous–Tertiary boundary at El Haria, northwest Tunisia. *Marine Micropaleont.* **13**, 153–191.

Connor, E. F. & McCoy, E. D. 1979 The statistics and biology of the species–area relationships. *Am. Nat.* **113**, 791–833.

Chamberlin, T. C. 1909 Diastrophism as the ultimate basis of correlation *J. Geol.* **17**, 689–693.

Coope, G. R. 1979 Late Cenozoic fossil Coleoptera: evolution, biogeography and ecology. *A. Rev. Ecol. Syst.* **10**, 247–267.

Corliss, B. H., Aubry, M.-P., Berggren, W. A., Fenner, J. M., Keigwin, L. D. & Keller, G. 1984 The Eocene–Oligocene boundary event in the deep sea. *Science, Wash.* **226**, 806–810.

Diamond, J. M. 1984 'Normal' extinctions of isolated populations. In *Extinctions* (ed. M. H. Nitecki), pp. 191–246. University of Chicago Press.

Dockery, D. T. 1986 Punctuated succession of Paleogene molluscs in the northern Gulf Coastal Plain. *Palaios* **1**, 582–589.

Dockery, D. T. & Hansen, T. A. 1987 Eocene–Oligocene molluscan extinctions: comment and reply. *Palaios* **2**, 620–622.

Donovan, S. K. 1987 Iridium anomalous no longer? *Nature, Lond.* **326**, 331–332.

Elder, W. P. 1987 The paleoecology of the Cenomamian–Turonian (Cretaceous) stage boundary extinctions at Black Mesa, Arizona. *Palaios* **2**, 24–40.

Fagerstrom, J. A. 1987 *The evolution of reef communities*. New York: John Wiley.

Fischer, A. G. & Arthur, M. A. 1977 Secular variations in the pelagic realm. In *Deep–water carbonate environments* (ed. H. E. Cook & P. Enos) (*Spec. Publs. Soc. econ. Palaeont. Miner., Tulsa no. 25*), pp. 19–50.

Fitch, F. J. & Miller, J. A. 1984 Dating Karoo igneous rocks by the conventional K–Ar and ^{40}Ar–^{39}Ar age spectrum methods. *Spec. Publ. geol. Soc. S. Afr.* **13**, 247–266.

Flessa, K. W. & Sepkoski, J. J. 1978 On the relationship between Phanerozoic diversity and changes in habitable area. *Paleobiol.* **4**, 359–366.

Fortey, R. A. 1983 Cambrian–Ordovician trilobites from the boundary beds in western Newfoundland and their phylogenetic significance. *Spec. Pap. palaeont. no. 20*, 179–211.

Frakes, L. A. & Francis, J. E. 1988 A guide to Phanerozoic cold polar climates from high-latitude ice-rafting in the Cretaceous. *Nature, Lond.* **333**, 547–549.

Glass, B. P. 1988 Late Eocene impact events recorded in deep–sea sediments. *Lunar planet. Inst. Contrib.* no. 673, 63–64.

Goodfellow, W. D., Geldsetzer, H., McLaren, D. J., Orchard, M. J. & Klapper, G. 1989 Geochemical and isotopic anomalies associated with the Frasnian–Famennian extinction. *Histor. Biol.* 2, 51–72.

Goodwin, P. W. & Anderson, E. J. 1985 Punctuated aggradational cycles: a general hypothesis of episodic stratigraphic accumulation. *J. Geol.* 93, 515–533.

Hallam, A. 1978 How rare is phyletic gradualism? Evidence from Jurassic bivalves. *Paleobiology* 4, 16–25.

Hallam, A. 1981a *Facies interpretation and the stratigraphic record.* Oxford: W. H. Freeman.

Hallam, A. 1981b The end-Triassic bivalve extinction event. *Palaeogeog. Palaeoclimatol. Palaeoecol.* 35, 1–44.

Hallam, A. 1984 Pre-Quaternary sea-level changes. *A. Rev. Earth planet. Sci.* 12, 205–243.

Hallam, A. 1986 The Pliensbachian and Tithonian extinction events. *Nature, Lond.* 319, 765–768.

Hallam, A. 1987a Mesozoic marine organic-rich shales. In *Petroleum marine source rocks* (ed. J. Brooks & A. J. Fleet), pp. 251–261. Oxford: Blackwell Scientific.

Hallam, A. 1987b Radiations and extinctions in relation to environmental change in the marine Lower Jurassic of northwest Europe. *Paleobiology* 13, 152–168.

Hallam, A. 1987c End-Cretaceous mass extinction event: argument for terrestrial causation. *Science, Wash.* 238, 1237–1242.

Hallam, A. 1988 A reevaluation of Jurassic eustasy in the light of new data and the revised Exxon curve. In *Sea-level changes – an integrated approach* (ed. C. Wilgus *et al.*) (*Spec. Publs. Soc. econ. Palaeont. Miner., Tulsa* no. 42, 261–273.

Hallam, A. & Bradshaw, M. J. 1979 Bituminous shales and oolitic ironstones as indicators of transgressions and regressions. *J. geol. Soc. Lond.* 136, 157–164.

Hansen, T. A. 1987 Extinction of late Eocene to Oligocene molluscs: relationship to shelf area, temperature changes and impact events. *Palaios* 2, 69–75.

Haq, B. U., Hardenbol, J. & Vail, P. R. 1987 Chronology of fluctuating sea levels since the Triassic. *Science, Wash.* 235, 1156–1167.

Holser, W. T. & Margaritz, M. 1987 Events near the Permian–Triassic boundary, *Modern Geol.* 11, 155–180.

House, M. R. 1985 Correlation of mid-Palaeozoic ammonoid evolutionary events with global sedimentary perturbations. *Nature, Lond.* 313, 17–22.

Hubbard, R. J. 1988 Age and significance of sequence boundaries on Jurassic and early Cretaceous rifted continental margins. *Bull. Am. Ass. Petrol. Geol.* 72, 49–72.

Hut, P., Alvarez, W., Elder, W. P., Hansen, T. A., Kauffman, E. G., Keller, G., Shoemaker, E. M. & Weissman, P. R. 1987 Comet showers as a cause of mass extinctions. *Nature, Lond.* 329, 118–126.

Jablonski, D. 1985 Marine regressions and mass extinctions: a test using the modern biota. In *Phanerozoic diversity patterns* (ed. J. W. Valentine), pp. 335–354. Princeton University Press.

Jablonski, D. 1986a Causes and consequences of mass extinctions. In *Dynamics of extinction* (ed. D. K. Elliott), pp. 183–229. New York: John Wiley.

Jablonski, D. 1986b Evolutionary consequences of mass extinctions. In *Patterns and processes in the history of life* (ed. D. M. Raup & D. Jablonski), pp. 313–329. Berlin: Springer–Verlag.

Jarvis, I., Carson, G. A., Cooper, M. K. E., Hart, M. B., Leary, P. N., Tocher, B. A., Horne, D. & Rosenfeld, A. 1988 Microfossil assemblages and the Cenomanian–Turonian (late Cretaceous) oceanic anoxic event. *Cretac. Res.,* 9, 3–103.

Jenkyns, H. C. 1988 The early Toarcian (Jurassic) anoxic event: stratigraphic, sedimentary, and geochemical evidence. *Am. J. Sci.* 288, 101–151.

Johnson, J. G. 1974 Extinction of perched faunas, *Geology* 2, 479–482.

Johnson, J. G., Klapper, G. & Sandberg, C. A. 1985 Devonian eustatic fluctuations in Euramerica. *Bull. geol. Soc. Am.* 96, 567–587.

Johnson, M. E. & Colville, V. R. 1982 Regional integration of evidence for evolution in the Silurian *Pentamerus–Pentameroides* lineage. *Lethaia* 15, 41–54.

Keith, M. L. 1982 Violent volcanism, stagnant oceans and some inferences regarding petroleum, strata-bound ores and mass extinctions. *Geochim. cosmochim. Acta* 46, 2621–2637.

Keller, G. 1988a Biotic turnover in benthic foraminifera across the Cretaceous–Tertiary boundary at El Kef, Tunisia. *Palaeogeog. Palaeoclimatol. Palaeoecol.* 66, 153–172.

Keller, G. 1988b Extinction, survivorship and evolution of planktic foraminifera across the Cretaceous–Tertiary boundary at El Kef, Tunisia. *Marine Micropaleant.* 13, 239–263.

Keller, G., D'Handt, S. L., Oat, C. J., Gilmore, J. S., Oliver, P. Q., Shoemaker, E. M. & Molina, E. 1987 Late Eocene impact microspherules: stratigraphy, age and geochemistry. *Meteoritics* 22, 25–60.

Klapper, G. & Johnson, J. G. 1980 Endemism and dispersal of Devonian conodonts. *J. Paleont.* 54, 400–455.

Logan, A. & Hills, L. V. (eds.) 1973 The Permian and Triassic Systems and their mutual boundary. *Mem. Canad. Soc. Petrol. Geol.* no. 2.

Loper, D. E., McCartney, K. & Buzyna, G. 1988 A model of correlated periodicity in magnetic-field reversals, climate, and mass extinctions. *J. Geol.* 96, 1–15.

[213]

Lutz, T. M. 1987 Limitations to the statistical analysis of episodic and periodic models of geologic time series. *Geology* 15, 1115–1117.

McLaren, D. J. 1983 Bolides and biostratigraphy. *Bull. geol. Soc. Am.* 94, 313–324.

McLaren, D. J. 1989 Detection and significance of mass killings. *Histor. Biol.* 2, 5–16.

Martin, T. E. 1981 Species–area slopes and their coefficients: a caution on their interpretation. *Am. Nat.* 118, 823–827.

Miller, K. G., Fairbanks, R. G. & Mountain, G. S. 1987 Tertiary oxygen isotope synthesis, sea-level history, and continental margin erosion. *Paleoceanography* 2, 1–19.

Moore, R. C. 1954 Evolution of late Paleozoic invertebrates in response to major oscillations of shallow seas. *Bull. Mus. comp. Zool. Harv.* 122, 259–286.

Murray, J. 1895 A summary of the scientific results. In *Challenger report summary*, 2 vols. London.

Newell, N. D. 1967 Revolutions in the history of life. *Spec. Pap. geol. Soc. Am.* no. 89, 63–91.

Officer, C. B., Hallam, A., Drake, C. L. & Devine, J. D. 1987 *Nature, Lond.* 326, 143–149.

Orth, C. J., Quintana, L. R., Gilmore, J. S., Grayson, R. C. & Westergaard, E. H. 1986 Trace-element anomalies at the Mississippian–Pennsylvanian boundary in Oklahoma and Texas. *Geology* 14, 986–990.

Orth, C. J. & Attrep, M. 1988 Iridium abundance measurements across the bio-event horizons in the geologic record. *Lunar planet. Inst. Contrib.* no. 673, 139–140.

Orth, C. J. 1989 Geochemistry of the bio-eventhorizons. In *Mass extinctions: processes and evidence* (ed. S. K. Donovan), pp. 37–72. London: Belhaven Press.

Palmer, A. R. & James, N. P. 1980 The Hawke Bay Event: a circum-Iapetus regression near the Lower–Middle Cambrian boundary. In *The Caledonides in the USA* (ed. D. R. Wones) (*Dep. geol. Sci. Virgin. Polytech. Inst. & State Univ. Meon.* no. 2), pp. 15–18.

Powell, C. McA. & Veevers, J. J. 1987 Namurian uplift in Australia and South America triggered the main Gondwanan glaciation. *Nature, Lond.* 326, 177–179.

Prinz, P. 1985 Stratigraphie und Ammonitenfauna der Pucara-Gruppe (Obertrias–Unterjura) von Nord Peru. *Palaeontographica* 188, 153–197.

Prothero, D. R. 1985 North American mammalian diversity and Eocene–Oligocene extinctions. *Paleobiology* 11, 389–405.

Quinn, J. F. 1987 On the statistical detection of cycles in extinctions in the marine fossil record. *Paleobiology* 13, 465–478.

Rampino, M. R., Self, S. & Stothers, R. B. 1988 Volcanic winters. *A. Rev. Earth plant. Sci.*, 16, 73–100.

Ramsbottom, W. H. C. 1981 Eustatic control in Carboniferous ammonoid biostratigraphy. In *The Ammonoidea* (ed. M. R. House & J. R. Senior) (*System Ass. Spec. Vol.* no. 18), pp. 369–388. London: Academic Press.

Roberts, J. 1981 Control mechanisms of Carboniferous brachiopod zones in Australia. *Lethaia* 14, 123–134.

Rollins, H. B., Carothers, M. & Donohue, J. 1979 Transgression, regression and fossil community succession. *Lethaia* 12, 89–104.

Sandberg, C. A., Ziegler, W., Dreesen, R. & Butler, J. L. 1988 Late Frasnian mass extinction: conodont event biostratigraphy, global changes, and possible causes. *Courr. Forsch.-Inst. Senckenberg* 102, 263–307.

Saunders, W. B. & Ramsbottom, W. H. C. 1986 The mid-Carboniferous eustatic event. *Geology* 14, 208–212.

Saunders, W. B. & Swan, A. R. H. 1984 Morphology and morphologic diversity of mid-Carboniferous (Namurian) ammonoids in time and space. *Paleobiology* 10, 195–228.

Schlanger, S. O., Arthur, M. A., Jenkyns, H. C. & Scholle, P. A. 1987 The Cenomamian–Turonian oceanic anoxic event, I. Stratigraphy and distribution of organic carbon-rich beds and the marine ^{13}C excursion. In *Marine petroleum source rocks* (ed. J. Brooks & A. J. Fleet) (*Geol. Soc. spec. Publ.* Vol. no. 36), pp. 371–400. Oxford: Blackwell Scientific.

Schopf, T. J. M. 1974 Permo-Triassic extinctions: relation to sea-floor spreading. *J. Geol.* 82, 129–143.

Schopf, T. J. M. 1979 The role of biogeographic provinces in regulating marine faunal diversity through geologic time. In *Historical biogeography, plate tectonics, and the changing environment* (ed. J. Gray & A. J. Boucot), pp. 449–457. Oregon State University Press.

Sepkoski, J. J. 1976 Species diversity in the Phanerozoic: species–area effects. *Paleobiology* 2, 298–303.

Sepkoski, J. J. 1986 Phanerozoic overview of mass extinction. In *Patterns and processes in the history of life* (ed. D. M. Raup & D. Jablonski) (*Life Sciences Research Report* 36), pp. 259–276. Berlin: Springer-Verlag.

Shackleton, N. J. 1986 Paleogene stable isotope events. *Palaeogeog. Palaeoclimatol. Palaeoecol.* 57, 91–102.

Sheng, J.-z., Chen, C. z., Wang, Y.-g., Rui, L., Liao, Z.-t., Bando, Y., Ishii, K., Nakazawa, K. & Nakamura, K. 1984 Permian–Triassic boundary in Middle and Eastern Tethys. *J. Fac. Sci. Hokkaido Univ.* Series IV, 21, 133–181.

Simberloff, D. 1974 Permo-Triassic extinctions: effects of an area on biotic equilibrium. *J. Geol.* 82, 267–274.

Sliter, W. V. 1976 Cretaceous foraminifers from the southwestern Atlantic Ocean, leg 36, deep-sea drilling project. In *Init. Reps. DSDP* 36 (ed. P. F. Barker, I. W. D. Dalziel, *et al.*), pp. 519–537. Washington: U.S.A. Government.

Smit, J., Groot, H., de Jonge, R. & Smit, P. 1988 Impact and extinction signatures in complete Cretaceous–Tertiary boundary sections. *Lunar planet. Inst. Contrib.* no 673, 182–183.

Stanley, S. M. 1984 Marine mass extinction: a dominant role for temperature. In *Extinctions* (ed. M. H. Nitecki), pp. 69–117. University of Chicago Press.

Stanley, S. M. 1986 *Extinction*. New York: Scientific American Library.

Stanley, S. M. 1988 Paleozoic mass extinctions: shared patterns suggest global cooling as a common cause. *Am. J. Sci.* **288**, 334–352.

Suess, E. 1906 *The face of the earth*, vol. 2. Oxford: Clarendon Press.

Surlyk, F. & Johannsen, M. B. 1984 End-Cretaceous brachiopod extinctions in the Chalk of Denmark. *Science, Wash.* **223**, 1174–1177.

Thickpenny, A. & Leggett, J. K. 1987 Stratigraphic distribution and palaeo-oceanographic significance of European early Palaeozoic organic-rich sediments. In *Marine petroleum source rocks* (ed. J. Brooks & A. J. Fleet) (*Geol. Soc. spec. Publ.* no. 26), pp. 231–248. Oxford: Blackwell Scientific.

Vail, P. R., Mitchum, R. M., Todd, R. G., Widmier, J. M., Thompson, S., Sangree, J. B., Bubb, J. N. & Hatlelid, W. G. 1977 Seismic stratigraphy and global changes of sea level. *Am. Ass. Petrol. Geol. Mem.* no. 26, 49–212.

Veevers, J. J. & Powell, C. McA. 1987 Late Paleozoic glacial episodes in Gondwanaland reflected in transgressive–regressive depositional sequences in Euramerica. *Bull. geol. Soc. Am.* **98**, 475–487.

Westrup, S. R. & Ludvigsen, R. 1987 Biogeographic control of trilobite mass extinction at an Upper Cambrian 'biomere' boundary. *Paleobiology* **13**, 84–99.

Wiedmann, J. 1986 Macro-invertebrates and the Cretaceous–Tertiary boundary. In *Global bio-events* (ed. O. H. Walliser) (*Lecture Notes in Earth Sciences*, vol. 8), pp. 397–409. Berlin: Springer–Verlag.

Wilde, P. & Berry, W. B. N. 1984 Destabilisation of the oceanic density structure and its significance to marine 'extinction events'. *Palaeogeog. Palaeoclimatol. Palaeoecol.* **48**, 142–162.

Wise, K. P. & Schopf, T. J. M. 1981 Was marine faunal diversity in the Pleistocene affected by changes in sea level? *Paleobiology* **7**, 394–399.

Wyatt, A. R. 1987 Shallow water areas in space and time. *J. geol. Soc. Lond.* **144**, 115–120.

Yanshin, A. L. 1973 On so-called world transgressions and regressions. *Byull. mosk. Obshch. Ispyt. Prir.* (In Russian.) **48**, 9–44.

Discussion

J. M. COHEN (39 *Greenhill, Blackwell, Worcestershire, U.K.*). How could such organisms as sharks, mid-water cephalopods and coelacanths have survived your universally anoxic deep-seas? Surely any such black-shale producing anoxias must have been local rather than global?

A. HALLAM. I fear that Dr Cohen obtained an exaggerated impression from my lecture as to the extent to which I thought the deep ocean was anoxic in the past. It would be more accurate to state that there was a greater tendency towards anoxia in the past, at times when the earth's climate was more equable. Dr Cohen's point is, nevertheless, an interesting one, because it implies that at least limited refugia must have existed at times of mass extinction. Otherwise it is difficult to explain the survival of so-called Lazarus taxa. He could well be right, the anoxic events recorded from the stratigraphic record could perhaps signify relatively local phenomena. I speculate in my paper that many such events may have been confined to epicontinental seas.

Orkney (Sharrock 1976). In fact there were always a few wintering birds in Shetland, but these may have come from Scandinavia (Lack 1986), and a Shetland breeding population was re-established in 1976 (Berry & Johnston 1980). Another example is shown in figure 1. Remarkably full counts of the breeding birds of Skokholm Island, Wales, were made from 1928 to 1979, except for the war years. This data set shows many local extinctions and immigrations,

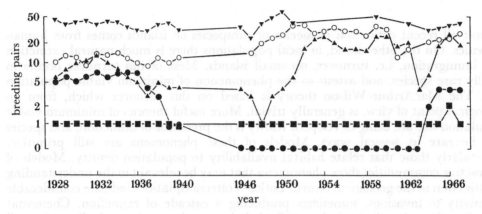

FIGURE 1. Numbers of breeding pairs for five species of land birds on Skokholm Island, Wales. The ordinate is scaled by arcsinh y, essentially the same as a logarithmic transformation for numbers greater than two. Counts for these species up to 1979 are in Williamson (1983). (▼) *Anthus pratensis*, meadow pipit; (○) *Oenanthe oenanthe*, wheatear; (▲) *Vanellus vanellus*, lapwing; (●) *Prunella modularis*, dunnock; (■) *Corvus corax*, raven. Modified from Williamson (1981).

mostly relating to casual breeding by single pairs (Williamson 1981, 1983). A few slightly more common species, such as *Prunella modularis*, the dunnock, also show turnover. In figure 1 it can be seen that a population of about five pairs went extinct on the island in the early 1940s, and that a population was not re-established until the 1960s. Dunnocks are primarily birds of scrubland and hedgerows; the bleak island of Skokholm is a distinctly marginal habitat for them. The island is almost exactly 1 km² in area, so the density of five pairs per square kilometre can be compared with the figure of 28 pairs per square kilometre for English farmland in general and 52 pairs per square kilometre in favourable habitats (Sharrock 1976) to show how unfavourable Skokholm is for this particular species. Indeed, there may be a lack of genetic adaptation as a result of this sort of turnover. Some permanently established dunnock populations in Skokholm-like habitats are in the subspecies *P.m. hebridium* (Sharrock 1976).

That most of the local extinctions associated with turnover occur in thin, marginal populations can be seen in another bird community, that in the 16 ha† oak wood at Bookham Common, Surrey, England. Figure 2 is a diagram of the breeding records over 29 years. The species are ordered by their arithmetical mean over the period. About one third of them have an average of less than one pair a year, which means they are, in this survey, normally locally extinct. It is evident that in this relatively small patch of habitat most species occur at so low a density that they are not recorded as breeding in every year, i.e. they regularly become extinct locally. This sort of extinction can be called demographic extinction. There is also environmental extinction, as shown by two warblers that became extinct in the wood in 1971.

† 1 ha = 10⁴ m².

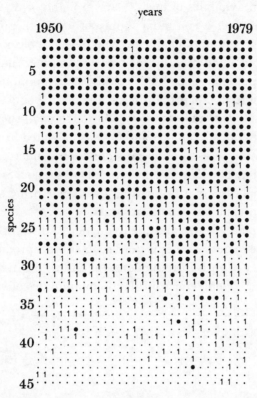

FIGURE 2. A diagram of the breeding community of birds at Eastern Wood, Bookham Common, England. The columns are years. There was no count in 1957. Each row is a species. (●) Two or more breeding pairs; (1) one breeding pair; (·) not breeding. The full data can be found in Williamson (1981, 1987). Modified from Williamson (1987).

These are the ninth species, *Phylloscopus trochilus*, the willow warbler, and the twentieth species, *Sylvia borin*, the garden warbler. Both were affected by the management policy of allowing the rides through the wood to become more overgrown, changing the habitat to their disadvantage (Williamson 1981).

An extinction rate can be calculated from the organisms present originally, and those at some later time. Figure 2 shows how surprisingly difficult it can be to define what is meant by being present in an area. For the whole arbitrary period of 29 years, 45 species are recorded as breeding. In any one year, the number varies between 28 and 36 species (Williamson 1987). Only 16 species bred in all 29 years, and a mere five of these had more than one pair each year. So if the wood by some geological process became isolated and only these last five species survived, would the extinction rate be 89, 86, 82, 69 or 0%? Apparent high extinction rates on small islands (Richman *et al.* 1988) show the importance both of immigration and minimum viable populations in the study of community dynamics.

My final example of natural extinction on real islands is the human disease measles. This is caused by a virus closely related to rinderpest and canine distemper and well known for producing lethal epidemics in primitive peoples. In its present genetic form it is widely regarded as having evolved recently, as it requires large urban populations to sustain it. Even more than the bird data, it demonstrates how a species will become extinct if its population size falls too low.

[219]

The stochastic population dynamics of measles are probably as well understood as that of any organism. The virus requires a continual supply of new susceptibles, that is a constant supply of children old enough to catch the disease who have not been infected. Infection leads to an almost universal life-long immunity. From an early computer simulation, Bartlett (1957) deduced that an urban population of 200000 would be the minimum needed to provide a sufficient supply of susceptibles to keep the disease going. In populations smaller than that measles would go extinct. He was able to show (Bartlett 1957, 1960) empirically that his estimate was a little low; a figure of 250000–300000 was indicated from public health returns from both sides of the Atlantic. In a more recent simulation, Anderson & May (1986) found that in a population of less than 6500 an epidemic would fail to start. Such communities would, of course, still have occasional cases of measles, from the disease being brought in from outside, but the chain reaction of an epidemic would not start.

The relation between population size and the frequency and strength of measles epidemics is bound to be fuzzy, because there are other factors, such as the reproductive rate of the human population and the frequency of virus introductions, that affect the dynamics. However, figure 3, which shows the frequency of epidemics in island communities of various sizes, shows how

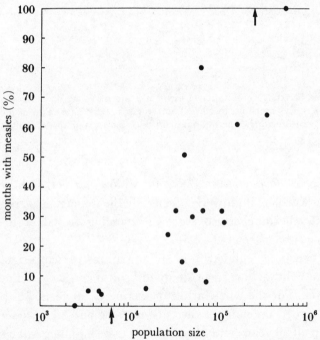

FIGURE 3. Extinction of measles in small island communities. Data from Black (1966). The upper arrow is Bartlett's (1957) estimate of the minimum population size for endemic measles, i.e. for no extinction. The lower arrow is Anderson & May's (1986) estimate of the minimum population size for measles to invade. Modified from Williamson (1981).

accurate the two predictions are. Note that this result can only be brought out graphically with a suitable scale on the abscissa, and that plotted this way the location and uncertainty of the rate of increase of probability of epidemics is brought out neatly.

Measles shows most clearly that rare populations are in danger of extinction, because they may by chance fall below a minimum viable population density. So I next consider some features of rarity.

RARITY

As Darwin said (1859, p. 109) 'Rarity, as geology tells us, is the precursor of extinction', though we might nowadays wish to rest this conclusion on ecological studies as well. Because of its importance in conservation in work aiming to avoid extinction, rarity has been much studied recently. One major conclusion is that there are many ways in which an organism can be rare. Perhaps the neatest classification of these is by Rabinowitz (1981), though she notes the other classifications put forward in the same symposium. Her seven forms of rarity come from a dichotomous contrast in three characters. These are the geographical range, which may be either large or small, the habitat specificity, which may be wide or narrow, and the local population size, which may be large or small. Her classification, of course, produces eight categories, but the eighth is the set of common species. In her view, a rare species may have any or all of the three characters of restricted range, restricted habitat or restricted population density.

Earlier, I distinguished environmental extinction from demographic extinction. Rabinowitz, unlike Harper (1981), was not concerned with the change of species characters in time, or indeed in space (Schoener 1987). However, environmental extinction in general implies that there has been a loss of habitat, which will have been reflected, depending on circumstances, in a loss of geographical range or an apparent increase in habitat specificity or both. The characters of environments, as geologists well know, are changing all the time. In figure 4 I have shown an indirect measure of this. Figure 4 is an ordination of the land-bird community of Skokholm (Williamson 1983), and the track of the community looks remarkably like a random walk. The first axis of the ordination is related to the total number of pairs (of all species) breeding, the second to the number of species breeding (Williamson 1987). It can be seen that the total number of pairs has been increasing fairly steadily since records began, whereas the number of species first declined and then increased again. If the course of community change is a reflection of environmental change, and the details of the changes in the individual species that give rise to the ordination strongly suggest that it is, then figure 4 is also related to the important phenomenon that environmental heterogeneity has a reddened

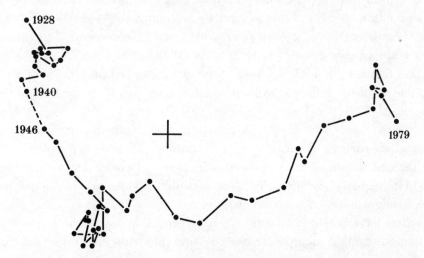

FIGURE 4. A 'step-along' ordination of the land-bird community of Skokholm Island, Wales. The points are connected in time sequence. There were no counts between 1940 and 1946. Modified from Williamson (1983).

spectrum (Williamson 1981, 1983, 1987). Over time, environments will become more and more different from the way they were, and this in itself will lead to the environmental extinction of populations. The fate of the species as a whole will depend on the distribution of its populations, and on its ability to adapt by genetic change to the environmental changes.

Another point that Rabinowitz (1981) brought out is that a population can adapt to the state of being rare. This additional genetic dimension of the problem of understanding demographic extinction certainly makes prediction more difficult. But again, as Darwin (1859, p. 110) was the first to say, rare species will usually evolve more slowly than common ones.

Perhaps the most interesting, and most puzzling, of Rabinowitz' seven types is that of species that have, for their life-form, low population densities yet quite wide geographical ranges and that occur in a variety of habitats. Salisbury (1942) suggested that some such cases could be related to demographic features, particularly to low reproductive output. Whatever the cause, those species are both those most likely to be adapted genetically to rarity and to be most at risk of demographic extinction. Can mathematical models be developed that allow us to quantify this risk?

MATHEMATICAL MODELS WITH EXTINCTION

The most famous theory involving extinction is the MacArthur & Wilson (1967) equilibrium theory of the number of species on an island. This theory postulates a balance between immigration and local extinction, leading to an equilibrium maintained by turnover. Data such as those in figures 1 and 2 show that the central part of the theory is true but ecologically trivial (Williamson 1989b). There is turnover, but most of the species turning over are marginal, casual species. Schoener & Spiller (1987) make the same point for spiders on Bahamian islands. The species turning over are rare, and it is often the same species that re-immigrates after going locally extinct. Schoener (1987), working with data for Australian birds, shows that most such species are at the edge of their habitat and geographical range, pseudo-rare in Rabinowitz's (1981) phrase, and common somewhere else, in habitats more favourable to them.

As well as dealing with an essentially trivial ecological phenomenon, the relation of local extinction and of immigration to species number postulated by MacArthur & Wilson is much fuzzier than they assumed (Williamson 1981, 1983), and developments of the theory to predict species–area relations give unsatisfactory answers (Williamson 1988, 1989b). More seriously, the theory fails to catch the main features of the change in communities with time, as shown in figures 2 and 4, or to predict or explain which species will form the bulk of the ecological community.

Another related weakness in the MacArthur–Wilson theory is that it is purely a phenomenological theory. It does not 'incorporate detailed demographic mechanisms permitting the calculation of rates of local extinction and colonization from the life history parameters of individuals' (Lande 1987). Nor does it include the environmental variation that drives some immigrations and some extinctions.

Recently there have been several attempts to remedy this, by building models of minimum viable population size that incorporate both demographic and environmental extinction and extinction that involves an interaction between these two processes (Belovsky 1987; Lande 1987). Theories involving purely stochastic change in demographic variables, that is assuming a constant environment, can be remarkably successful. Measles, discussed above, is an example. Incorporating environmental change, which will act by altering the mean and possibly the variance of demographic parameters, is much more difficult. One problem is knowing how to relate environmental change to biological change. A very great deal is known about the pattern of some environmental variables; practically nothing is known about how such changes work through to produce changes in schedules of birth and death rates.

At the end of the Pleistocene, many real and habitat islands were formed as climates changed and sea levels rose. The populations isolated on such patches could, in some cases, no longer be reinforced by immigration. Brown (1971) and Patterson (1984) for mammals on habitat islands in the Rocky Mountains, and Richman *et al.* (1988) for lizards on real islands near Baja California and in South Australia, have estimated local extinction rates, having first estimated the original number of species present from species–area curves. It is unfortunately true, as was shown with the birds of Bookham Common, that the species recorded in a patch of more or less continuous habitat, from range maps and the like, will include many that are not maintaining viable populations in that patch. So it could be said that many of the species apparently going extinct were not there, in the sense of not maintaining permanent populations, in the first place. This reinforces the current conservationist view that reserves will need to be appreciably larger than was thought a few years ago if extinction is to be prevented. Newmark (1987) presents historical records that support this.

Times to extinction are much shorter in minimum viable population models with environmental variation than in purely demographic ones. They are shorter still if catastrophes are included (Shaffer 1987). By making some bold assumptions, Belovsky (1987) calculates extinction rates for relic populations of mammals in the Rocky Mountains, and finds that his environmental extinction models predict too low a rate. He has to assume a rather precise relation between size and demographic parameters to get his result. It is well known, for instance, that the intrinsic rate of natural increase, r, decreases with increasing size across orders. It is less well known that the relation is the other way round within genera and families (Williamson 1989a) and even sometimes between families (Stemberger & Gilbert 1985). What is undoubtedly true is that small, rather random, environmental fluctuations increase the chances of small populations going extinct.

One important conclusion to come from these rather gloomy studies of conservation problems is that closely related species may have quite different minimal area requirements. Thomas (1984) from field studies of the persistence of populations, gives the data in table 1 for the minimal breeding area of non-migratory British butterflies. There are a further 12 migratory species, and 12 species where minimal area is not known, but it is mildly encouraging that so many require such small areas.

TABLE 1. MINIMUM AREA OF VIABLE COLONIES OF BRITISH BUTTERFLIES

(Simplified from Thomas (1984).)

area/ha[a]	0.5–1	1–2	2–5	5–10	10–50	> 50
no. of species of butterflies	15	11	2	2	4	1

[a] 1 ha = 10^4 m^2

Interactive and community models

If a general theory of extinction is to be developed, a general theory of rarity is needed. One point that comes through strongly from the SCOPE programme on the Ecology of Biological Invasions, is that there is no general theory that allows us to predict which species can invade and which cannot (Williamson 1989 a), and part of the reason for this is that there is no general theory of the distribution of commonness and rarity between species in the same trophic level.

The fact that some species are common and some rare could be due either to some habitats or environments being commoner than others, or to interactions between species leading to different equilibrium population densities. It has certainly long been a general view that subtle biological interactions are involved in many extinctions. For instance Darwin (1859, p. 319) said that 'unperceived injurious agencies... are amply sufficient to cause rarity, and finally extinction'. What some models of species interactions show is that quite simple, fairly strong, ecological interactions may lead to quite surprising results. I shall give two examples.

The first is a large computer-simulation model developed by Drake (1985). This is a Lotka–Volterra type model, which means that the equations are the simplest and most linear that can be reasonably postulated. He modelled only feeding relations as between carnivore and herbivore; all competitive effects were indirect through share resources or shared predators. He had a pool of 100 species, some primary producers, some herbivores and so on, each with fixed parameters. He started each run with two producers and one herbivore, and then introduced other species randomly from his pool. He calculated the stability of the system after each introduction and considered that species went extinct if the system was unstable. These simulations require much computer time, and he only completed ten runs, ending with ten different communities.

One of Drake's runs is shown in figure 5; I have chosen it because it has the largest extinction of any of his runs. After rather more than 1100 introductions, which means that each species has already been introduced about 11 times, one particular introduction produces a mass extinction, reducing the number of species present in stable equilibrium from 19 to nine. Other cascades of extinction can be seen in the graph, and Drake was unable to find any general rules about when such cascades would occur and what species would be involved.

It is possible that Drake's results come from using over-simplified dynamical models, or from using an unnatural distribution of interaction strengths between species. Opinions differ, relevant data are rare. My second example is more widely agreed to be realistic. Equations that govern continuous culture systems, otherwise known as chemostats, incorporate equations and functional forms that have been established experimentally by microbiologists (Williamson 1972).

The simplest chemostat involves one species of bacterium limited by one particular substrate, such as a carbon source. Such a system is stable at all dilution rates almost up to the maximum growth rate of the bacterium. Adding a predator, a ciliate protozoan, produces quite a different result. Figure 6 summarizes the pattern found by a simulation study of such a system.

The parameters used in this simulation system are those of Curds (1971) and are based on those found experimentally for real species of bacteria and ciliates. The parameters are a maximum specific growth rate, a saturation constant and a yield coefficient for each of the two simulated species. As is natural, the maximum growth rate of the bacterium is set at a greater value than that of the ciliate. Consequently on the right of figure 6 the ciliate becomes extinct,

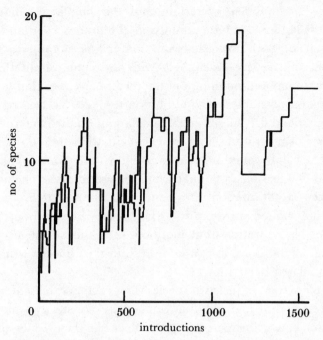

FIGURE 5. Immigration and extinction in a theoretical Lotka–Volterra community. The species are drawn from 100 species defined by their parameters. There is extinction of over 50 % of the community after 1150 invasions. The community stabilises after 1500 invasions. From Drake (1985), with permission.

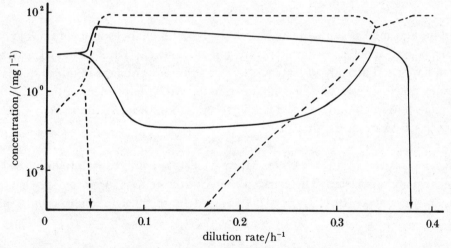

FIGURE 6. The theoretical behaviour of a bacterium–ciliate chemostat system. The lines show the maximum and minimum of the limit cycle oscillations at different dilution rates. Parameters of the system as in Curds (1971). Ciliates go extinct at high dilution rates. There are stable states at very low dilution rates and at dilution rates just less than the critical point for ciliate extinction. The oscillations in the middle range of dilution are so large that extinction of the bacterium, followed by extinction of the ciliate, would be expected.

or cannot invade, because the dilution rate is more than its growth rate can cope with. The transition from a stable community of predator and prey to the extinction of the predator by wash out occurs very rapidly with change of dilution, as can be seen by the almost vertical line in the ciliate graph at the right-hand side.

Below the critical dilution rate for ciliates, about 0.38 h^{-1}, there is a small range of dilution rates in which both the simulated bacterium and the simulated ciliate come to stable equilibrium points. At 0.34 there is a bifurcation, and at dilutions less than this the system goes into limit cycles. That is, both the bacterium and ciliate numbers converge to stable oscillations, oscillations that repeat indefinitely with the same amplitude and wavelength. Figure 6 shows the maximum and minimum of the oscillations at each dilution. Drake's rules would regard the system as unstable below a dilution rate of 0.34, and record an extinction. Certainly for dilutions around 0.1 the size of the oscillations on the bacterial population are so vast, six to eight orders of magnitude, that in a real situation the population would almost certainly go extinct. The ciliate would inevitably go extinct afterwards. Taking three orders of magnitude as the limit of a feasible cycle that would persist in a real chemostat, extinction of both species would occur at all dilution rates between 0.23 and 0.045.

There is a surprise, only hinted at by Curds (1971), at very low dilutions. The system is once again in a steady state. The transition from enormous oscillations to steady state again occurs over a very small range of dilutions, as can be seen from the almost vertical curve at the left of the graph of the simulated bacterium.

Altogether this simple system of just two species shows remarkably complex behaviour. In two zones, at the lowest and at high dilution rates, there are steady states. Each zone ends abruptly as the dilution rate is increased slightly. Between these zones the system shows a remarkable range in the amplitude of oscillations. With such complexities in an extremely simple system, is it surprising there is still much to learn about real ecosystems?

Conclusions

All these data and models show that with relatively small changes in conditions extinction may suddenly occur, and for reasons that are not obvious. In Drake's model, cascades of extinction can happen. There are at least two large and important gaps in our understanding. The first is being able to say for any particular extinction why it happened, rather than to show a range of reasons that might have caused it. The second is knowing, in almost all communities, why some species are rare in the first place.

I thank Kevin Gaston, Brian McArdle, Moya McCloskey and Charlotte Williamson for much advice, help and discussion. The work on chemostat systems is supported by Department of the Environment contract PECD7/8/107. I am grateful to Jim Drake for permission to use figure 5.

Discussion

C. B. Goodhart (*Gonville and Caius College, Cambridge, U.K.*). The examples given here of the extinction of island populations are mostly no more than 'pseudo-extinctions', where there are plenty of other flourishing populations living elsewhere apart from the one going extinct. This is quite different from the historical extinction of whole species as discussed in the other contributions to this symposium, and it is important to make a clear distinction between the two.

Nowadays, whenever a much-needed bypass is proposed, for example, conservationists can be relied upon to discover some obscure moth or moss, or whatever, whose only known home in Britain will be destroyed unless the road is diverted, often at a cost of millions of pounds. But it usually turns out that the creature concerned is rare in Britain only because it is here at the edge of its distribution, and it is common enough on the Continent. So its loss from this country, however regrettable, would really only be a pseudo-extinction, and not the irrevocable biological disaster that the word 'extinction' might be thought to imply.

So professional ecologists would be wise not to cry 'wolf' too often in such circumstances, lest they are ignored when there really is something worth saving. Indeed, they have a professional duty to look at each case objectively on its merits, taking into account also the costs involved, in any advice they may feel called upon to offer. However desirable something may be, it is always possible that the cost could be too high, even when it is somebody else who is going to have to pay it.

M. H. WILLIAMSON. In reply to Dr Goodhart's first point, I do indeed discuss the extinction of local populations rather than the extinction of species, but these are real extinctions for the populations concerned, not pseudo-extinctions. Species extinction occurs when all local populations have become extinct, so the study of local extinctions is relevant to the study of species extinction.

Dr Goodhart's second point is valid, though I would have preferred a more objective statement of it.

D. A. WEBB (*Trinity College, Dublin, Republic of Ireland*). I am always rather suspicious of models, and I can't help wondering whether, if Professor Williamson instead of using a chemostat had built a vast vivarium and had introduced owls and mice at recorded rates the curves would have come out the same. I have no doubt that his model would apply well to certain lower organisms, but when you move to higher organisms the differences in behaviour and limitations of habitat and so on are so vast that I am sceptical as to the predictions derived from one field being applicable in the other.

M. H. WILLIAMSON. Many ecologists and field biologists are suspicious of models, and this may have slowed the agreement about valid general statements in ecology. In Professor Webb's example, the curves would be similar but not the same. The detailed mathematical formulation would certainly be different, but almost any predator–prey system would have a tendency to oscillations. Local extinctions following a boom and crash (a single oscillation) are well known for vertebrates.

J. COHEN. Dr Williamson's use of the word 'extinction' seems rather different from that of other speakers at this meeting. For most of us, 'temporary extinction' resembles usage like that on the Vapona packet: 'Kills insects for up to three months'.

M. H. WILLIAMSON. I do not use the phrase 'temporary extinction', but Dr Cohen is perhaps trying to make the point made by Dr Goodhart in his first paragraph.

REFERENCES

Anderson, R. M. & May, R. M. 1986 The invasion, persistence and spread of infectious diseases within animal and plant communities. *Phil. Trans. R. Soc. Lond.* B **314**, 533–570.

Bartlett, M. S. 1957 Measles periodicity and community size. *Jl R. statist. Soc.* A **120**, 48–70.

Bartlett, M. S. 1960 The critical community size for measles in the United States. *Jl R. statist. Soc.* A **123**, 37–44.

468   M. WILLIAMSON

Belovsky, G. E. 1987 Extinction models and mammalian persistence. In *Viable populations for conservation* (ed. M. E. Soulé), pp. 35–57. Cambridge University Press.

Berry, R. J. & Johnston, J. L. 1980 *The natural history of Shetland*. London: Collins.

Black, F. L. 1966 Measles endemicity in insular populations: critical community size and its implications. *J. theor. Biol.* **11**, 207–211.

Brown, J. H. 1971 Mammals on mountaintops: non-equilibrium insular biogeography. *Am. Nat.* **105**, 467–478.

Curds, C. R. 1971 A computer-simulation study of predator–prey relationships in a single-state continuous-culture system. *Water Res.* **5**, 793–812.

Darwin, C. 1859 *On the origin of species*. London: Macmillan.

Drake, J. A. 1985 Some theoretical and empirical explorations of structure in food webs. Ph.D. thesis, Purdue University, Indiana, U.S.A.

Harper, J. L. 1981 The meanings of rarity. In *The biological aspects of rare plant conservation* (ed. H. Synge), pp. 189–203. Chichester: John Wiley.

King, W. B. 1985 Island birds: will the future repeat the past? In *Conservation of island birds* (ed. P. J. Moors), pp. 3–15. Cambridge: International Council for Bird Preservation.

Lack, D. 1976 *Island biology illustrated by the land birds of Jamaica*. Oxford: Blackwell Scientific Publications.

Lack, P. 1986 *The atlas of wintering birds in Britain and Ireland*. T. Calton & A. D. Poyser.

Lande, R. 1987 Extinction thresholds in demographic models of territorial populations. *Am. Nat.* **130**, 624–635.

MacArthur, R. H. & Wilson, E. O. 1967 *The theory of island biogeography*. Princeton University Press.

Newmark, W. D. 1987 A land-bridge island perspective on mammalian extinctions in western North American parks. *Nature, Lond.* **325**, 430–432.

Patterson, B. D. 1984 Mammalian extinction and biogeography in the Southern Rocky Mountains. In *Extinctions* (ed. M. H. Nitecki), pp. 247–293. Chicago University Press.

Rabinowitz, D. 1981 Seven forms of rarity. In *The biological aspects of rare plant conservation* (ed. H. Synge), pp. 205–217. Chichester: John Wiley.

Raffaele, H. 1977 Comments on the extinction of *Loxigilla portoricensis grandis* in St. Kitts. *Condor* **97**, 389–390.

Richman, A. D., Case, T. J. & Schwaner, T. D. 1988 Natural and unnatural extinction rates of reptiles on islands. *Am. Nat.* **131**, 611–630.

Salisbury, E. J. 1942 *The reproductive capacity of plants*. London: Bell.

Schoener, T. W. 1987 The geographical distribution of rarity. *Oecologia* **74**, 161–173.

Schoener, T. W. & Spiller, D. A. 1987 High population persistence in a system with high turnover. *Nature, Lond.* **330**, 474–477.

Sharrock, J. T. R. 1976 *The atlas of breeding birds in Britain and Ireland*. Berkhamsted: Poyser.

Shaffer, M. 1987 Minimum viable populations: coping with uncertainty. In *Viable populations for conservation* (ed. M. E. Soulé), pp. 69–86. Cambridge University Press.

Stemberger, R. S. & Gilbert, J. J. 1985 Body size, food concentration, and population growth in planktonic rotifers. *Ecology* **66**, 1151–1159.

Thomas, J. A. 1984 The conservation of butterflies in temperate countries: past efforts and lessons for the future. *Symp. R. ent. Soc. Lond.* **11**, 333–353.

Williamson, M. 1972 *The analysis of biological populations*. London: Edward Arnold.

Williamson, M. 1981 *Island populations*. Oxford University Press.

Williamson, M. 1983 The land-bird community of Skokholm: ordination and turnover. *Oikos* **41**, 378–384.

Williamson, M. 1987 Are communities ever stable? *Symp. Br. ecol. Soc.* **26**, 353–371.

Williamson, M. 1988 Relationship of species number to area, and other variables. In *Analytical biogeography* (ed. A. A. Myers & P. S. Giller), pp. 91–115. London: Chapman & Hall.

Williamson, M. 1989*a* Mathematical models of invasion. In *The ecology of biological invasions: a global perspective* (ed. J. A. Drake, F. di Castri, R. H. Groves, F. J. Kruger, H. A. Mooney, M. Rejmanek & M. Williamson), pp. 329–350. Chichester: John Wiley.

Williamson, M. 1989*b* The MacArthur and Wilson theory today: true but trivial. *J. Biogeogr.* **16**, 3–4.

Phil. Trans. R. Soc. Lond. B **325**, 469–477 (1989)

Printed in Great Britain

The present, past and future of human-caused extinctions

By J. M. Diamond

*Department of Physiology, University of California Medical School, Los Angeles,
California 90024-1751, U.S.A.*

This paper re-evaluates whether we are really at the start of a mass extinction caused by humans. I consider the present, past and future of human-caused extinctions.

As regards the present, estimates of extinction rates based on *Red Data Books* underestimate real values by a large factor, because the books evaluate only those species that have attracted specific attention and searches. Especially in tropical areas with few resident biologists, many poorly known species go extinct without having been the object of specific attention, and others disappear even before being described. A 'green list' of species known to be secure is needed to complement 'red books' of species known to be extinct.

As regards the past, it is now clear that the first arrival of humans at any oceanic island with no previous human inhabitants has always precipitated a mass extinction in the island biota. Well-known victims include New Zealand's moas, Madagascar's giant lemurs, and scores of bird species on Hawaii and other tropical Pacific islands. Late-Pleistocene or Holocene extinctions of large mammals after the first arrival of humans in North America, South America and Australia may also have been caused by humans. Hence human-caused mass extinction is not a hypothesis for the future but an event that has been underway for thousands of years.

As regards the future, consideration of the main mechanisms of human-caused extinctions (overhunting, effects of introduced species, habitat destruction, and secondary ripple effects) indicates that the rate of extinction is accelerating. The basic reason is that there are now more humans than ever before, armed with more potent destructive technology, and encroaching on the world's most species-rich habitats: the continental tropical rainforests.

INTRODUCTION

Are we now at the start of another mass extinction, similar in magnitude to the greatest ones of the past, but differing in being caused by humans? I am not referring to the risk of a nuclear war, but instead to the risk resulting from effects (such as habitat destruction) that are already operating and that have already caused extinctions.

The answers offered to this question vary greatly. The *Red Data Book* of the International Council for Bird Preservation (ICBP) lists 88 Recent bird species as extinct and 283 as extant but endangered, about 1 and 3% of all Recent bird species, respectively. Many conservation biologists consider these numbers as gross underestimates, whereas many economists and developers consider them gross overestimates. How many species really have already become extinct through human causation? How many more are likely to become extinct in the next 50 years? If it is true that humans cause extinctions, so what? Aren't extinctions part of normal evolutionary history, with all species destined sooner or later for extinction anyway? Is an impending mass extinction crisis a hysterically exaggerated hypothesis, an event already underway or an already partly accomplished act?

To answer these questions, I consider the present, past and future of human-caused extinctions. I begin with an assessment of claims about the number of extinctions that have already occurred in modern times (i.e. since 1600). I then assess the number of extinctions that humans might have caused in recent millenia but were not witnessed by literate observers. Next, I discuss the number of extinctions likely to occur in the near future (e.g. by the year 2050), based on knowledge of mechanisms of human-related extinctions and on extrapolation of current trends. Finally, I return to the initial question of whether we are now in the middle of a mass-extinction wave.

THE PRESENT

As a start, consider the claim in the ICBP Bird *Red Data Book* that about 88 full species of birds plus an additional 83 subspecies have become extinct since 1600 (King 1981). For a bird species to get on this list requires that it has been specifically searched for and not found for many years, despite ornithologists knowing where it previously occurred and having looked there. For some species and races that became extinct recently, ornithologists have watched a population dwindle down to the last few individuals and then followed the fates of those individuals in their last years. Thus it is very unlikely that species that pass the tests for inclusion on the ICBP list of extinct species actually still exist.

The occasional, well-publicized 'rediscoveries of extinct species' have instead involved birds that did not make it onto the ICBP list of species considered extinct. For example, when I rediscovered New Guinea's long-lost yellow-fronted gardener bowerbird (*Amblyornis flavifrons*) in 1981, the media cited it as an example of a species formerly considered extinct. In reality, no specimen had been collected of that bowerbird since its description in 1895, and there had been some speculation that it was possibly extinct, but the fact is that ornithologists did not know where the sole specimens had been collected. When I happened by chance on the bowerbird's range, New Guinea's previously unexplored Foja Mountains, the bowerbird proved reasonably common there (Diamond 1982).

The lengthy evaluation required for each species considered for inclusion in the *Red Data Book* meant that the books were widely recognized as out of date at the moment of publication. As Collar & Andrew (1988) of the ICBP aptly express it, 'With only limited resources to finance the evaluation and documentation of species at risk, there is inevitably tension between the urgency imposed by their plight and the responsibility of making the analyses as thorough and hence as truthful as possible'. The inevitable incompleteness of the *Red Data Books* attracted criticism that resulted in a new ICBP volume listing bird species considered to be at risk of global extinction and entitled *Birds to Watch* (Collar & Andrew 1988). This book lists 1029 bird species at risk of global extinction, and an additional 637 species considered near-threatened. Adding these two catagories to the 88 species that had already passed rigorous evaluation as being extinct yielded a total of 1754 bird species, or about 20% of the world's approximately 9000 Recent bird species, at risk of extinction or already extinct.

However, in the areas that I know best as a field ornithologist – New Guinea, Melanesia and Indonesia – some of those species that *Birds to Watch* lists as threatened are in fact surely extinct, because they occur on small islands and have not been seen in 50 years by residents of those islands who knew the species well. How could this renewed ICBP effort list extinct species as merely threatened? The basic problem is that even this latest effort continues to place the burden of proof on documenting that a species is extinct or threatened. Those species that get into the book are ones for which someone suspected danger and gathered supporting evidence. It is not the case that the status of all the world's bird species was evaluated and that those species found to be in danger were then listed.

This distinction between 'probably extinct' and 'positively known to be extinct' means little in Europe or North America, where there are millions of fanatically devoted bird-watchers. All species of European and North American birds are repeatedly monitored every year. The rarer they are, the more diligently they are monitored. No European or North American bird species could possibly become extinct without having been known to be in danger, and having been sought after, for many years.

However, most of the world's species occur in the tropics, where there are few resident trained naturalists and few specific searches for possibly endangered species. We don't even know how best to identify many tropical species alive, and we don't know where they were collected. All that we have may consist of specimens, sometimes from an uncertain locality, dating back to the 19th century. Thus if the burden of proof were on showing that a species is still extant in healthy numbers in a non-threatened habitat, many more species would fail that test than the ICBP now lists as extinct or threatened.

Some recent studies emphasize the gulf between 'proved extinct' and 'not proved extant' (Diamond 1987). In the Solomon Islands, a tropical southwest Pacific archipelago where I have done much fieldwork, the *Red Data Book* (King 1981) listed one extinct bird species. However, when I tabulated what species had actually been reported alive since 1953, it turned out that there were no definite records for 12 of the Solomon's 164 bird species, even though some of these species were formerly described as common. Most of these missing species are ground birds susceptible to introduced predators, and Solomon islanders specifically told me that some of them had been exterminated by introduced cats.

Still, the Solomon Islands are a mild example, because they have fewer native species to begin with, fewer introduced predators and more intact natural habitat than most other tropical areas. More typical of the world's tropics is lowland Malaysia, which is rich in species and has been extensively deforested. A recent four year search for the 266 species of primary freshwater fishes that had been described from the Malay Peninsula was able to find less than half (only 122) of those species (see Mohsin & Ambok 1983). The remainder have either become extinct, endangered or rare as a result of habitat changes.

Many other tropical habitats rich in endemic species, like Peninsular Malaysia, have already been extensively damaged or destroyed, but differ from Peninsular Malaysia in that no biologist has troubled to look for the habitat's former species. Other habitats have been destroyed even before they were surveyed by biologists. For example, a botanical survey of one forested ridge in Ecuador discovered numerous previously undescribed species of plants, just before those species were exterminated by logging at the ridge (Gentry 1987). That ridge's insects, and the plants of innumerable other logged ridges, were never described before being exterminated.

For these reasons the ICBP is shifting its emphasis from 'red books' of species proven to be extinct towards 'green lists' of species proven to be extant and secure (Imboden 1987). When one considers that most of the world's species live in tropical habitats now under siege and rarely visited by biologists, it becomes unlikely that as many as half of the world's Recent species will qualify for the Green List.

The past

So far, I have confined my discussion to extinctions that occurred in modern times and that were recorded by literate observers. But many other Recent species are known only from fossil or subfossil bones and were never recorded by literate observers. Famous examples include the moas of New Zealand, the elephant birds of Madagascar, the flightless geese of Hawaii, and the mammoths of the Americas and Eurasia. There has been a long-standing debate over whether these extinctions too were caused by humans.

Twelve or thirteen species of moas – large, flightless, ostrich-like birds confined to New Zealand – have been described from subfossil bones. It was long argued whether the moas died out before or after the Maori colonization of New Zealand around 1000 A.D., hence whether the extinctions of the moas were from natural or human causes. The argument has now been settled, because bones of almost all moa species have been found in close association with humans: namely, in Maori ovens and butchering sites, where moas were cut up and eaten (Anderson 1984, 1989). The estimated number of moa skeletons at the sites exceeds 100 000. Radiocarbon dates show that the moas became extinct within about 500 years of human arrival. Hence there is now no doubt that the moas were exterminated as a result of human activity. Many other New Zealand bird species, as well as lizards and frogs, became extinct at the same time because of overhunting or habitat destruction, or commensal mammals that arrived with the Maoris.

Madagascar, like New Zealand, had giant flightless birds known as elephant birds, abundantly attested by subfossils and especially by eggshells. More recently, archaeologists have unearthed on Madagascar the bones of a dozen large species of lemurs up to the size of a gorilla, plus giant land tortoises and a hippopotamus. Madagascar, too, suffered the extinction of this whole megafauna soon after humans arrived around 500 A.D. (Dewar 1984).

Since 1982 more than 50 species of now-extinct subfossil birds have been found in Hawaii at sites postdating colonization by Polynesians, including a radiation of flightless geese. This mass extinction removed more than 50 % of Hawaii's original avifauna (see Olson & James 1982; James et al. 1987). Similar reports of extinct birds are coming in from archaeological sites on all other investigated Pacific islands, including Tonga, the Marquesas, Cooks, Chathams, Bismarcks, Tahiti, Henderson, Fiji, Tikopia and New Caledonia (Steadman 1989).

All these oceanic islands – New Zealand, Madagascar, Hawaii and other Pacific islands – have in common the fact that they were initially not inhabited by humans. Their faunas collapsed quickly after human arrival, because the island species had evolved in the absence of human hunters and of introduced predators such as cats and rats. In particular, virtually all oceanic islands studied by palaeontologists have yielded one or more species of flightless rail, mostly now extinct. When one extrapolates from the studied islands to unstudied islands, one estimates that about one fifth of the species of birds that existed in the world a few thousand years ago have disappeared as a result of human activities on oceanic islands (Olson 1989).

These recent extinctions of 'naïve' island species (i.e. ones without experience of humans) are now widely accepted as human caused, because the extinctions are recent enough to be accurately datable, because of their close coincidence in time with human arrival, and because no natural environmental change even remotely adequate to account for them has been found. Still controversial, however, are extinctions of the naïve species that also existed on the continents not occupied by humans until the late Pleistocene: North America, South America and Australia. In North America and South America, respectively, 73 and 80% of large mammal genera became extinct around the time of appearance 11000 years ago of Clovis hunters, who are considered by many to be the first human occupants of the New World (Martin 1984). Some of the extinctions are dated to within a century or two of the arrival of the Clovis hunters (see Mead *et al.* 1986). For a few of the extinct species, notably mammoths, evidence of human hunting is available in the form of butchered kills. The large mammals of the Americas may thus have been exterminated by humans, and Mosimann & Martin (1975) proposed that this occurred by a rapid 'blitzkrieg' that decimated naïve prey, as now widely accepted for the megafauna of oceanic islands. However, the cause of these late-Pleistocene extinctions of large American mammals is still debated, because there were major changes in climate around the same time; the exact time relation between extinction and human arrival is not established for most species, and evidence of human hunting has been obtained for only a few species.

The remaining continent that humans occupied only late in the Pleistocene, Australia, also lost most (86% (Martin 1984)) of its large mammal genera, plus some large flightless birds and giant lizards and snakes, after human arrival around 50000 years ago. Many native mammals of Mediterranean and West Indian islands, including all large species, also became extinct in the late Pleistocene or Holocene, though the exact time relation to arrival of humans is uncertain. Eurasia and Africa, where humans coevolved with large mammals for a million years or more, did not suffer such mass extinctions. However, a few species of large mammals, such as Eurasia's mammoths and woolly rhinoceroses, and Africa's giant buffalo, did disappear in the late Pleistocene as human hunting skills improved (Klein 1983). In all these cases, the debate continues as to whether these extinctions of large mammals were due to effects of humans or of climate.

My own evaluation of the evidence is that most of these debated late-Pleistocene extinctions of large mammals were indeed caused by humans. Proponents of climate-based theories have failed to explain why large species became extinct in each part of the world after the first arrival of humans but did not become extinct in other parts of the world with similar climate changes at the same time or in the same part of the world when similar climate changes had occurred in the past (Martin 1984; Diamond 1989).

Thus at least the evidence from oceanic islands makes clear that human-caused mass extinctions are not a hypothesis about something that could happen in the future, but instead a proven event that has already overtaken one fifth of the world's Recent avifauna. Depending on one's evaluation of the continental evidence, human-caused mass extinction may also have already befallen half of the world's Recent large mammals.

THE FUTURE

Is the peak of the current human-caused extinction already passed, or is most of it still to come? Let us try to answer this question by considering the possible mechanisms by which humans exterminate species, and the possible remaining victims. Four mechanisms stand out (Diamond 1984a).

The first and most obvious mechanism of extermination is overhunting. This is the mechanism that played a large role in the extinctions of New Zealand's moas, the 50 species or subspecies of large mammals exterminated since 1600, and possibly the large mammals that became extinct in the late Pleistocene. Hence one might wonder whether, after tens of thousands of years, we have already hunted out any species that we are likely to be able to hunt out. However, many species of large mammals survive in Africa, Eurasia and the oceans, and a smaller number survive in the Americas and Australia. Improved hunting technology is rapidly reducing populations of surviving large mammal species. Obvious candidates to disappear in the near future are most large mammals of Africa and southeast Asia (at least outside of zoos and game parks) and the larger cetaceans.

A second mechanism of human-caused extinction involves effects of introduced species on native species: effects as predators, grazers, browsers, competitors and vectors of disease. Introduced predators such as cats and rats have been the dominant cause of bird extinctions on oceanic islands, whereas introduced mammalian herbivores such as goats have been the main cause of reduction of island vegetation. Famous dramatic examples include the extinctions of native birds within a few months or years after rats arrived at Lord Howe Island in 1918 and at New Zealand's Big South Cape Island in 1964 (Atkinson & Bell 1973; Diamond 1984a). The reason for the susceptibility of island species is of course that they evolved in the absence of species functionally equivalent to the introduced ones. Hence island birds had evolved no behavioural defences against mammalian predators, whereas island plants had evolved no chemical and mechanical deterrants to mammalian herbivores.

Again, one might wonder: hasn't most of the damage already been done, and haven't cats and rats and goats spread around the world? Perhaps surprisingly, the answer is 'no'. The decimation of Australia's small native marsupials and rodents by introduced cats and foxes is sufficiently recent that some native species still survive to provide likely future victims. Oceanic islands still rich in endemic birds, lacking *Rattus rattus* and *R. norvegicus*, and hence candidates for a rat-caused extinction wave, include Rennell and Little Barrier (Diamond 1984b). What may ultimately prove to be the biggest modern extinction wave caused by an introduced predator has just started in Africa's Lake Victoria, whose hundreds of endemic species of cichlid fish are now in the process of being exterminated by a predatory fish, the Nile perch, introduced in a misguided effort to establish a new fishery.

A third mechanism is habitat destruction, which is now becoming the leading mode of human-caused extinctions. In the past, habitat destruction has accounted for about half of the extinctions of bird species on continents (King 1980). In this case the worst is surely yet to come because of accelerated destruction of the world's most species-rich habitats, the continental tropical rainforests (Myers 1980). At current rates of logging and forest destruction, rainforests that are rich in endemic species and that may be largely cleared within the coming decade include the Atlantic forest of Brazil, the forests of Madagascar, and those of lowland Malaysia, whereas those of Borneo and the Philippines are expected to follow soon thereafter. By the middle of the next century, large blocks of rainforest are unlikely to survive outside Amazonia and Zaire. Nor is it necessary to destroy a habitat completely to exterminate many of its species: reduction and fragmentation of habitats also cause extinctions (Diamond 1984c).

I have reasoned so far about human-related factors impinging directly on target species. However, almost every species is dependent on other species for food, habitat structure, pollination, seed dispersal or other necessities. Thus each extinction is likely to cause a cascade of secondary extinctions. For example, human removal of top predators (jaguars, pumas and harpy eagles) on Barro Colorado Island caused a population surge in medium-sized predators (monkeys and coatamundis) on which the top predators normally prey, and that surge of medium-sized predators then led to extinctions of ground-nesting birds on which humans had no direct effects (Terborgh & Winter 1980). Similarly, the disappearance of coyotes from southern Californian canyons turned out to result in declines rather than increases in populations of native birds, because coyotes prey on cats and foxes that prey more heavily on birds than do the coyotes themselves (Soule *et al.* 1988).

Finally, recall that our exclusive focus up to this point has been on extinctions of species. But there are other types of losses of biological diversity, such as extinctions of genetically distinct populations (subspecies) and losses of genetic diversity within a population. Some species that have recovered after passing through severe population bottlenecks, such as elephant seals (Bonnell & Selander 1974) and Cheetahs (O'Brien *et al.* 1983), are now virtually homozygous at all loci studied electrophoretically. In the long term, such species are less able to adapt to altered environmental conditions, and in the short term their survival may be jeopardized by inbreeding depression.

Coda

Among the questions with which I began this article was a common observation disparaging the significance of current human-caused extinctions. 'Extinctions are occurring normally all the time, and it is the fate of all species eventually to become extinct, so what is so different about human-caused extinctions?'

There are three things wrong with this reasoning. First, it is not the fate of all species to go extinct: the existence of tens of millions of species today shows that many species of the past survived to evolve into chronologically distinct species.

Second, the current rate of conservatively documented extinctions is far above the background rate: e.g. one or two bird species per year, sufficient to eliminate the world's entire avifauna in 4500–9000 years. My consideration of the main mechanisms of human-caused extinctions, and our extrapolation from current trends, led me to the conclusion that that current rate of human-caused extinctions will increase. Basically, this is because there are more humans alive now than ever before, armed with more potent means of destruction, and now beginning to assault the most species-rich areas on earth.

Third, in other spheres of life besides conservation biology humans do not simply accept whatever Nature or their own deeds deal to them. Instead, we make choices and alter the course of events around us. To dismiss the current extinction wave on the grounds that extinctions are normal events is like ignoring a genocidal massacre on the grounds that every human is bound to die at some time anyway.

References

Anderson, A. 1984 The extinction of moa in southern New Zealand. In *Quaternary extinctions* (ed. P. S. Martin & R. G. Klein), pp. 728–740. University of Arizona Press.

Anderson, A. 1989 Mechanics of overkill in the extinction of New Zealand moas. *J. archaeol. Sci.* **16**, 137–151.

Atkinson, I. A. E. & Bell, B. D. 1973 Offshore and outlying islands. In *The natural history of New Zealand* (ed. G. R. Williams), pp. 372–392. Wellington: Read.

Bonnell, M. L. & Selander, R. K. 1974 Elephant seals: genetic variation and near extinction. *Science, Wash.* **184**, 908–909.

Collar, N. J. & Andrew, P. 1988 *Birds to watch. The ICBP world checklist of threatened birds.* Cambridge: International Council for Bird Preservation.

Dewar, R. E. 1984 Extinctions in Madagascar: the loss of the subfossil fauna. In *Quaternary extinctions* (ed. P. S. Martin & R. G. Klein), pp. 574–593. University of Arizona Press.

Diamond, J. M. 1982 Rediscovery of the yellow-fronted gardener bowerbird. *Science, Wash.* **216**, 431–434.

Diamond, J. M. 1984a Historic extinctions: a Rosetta Stone for understanding prehistoric extinctions. In *Quaternary extinctions* (ed. P. S. Martin & R. G. Klein), pp. 824–862. University of Arizona Press.

Diamond, J. M. 1984b The avifaunas of Rennell and Bellona Islands. In *The natural history of Rennell Island, British Solomon Islands*, vol. 8, pp. 127–168. Copenhagen: Zoological Museum, University of Copenhagen.

Diamond, J. M. 1984c "Normal" extinctions of isolated populations. In *Extinctions* (ed. M. H. Nitecki), pp. 191–246. University of Chicago Press.

Diamond, J. M. 1987 Extant unless proven extinct? Or, extinct unless proven extant? *Conserv. Biol.* **1**, 77–79.

Diamond, J. M. 1989 Quaternary megafaunal extinctions: variations on a theme by Paganini. *J. archaeol. Sci.* **16**, 167–175.

Gentry, A. 1988 Changes in plant community diversity and floristic composition on environmental and geographical gradients. *Ann. Mo. bot. Gdn* **75**, 1–34.

Imboden, C. 1987 Green lists instead of red books? *Wld Birdwatch* **9** (2), 2.

James, H. F., Stafford, T. W. Jr., Steadman, D. W., Olson, S. L., Martin, P. S., Jull, A. J. T. & McCoy, P. C. 1987 Radiocarbon dates on bones of extinct birds from Hawaii. *Proc. natn. Acad. Sci. U.S.A.* **84**, 2350–2354.

King, W. B. 1980 Ecological bases of extinction in birds. In *Proceedings of the 17th International Ornithological Congress* (ed. R. Nöhring), pp. 905–911. Berlin: Deutsche Ornithologische Gesellschaft.

King, W. B. 1981 *Endangered birds of the world. The ICBP bird red data book.* Washington: Smithsonian.

Klein, R. G. 1983 The stone-age prehistory of southern Africa. *Rev. Anthropol.* **12**, 25–48.

Martin, P. S. 1984 Prehistoric overkill: the global model. In *Quaternary extinctions* (ed. P. S. Martin & R. G. Klein), pp. 354–403. University of Arizona Press.

Mead, J. I., Martin, P. S., Euler, R. C., Long, A., Jull, A. J. T., Toolin, L. J., Donahue, D. J. & Linick, T. W. 1986 Extinction of Harrington's mountain goat. *Proc. natn. Acad. Sci. U.S.A.* **83**, 836–839.

Mohsin, A. K. M. & Ambok, M. A. 1983 *Freshwater fishes of Peninsular Malaysia.* Kuala Lumpur: University Pertanian Malaysia Press.

Mosimann, J. E. & Martin, P. S. 1975 Simulating overkill by Paleoindians. *Am. Scient.* **63**, 304–313.

Myers, N. 1980 *Conversion of tropical moist forest.* Washington: National Academy of Sciences.

O'Brien, S. J., Wildt, D. E., Goldman, D., Merril, D. R. & Bush, M. 1983 The cheetah is depauperate in genetic variation. *Science, Wash.* **221**, 459–462.

Olson, S. L. 1989 Extinction on islands: man as a catastrophe. In *Conservation for the twenty-first century* (ed. M. Pearl & D. Western). New York: Oxford University Press. (In the press.)

Olson, S. L. & James, H. F. 1982 Prodromus of the fossil avifauna of the Hawaiian islands. *Smithson. Contr. Zool.* no. 365.

Soule, M. E., Boulger, D. T., Alberts, A. C., Sauvajot, R., Wright, J., Sorice, M. & Hill, S. 1988 Reconstructed dynamics of rapid extinctions of chapparal-requiring birds in urban habitat islands. *Conserv. Biol.* **2**, 75–92.

Steadman, D. W. 1989 Extinction of birds in Eastern Polynesia: a review of the record, and comparisons with other Pacific island groups. *J. archaeol. Sci.* **16**, 177–205.

Terborgh, J. & Winter, B. 1980 Some causes of extinction. In *Conservation biology* (ed. M. E. Soule & B. A. Wilcox), pp. 119–133. Sunderland: Sinauer.

Discussion

N. P. ASHMOLE (*Department of Zoology, University of Edinburgh, U.K.*). The picture that Professor Diamond has painted is a very gloomy one, but perhaps we can take a little comfort from the fact that what we have heard in this meeting, and know from previous studies of the fossil record, suggests that episodes of extinction are typically followed by rapid evolution and adaptive radiation. It seems likely that the mass extinction caused by humans will in due course result in exciting evolutionary changes and proliferation of new species: the only trouble is that we may not be here to see it!

P. E. PURVES (35 *Morgan House, Tachbrook Street, London, U.K.*). Professor Diamond, during your talk on man-related extinctions you referred to a number of marine mammals that have received your attention. I was surprised, however, that there was no mention of the near extermination of the largest animal that has ever existed on this earth, namely the blue whale, *Balaenoptera musculus*.

During the years immediately before World War II, blue whales were being killed at a rate of 37000 annually. By 1966 the southern blue whale stock was estimated at less than 1000 animals and therefore no longer commercially viable.

The whaling companies then turned their attention to the second largest of the baleen whales, the fin whale, *Balaenoptera physalus*. These in turn were systematically slaughtered until they too, became so rare as to be not worth hunting.

At present, only the smallest of the baleen whales, the minke whale, *Balaenoptera acutorostrata*, remains a viable resource. The large baleen whales have become so widely dispersed that recovery of the whale populations seems very doubtful. The Government of the U.S.A. has placed a total embargo on the hunting of Californian grey whales and thankfully the population of grey whales is increasing by approximately 12% per annum.

Phil. Trans. R. Soc. Lond. B **325**, 479–488 (1989)

Printed in Great Britain

The rise and fall of *Homo sapiens sapiens*

By C. Tudge

208 *Clive Road, London SE*21 8*BS, U.K.*

Human beings have broken the ecological 'law' that says that big, predatory animals are rare. Two crucial innovations in particular have enabled us to alter the planet to suit ourselves and thus permit unparalleled expansion: speech (which implies instant transmission of an open-ended range of conscious thoughts) and agriculture (which causes the world to produce more human food than unaided nature would do).

However, natural selection has not equipped us with a long-term sense of self-preservation. Our population cannot continue to expand at its present rate for much longer, and the examples of many other species suggests that expansion can end in catastrophic collapse.

Survival beyond the next century in a tolerable state seems most unlikely unless all religions and economies begin to take account of the facts of biology. This, if it occurred, would be a step in cultural evolution that would compare in import with the birth of agriculture.

I am going to address three issues. First, is there reason to suppose that human beings might be in danger of extinction; and in particular, will they be swept up in the present wave of mass extinction that they themselves have perpetrated, and which Jared Diamond (this symposium) has described? Second, if we don't become extinct, what is liable to happen to us, and to other living things? And third, will the mass extinction that now is going on provide the kind of kick that might promote our own evolution – rather than our extinction – and if so, what form would that evolution take?

I shall address these issues by playing devil's advocate, the method of discussion invented by the Catholic Church in the late sixteenth century, in which an appointed critic attacks ideas that are popularly held (and indeed are of a kind that people *want* to believe in) to see if they are true. The point is not to destroy the ideas, but to see what they're made of; to ensure that what survives the attack is robust. It's the method of science, in fact.

Issue one: are we ourselves in danger of extinction?

The possibility of human extinction has certainly been suggested of late, on several grounds, including nuclear winter, epidemic (such as AIDS), and – the matter that concerns us here – because of our own destruction of the planet. In particular, it has been suggested that we are sowing the seeds of our own destruction by destroying so many other species; that we need a planet that is in ecological 'balance'; and that that balance depends upon the multitude of other species, perhaps between 10 and 30 million, that the Earth is thought to contain.

If that argument were true, it would be very powerful from a conservationist point of view. I take it to be self-evident that human beings are important; even being exaggeratedly detached, we can hardly deny that our species is an interesting biological experiment, and it would be a pity if it were snuffed out before its time. But I take it also to be self-evident that ours is not the only important species; that other creatures have a 'right' to occupy this planet, and that we at times have to bow to their needs, even at cost to ourselves. Those self-evident

[239]

truths are the basis of 'Green' philosophy. But most people, I think, take only the first of those premises to be self-evident. Most people, if pressed, would probably maintain in a way that is not incompatible with much of the apparent teaching of the Bible, that other animals and plants were 'put on Earth' for our convenience, and that although we shouldn't be cruel to them, we may dispose of them at our will. In other words, the moral philosophy of the Greens is not exclusively anthropocentric, whereas that of most of humanity is.

If you are in a minority, of whatever kind, then it pays as far as possible to demonstrate that your philosophy is compatible, and preferably congruent, with that of the majority. Thus it is that Greens have been anxious to show, these past few years, that a moral philosophy that is not entirely anthropocentric is coincident in its effects with one that is exclusively anthropocentric. Specifically, to bring the discussion down to earth, they have tried to show that human beings benefit from the variousness of other creatures.

Well, do we? The answer, after we've run the gauntlet of devil's advocacy, is 'up to a point'; which is Evelyn Waugh's euphemism for 'not really'.

The arguments that affect to show that a wealth of other species is good for us are of two kinds, specific and general. Specifically, it's pointed out, for example, that new drugs might be found in the roots of plants as yet unexamined, or in the glands of tree frogs; or that the wild relatives of present-day crops – or even, in these days of genetic engineering, the non-relatives of crops – contain genes that may confer resistance to disease; or that people could derive income from wild animals, by attracting tourists, for example, or by allowing limited hunting of animals such as the black rhinoceros.

All these arguments are true. The examples abound, or at least make an impressive list. But none of them is critical. The human species is not dying for lack of drugs, and if you should say, 'what about AIDS?' we might answer 'does anyone believe that the best strategy for seeking an AIDS therapy is to search among the glands of tree-frogs? Wild ground nuts from South America recently supplied breeders at the International Crops Research Institute for the Semi-Arid Tropics (ICRISAT) in India with genes that protected the domestic crop against rust (Gibbons 1985). Very valuable, but not critical; and if it came to a toss-up between saving wilderness for its possible complement of genes, and planting that same wilderness with crops of known value, it would be perverse (if the extra food were really needed) to opt for the wild species. Some Africans do make money from elephants, but if oil is discovered beneath the reserves, what price the wildlife? Besides, we might argue that saving particular species may itself help to perpetrate mass extinction. True, the coat-tail effect is well known; a reserve designed to harbour some particularly charismatic species will also contain a huge number of hangers-on, just as some of the tiger reserves in India also provide homes for jungle cats. But this can work the other way. The bontebok of South Africa, a rare subspecies of the blesbok, very properly has its own small national park. It is good for the bontebok, but the park was established on land that once was fynbos, with its fabulous assemblage of species based upon proteas and ericas. But the fynbos has been banished locally, because bontebok prefer grass.

The more general argument in favour of natural variety is that human beings in some way depend upon the natural food webs that almost invariably are highly complex and rich in species. For example, it is commonly argued – in essence – that if tropical forest is removed or decimated so that the number of species is reduced, then what remains degenerates into desert, which is of no use to anyone. But this argument simply isn't true. A greatly simplified forest, dominated by commercial species of *Eucalyptus*, dipterocarp or *Aralcaria*, stands up just as well, and as far as we know for just as long, as pristine tropical forest that contains hundreds of species of tree. True, if you replace tropical forest with grassland and then overgraze it, the grass is liable to degenerate. But it's not the loss of species that counts, it is the change of habit; that and a level of husbandry that probably isn't properly matched to the demands of the tropics.

Mangroves seem to provide a cast-iron example of natural variety leading intricately but nonetheless inexorably to human benefit. Mangroves contain several species of trees which, in Queensland at least, according to studies by Tom Smith at the Australian Institute of Marine Science (T. Smith, personal communication), in turn depend oddly enough upon unprepossessing crabs to spread their propagules; there are algae in there, and detritus, and a host of insect larvae and Protozoa; all providing food and shelter, eventually, for the larvae of fish that grow into the kind that people love to eat. Take the mangrove away – or indeed, take individual elements away, such as the crabs – and the edible fish disappear as well.

There can be no argument with this. Yet a conscientious devil's advocate would point out that the fish that are nurtured in mangroves are for the most part eaten by rich people who are over-fed to start with; and indeed might point out that fish as a whole, including the apparently vital tilapias of Africa and the enormous yields of cod and the like from high latitudes, contribute a remarkably small proportion of the total protein and energy intake of human beings, and that most of what is consumed is indeed consumed by people who don't need it. An average monetarist – nothing so grand as a devil's advocate, which is a sacred office – could point out that most of the luxury species that Queenslanders or Floridians love to eat can perfectly well be farmed (salmon, turbot, catfish, abolone, giant clams, oysters, and numerous prawns are among the animals that take well to life in a pond or a cage); and if they are farmed they can be fed on ground beef, raised in Illinois. The mangroves can then be given over to hotels, as in Miami; and the tourists will pay to visit the fish farms, which can easily be turned into theme parks, and generate far more wealth, with far more human comfort, than miles of pristine and singularly inhospitable mangrove.

Indeed, when you think about it, it is obvious that the people-need-natural-variety argument is false, on two grounds. The first is that cultivated systems, whether of intensive grain or for fish, are always more productive than wild systems because they absorb a much higher level of nutrient, and process it much more efficiently into human food. Most wild plants hate being over-nourished; and indeed, fertilizer escaping from arable farms, even in small amounts, is in many places the greatest single threat to the marvellous natural variety of the Australian bush. But because they prefer infertile conditions, the output of wild plants is bound to be relatively meagre. Indeed, cultivated systems often out-produce wild systems by 100-fold or more. But cultivated systems are inevitably simplified. They should not, of course, be monocultures, but there is no deep ecology in that; it's just a matter of sensible husbandry. But few cultivated systems contain more than a dozen or so species; orders of magnitude fewer than the wild environment.

Secondly, the argument that humans need the variety of other species is, when you think about it, a theological one. It would be likely to be true only if the Lord had indeed created the world for our express benefit. If we reject that notion, as Green thinkers do on moral grounds and as post-Darwinian scientists are bound to do, then we must concede that other species are for the most part totally detached from any consideration of human welfare, and that the loss of most of them would do us no demonstrable harm, while the loss of several –

including many of the genus *Anopheles* – would be a definite plus. The loss of the Large Copper butterfly from the English Fens has done the British people no material harm at all, and unless the Fens had been drained they could not have become one of the world's most intensive foci of arable farming. Most societies through most of history have persecuted the wolf, and it is impossible to show that the demise of dozens of subspecies, and one or two full species, of wolf-like animals, has had the slightest adverse effect on human material wellbeing. I wish it were not so. I wish we could demonstrate that people need Large Coppers and wolves. But we cannot.

Thus my first conclusion in this diabolically adversarial role is that the elimination of all but a tiny minority of our fellow creatures does not affect the material wellbeing of humans one iota; and indeed, that if human beings really want to take over the world, then they are obliged to tidy most other living creatures away. This is what the European colonialists set out to do when they first encountered the fauna of Africa, and it is what all farmers have done, assiduously and deliberately, since the neolithic revolution began around 10000 years ago. In fact, if we were to appoint a committee to make a short list of creatures that truly contributed to human wellbeing, then I doubt if it would contain more than 10000 species; one tenth of one per cent of the number conservatively estimated now to be on Earth. And that list would include the black rhino for millionaires to hunt, and the Lady Amherst pheasant for ordinary people to look at. There has never been such a mass extinction; but if human beings care only about their material wellbeing and a little sport, they would not need to worry about it at all.

Indeed the only concern that human beings need have about their fellow creatures, a competent devil's advocate would point out, is whether there are enough. Never mind the species, what's the biomass? Provided we can produce enough cellulose, then in an age of biotechnology we can feed ourselves. And here there are two questions that are linked but are none the less separate, and should be treated separately. First, there is the matter of human numbers; can the world as it now is, or as we may contrive to make it, support all the people there are liable to be in the next few decades and centuries? Secondly, are we by our activities reducing the capacity of the world to provide biomass, and is this putative reduction irredeemable? The two issues of course compound each other, but they are separate issues nonetheless.

Human numbers are, of course, staggering. There is an ecological law – a simple extrapolation of bedrock physics – which says that large, predatory animals are rare. We break that law: we are large and have a penchant for predation, yet our population now stands at five billion; and of all feasible demographic projections the one that comes nearest to consensus says that this will double to around 10 billion by the middle of the 21st century, that it will remain at such a figure for several centuries, and that it will then begin to decline, in theory to some figure that our distant descendants feel is appropriate. Nuclear war or some form of super-AIDS could of course make nonsense of such figures. But these figures do represent the ground state.

If the fabric of the Earth stayed as it is throughout that time, and if we add a little more science (as we will), and organize the world a bit better, reducing some of the awful inequities between north and south, for example, then there is no doubt that the world could accommodate such numbers without difficulty. Britain's farming is as intensive as any in the world, but agricultural scientists agree that with present technologies, and without claiming more land, output could easily be increased by at least 25%. Along with most western countries, we give the greater proportion of our home-grown cereal and pulses to livestock. So if we farmed competently and ate less meat we could probably feed around 200 million people in Britain alone. Much of the rest of the world is incapable of such intensive output, but on the other hand, most of the rest makes a far worse job of realizing whatever potential it has, than we do here. If the world really pulled its socks up and if some of us were less greedy, then even with present techniques we could probably feed not 10 but 20 billion people fairly comfortably.

This, however, is where we run into the second consideration; whether the world can *continue* to be as productive as it is now. The issues are not simple. It isn't true to argue, for example, as some Green philosophers like to, that intensive food production inevitably and invariably leads to soil degradation. There are fields at Rothamsted, in Hertfordshire, that have produced cereal every year for 140 years, without added manure, and they are in better heart now than at the beginning. Though the straw and grain have been harvested, organic matter has been maintained by the rotting roots. You cannot treat heath in this way, but any soil can go on being productive, and indeed improve in agricultural terms, provided you stay within its limits; and the limits of some soils are very high indeed.

On the other hand, we cannot ignore the general argument of Paul Ehrlich, of Stanford (Ehrlich & Ehrlich 1987), that much of present-day food production depends not upon sustaining soil but on mining it; that in many soils, if not most, there is a steady loss of 'heart', and indeed of the soil itself, as it washes or blows into the sea; that there is a net increase of undesirables, such as soil salinity, which can be very hard to correct; that some useful commodities such as fossil fuels are being destroyed forever, while others, such as phosphorus and many metals, are being spread around the planet and will become increasingly difficult to harvest. Overall, there is a degradation of the planet's fabric. To a large extent this could be arrested, or circumvented: soil salination can be reversed, as is happening in places in Australia; the loss of fossil fuels need not matter, as there is enough energy in surplus straw to run a tractor and fix nitrogen. But it is clear that the technologies to correct the ill effects of over-farming are not being applied fast enough, and won't be in the foreseeable future.

It is obvious, then, that human numbers would have had to stop increasing at some point; and Professor Ansley Coale at Princeton has pointed out that our population would have reached 17 trillion (10^{18}) within 700 years if the rate of increase of the 1960s had been maintained (Coale 1974, 1987). It is clear, too, that the numbers will level out sooner than optimists might have hoped, as the planet's capacity to produce is underminded. Exactly where the cut-off will be, and when we will reach it, is not clear. What does seem to me extremely likely is that the monetarist argument that the human species will back away from disaster for economic reasons – that as production becomes difficult so demand will reduce – is simply nonsense. Human beings are just as capable as any other species of breeding their way into trouble; and in fact they are more so because of the principle of momentum, which says that in a species with a generation time as long as ours the effects of overbreeding at any one time are not felt until 30 years later, by which time the fabric of the planet could have changed dramatically for the worse (Coale 1974, 1987).

The general point, then, is that we cannot say that disaster for the human species and for the planet as a whole is inevitable; the tragedy of Ethiopia in the 1980s will not necessarily be rehearsed on a global scale. But as Paul Ehrlich has pointed out, it is simply feeble-minded to dismiss out of hand the possibility that at some time in the next few hundred years – in a very short time, indeed – human numbers will exceed the capacity of the world to provide support

(Ehrlich 1987). What happens at that point really is anybody's guess. Mathematicians versed in the intricacies of chaos are perhaps best qualified to comment.

In fact, the likely fate of the human species over the next few hundred years might profitably be modelled mathematically, as has been done for nuclear winter. Every known factor that might influence our material wellbeing, and every known interaction, would be fed into a computer, to see what turns up. In practice the models would be far more complicated than those for nuclear winter, partly because there are more material factors to feed in, but partly because there are other dimensions to take into account. The nuclear-winter models are purely physical; they attempt to assess what will happen *after* the bombs have fallen, and after human beings have done their worst. If we modelled the fate of the human species and our fellow creatures, we would also have to take into account future *intentions*: what kind of a world do we, and our immediate descendants, want to create; and also human fallibility: to what extent are we capable of achieving the end results we find desirable?

The physical factors to be fed into the human future model are complicated, as I have already said, but they are to some extent quantifiable. But it is a sad fact, a reflection on the discipline of sociology, that to my knowledge we have no information at all on the second set of factors we would need to feed in: information on human intention. We don't know what kind of a world human beings want. We may guess in a general way that people nowadays are saddened by the poaching of rhinoceroses, and wish it didn't happen; but it is doubtful if many people know that there are two distinct races of white rhino, for instance, or indeed that there's any difference between the African species and the Asian. And when the Javan tiger was officially declared extinct only a few years ago, the matter hardly featured in national newspapers, though it did feature – significantly – on children's television. It is doubtful if anyone cares, in any positive way, about the reduction in species in tropical forest; secondary forest, or even a plantation, tastefully laid out, looks much the same as a natural wood to the untrained eye. Indeed I suspect that when politicians – Margaret Thatcher, Neil Kinnock, George Bush – use the word 'environment', as now is mandatory in all campaigns, that all they have in mind is generalized green-ness, a golf-course and a bit of Repton-style landscaping, or even a Disney-style theme park with, to quote the blurb of Disney-World, 'clownish baboons and madcap macaws'. It's one thing to get politicians ostensibly on the side of environment, but it's another thing again to determine what actually goes on inside their heads. But what does go on inside their heads, and those of the electorate, matters; and we just don't know what kind of a world people think is desirable.

However, the point of nuclear winter models is not that they unequivocally predict the future, as a soothsayer would do, but that they show a range of possibilities. More specifically, they differentiate the possible from the impossible, and the likely from the less likely. In fact, present nuclear-winter models show that nuclear war is likely to have some effect on climate, and that this could be disastrous if, for example, it led to midsummer frosts in the north, and delayed monsoons in the south. Extreme scenarios – a new mini-Ice Age, as in the seventeenth century, or the total elimination of the human species – are shown to be on the cards, but very much at the extreme tips of the probability curve.

And if we made a model of future human possibilities, feeding in intention (if we knew it) and putting an arbitrary figure on fallibility, we too would finish up with a curve, or rather a three-dimensional curve, of possibilities. And I suspect – this being pure guess work, but I hope reasonably sensible guess work – that among the many scenarios on that curve would be the following six:

1. *Superabundance*. High human population; many other species; lush vegetation.
2. *Most people's ideal (the 'populist' scenario)*. High human population; small, select variety of other species; abundant vegetation.
3. *Fall-back position: the 'Crete' scenario*. Low but stable human population; small but select variety of other species; scenery devastated but acceptable, as in modern Crete.
4. *Failure*. Low human population, but unstable; small variety of other species, with many 'desirable' types already gone, and extinctions continuing; scenery devastated and continuing to degrade. Human extinction conceivable, though extremely unlikely.
5. *Green and pleasant*. Low, stable human population arrived at by voluntary means; high variety of other species, lush vegetation.
6. *Green and unpleasant*. The same as (5), but arrived at by coercion.

I should like to comment briefly on these points. I think we can say that (1) is extremely difficult and perhaps impossible to achieve. The growth of the human population is eliminating other species, and it is hard to see how that trend could immediately stop. Scenario (2) is the kind alluded to above; and probably what politicians have in mind, insofar as they have anything in mind, when they start pushing environmentalism. The select band of species envisaged in (2) would be the 10 000 that competent biologists might identify.

Scenario (3) represents the likely fall-back position if (2) fails. The proposal is that the world as a whole might come to resemble present-day Crete. Crete is stunningly beautiful. But it is, ecologically speaking, a mess. The Minoans finished off the devastation that the farmers of the neolithic began. In a hundred years time the hillsides of Malaysia might look like those of Crete, and we may draw comfort – cold comfort – from the fact that they will be beautiful; bare rock, after the soil is gone, shining in the sun; not so much like Crete, perhaps, as Utah. Clearly, if we treat all the world as the Minoans treated Crete, then we will perforce have a much smaller population than now (and Crete's population is only half what it was in its heyday) but life for those that are left could be highly agreeable, even though their lifestyle was arrived at by insouciance.

On the other hand if things go very badly wrong – in the way that Paul Ehrlich suggests is easily to be envisaged – then we would finish up with scenario (4). Human extinction seems unlikely even in this, the worst conceivable scenario, because even though extinction is very difficult to predict (Jablonsky, this symposium) we can make commonsense observations. And a species like ours that is numerous, ubiquitous, heterogeneous and individually adaptable, and yet shares a common gene pool so that different surviving bands can swap genes, must be a very strong candidate for survival. But if we reach the stage of (4), then we will never be the same again. As Paul Ehrlich has pointed out, recovery in a devastated world, with easily obtainable raw materials already gone, will not be possible; or at least it's very difficult to see how.

The Green scenario is (5). It has been described both by Paul Ehrlich and by Michael Soule (Ehrlich 1987; Soule 1987). Paul Ehrlich envisages a final human population of around one to two billion, while Michael Soule puts the figure much lower, at about 100 million, the likely world population at around the time of Christ; a time, as he points out, of flowering genius. Both Ehrlich and Soule are humanitarians, and envisage such low populations being achieved by voluntary means. The means need not be draconian; if married couples averaged two children, as people in rich countries generally seem happy to do, then the population would inexorably drop, given that some people will elect not to have children at all, and some will die before they have children. The only problem is that a non-draconian policy would take hundreds of years to bring about a significant decline in population, and would not prevent the rise that is imminent. Conservation thus would become a matter of tiding as many creatures as possible over the centuries of human populousness: a period that Michael Soule has called the 'demographic winter'. Ehrlich and Soule both argue that the diminution of human numbers is compensated by the increased quality of life of the people that are on Earth, and by the probable increased longevity of the human species as a whole; for (5) is almost undoubtedly the 'safest' of the scenarios here envisaged.

I agree with Soule and Ehrlich that (5) is the most desirable of the envisagable scenarios; and so, I suspect, do most people reading this paper. But although it's not known what people at large think, I'm sure that many people would not agree that (5) is good. Some feel that to contemplate reduction in human numbers is *ipso facto* inhumane, and others feel it's a kind of blasphemy. On a more secular level, people seeking public office in South Florida at this instant, in Everglades country, are arguing the case for growth and more growth; to quote from a political advertisement on Florida television in 1988, 'growth leads to greater consumer choice': Taco Bell as well as Kentucky Fried. Many people would argue, in short, that (2) is the most desirable scenario, one that has lots of people, albeit living dangerously; and that (3), which is probably more likely than (4), is not too bad as a fall-back. Crete is beautiful, after all; and so, for that matter, is Utah.

The burden of this paper, though, is that if we want (5) to come about – and this is the only realistic scenario that allows for a reasonable proportion of our fellow species to survive – then we have to persuade vast numbers of other people that this is worth aiming for. We cannot, however, simply rely on the materialist arguments that say that we should preserve our fellow creatures because they are of direct benefit to us, for three reasons. The first, as I suggested earlier, is that these arguments are, for the most part, simply untrue. The human species could survive just as well if 99.9% of our fellow creatures went extinct, provided only that we retained the appropriate 0.1% that we need. Secondly, if those who believe that our fellow creatures are important go down the monetarist road, and allow themselves to agree that other species are important only insofar as they bring material benefit, then they will have no arguments with which to resist the inevitable takeover of the national parks when the problems of tropical agriculture are finally solved, and none to resist the inexorable transformation of nature reserves into theme parks.

Thirdly, by couching their defence of our fellow creatures purely in materialist terms, conservationists and scientists – often the same people, of course – are simply failing to express what it is that they feel about nature; and if they fail to express what they feel, then they can hardly be surprised if they fail to make the impact they want. David Jablonsky said that he regarded fossils 'with love and respect' (Jablonsky, this symposium). Absolutely. That's exactly it. Another word, equally appropriate, would be 'reverence'. Such an attitude to nature is, I am sure, what motivates most creative scientists; not a desire to control nature, as Karl Marx supposed, or as Margaret Thatcher supposes. But scientists very rarely, except among themselves, express that reverence.

Because they are generally afraid to admit in public to anything so irrational, so 'unscientific', as an emotional response to nature, scientists reinforce the image that the public has of them: that they are cold, hyper-rational and therefore sinister individuals. Thus the scientists, who care deeply about nature, and who are best equipped to save what is left of it, have largely abandoned what British politicians now call the 'moral high ground' to people who lack the kind of reverence for nature that impels scientists to spend their lives studying it. In recent years, scientists have instead allowed themselves to be swept up in the monetarist fervour, and give the impression that they agree that the point of science, and indeed the point of nature, is materialist. Scientists must learn to argue their case in aesthetic terms – and indeed in religious terms. Religion doesn't have to be a lot of theological and mystical airy-fairiness. In essence, as J. B. S. Haldane averred, it is simply 'an attitude to the Universe' (Haldane 1985); and the proper attitude is that of respect. The point of saving the mangroves is not that they provide us with fish to eat, and that fish are big business, but that fish are good; and so are the skimmers, and herons, and dolphins, that feed upon them in the wild.

The final twist, is that if such a philosophy as this prevailed (a philosophy that may be called Green), and if indeed the human species as a whole, or as a majority, was persuaded that the Green scenario was worth aiming for – that human beings should reduce their numbers, largely in the interests of other species – then this would truly be an evolutionary shift. Human evolution is not a matter of genes these days, and hasn't been for some thousands of years; it's a matter of what goes on inside people's heads. We could argue that what goes on in people's heads has a physical basis, even though the proximate cause is not genetic; after all, the brain is a plastic organ, and takes different forms, depending on the stimuli and indeed on the attitudes to which it is exposed when young.

Since the Neolithic revolution, the transition from hunter-gathering into farming, which began 10 000 years ago, human beings have been weaned, and have weaned their children, on the notion that it was the destiny of humans, and indeed the God-given right of humans, simply to take over the world. Ever since then, that's what we've been doing. I think that the Book of Genesis can properly be read as a folk memory of the transition from late Palaeolithic hunter-gathering – when life in the Middle East must have been very good indeed – into the traumas of Early Neolithic farming, which, though hard, eventually prevailed. All the myths and the Godly admonitions of the first four chapters make perfect sense when viewed in that light. 'In the sweat of the face shalt thou eat bread' is what God said to Adam as He banished him from the easy pickings of the Garden of Eden, the hunting-gatheric Arcadia, and condemned him to a life of agriculture; and we've been sweating, very successfully, ever since. 'Be fruitful and multiply' said God to the sons of Noah, who survived the flood; and we have followed that injunction to the letter. But it is time to acknowledge that the Neolithic party is over.

Green philosophers have argued in recent years that we need to create a post-industrial society, but that is nonsense; future generations will need industry, and besides, to blame all present ills on industry is a serious mis-reading of history. Romantics have argued of late that we need a post-scientific society, which is an even grosser nonsense. Leaving aside the philistinism of such a sentiment, we can be sure that our chances of survival will be greatly compromised, and the death of our fellow creatures guaranteed, unless we practice science of a very high degree.

But we do need a post-Neolithic society, which in its attitudes, in its moral philosophy, in its religion, in its politics and economics and in its way of working would be quite different from all that we have been developing for the past 10 000 years. What exactly those attitudes should be, and how they should be expressed; what form the politics and economics should take; those, I suggest, are the most interesting questions now facing the human species.

REFERENCES

Coale, A. J. 1974 The history of the human population. *Scient. Am.* **231** (3), 40.

Coale, A. J. 1987 Interview in *Lost in the crowd*, BBC Radio 3, 11 October.

Ehrlich, A. & Ehrlich, P. 1987 *Earth*. London: Methuen.

Ehrlich, P. 1987 Interview in *Lost in the crowd*, BBC Radio 3, 11 October.

Gibbons, R. 1985 Interview in *Science now – in passing; from ICRISAT*, BBC Radio 4, 5 August.

Haldane, J. B. S. 1985 Science and theology as art-forms. In *On being the right size, and other essays* (ed. J. Maynard Smith). Oxford University Press.

Soule, M. 1987 Interview in *Lost in the crowd*, BBC Radio 3, 11 October.